Emerging Contaminants

Edited by Aurel Nuro

Published in London, United Kingdom

IntechOpen

Supporting open minds since 2005

Emerging Contaminants
http://dx.doi.org/10.5772/intechopen.87857
Edited by Aurel Nuro

Contributors
Maria Betânia Betânia Melo De Oliveira, Marlyete Chagas De Araujo, Rafael Artur De Queiroz Cavalcanti De Sá, Caio Rodrigo Dias De Assis, Jordany Gomes de Andrade, Bruno Veras, Kaline Catiely Campos Silva, Glauber Pereira Carvalho dos Santos, Vagne de Melo Oliveira, Juliana Ferreira dos Santos, Ana Lúcia Figueiredo Porto, Nosiri Ijeoma Chidi, Chukwuma Anyanwu, Joel Nwaogwugwu, Hugo Saldarriaga-Noreña, Pedro Reyes-Romero, Josefina Vergara-Sánchez, Mario Alfonso Murillo-Tovar, César Torres-Segundo, Luis Chávez-Almazán, Jose Gustavo Ronderos-Lara, Wan Norfazilah Wan Ismail, Siti Umairah Mohktar, Sovugaporumal Ganesan, Mathipriya Shanmugavelu, Maria Betânia Melo de Oliveira, Sivoneide Maria Silva, Lívia Caroline Alexandre de Araújo, Ana Vitoria Araujo Lima, Amanda Virginia Barbosa, Jaqueline dos Santos Silva, Kaleen Massari Leite, Wellenilton Jose do Nascimento Júnior, Vladimir da Mota Silveira-Filho, Carina Lucena Mendes-Marques, Francisco Henrique da Silva, Abdulrazaq Yahaya, Abdulsalam Abdulkareem, Rotimi Larayetan, Abdulbasit Anoze Aliyu, Racheal Olufunke Nayo, Joy A Akor, Victor O Omale, Clifford Okpanachi, Johannes Mbese, Muhammad Bilal, Abdulsalam O Abdulazeez, Riaz Shah, Cemile Ozcan, Barış Can Körükçü, Elizabeth N. Ndunda, Japheth M. Nzangya, Geoffrey O Bosire, Bice S. Martincigh, Vincent O. Nyamori, Pooja Chadha, Jihoon Seo, Sadaf Zaidi, Faizan Ahmad, Mauricio Corredor, Amalia Muñoz-Gómez, Milena Roberta Freire Silva, Karolayne Silva Souza, Flávia Steffany Leite Miranda, Marcilio de Oliveira Santos, Kelma Sirleide de Souza, Sameh Boudiba, Merzoug Benahmed, Karima Hanini, Lana AlShalati, Sukanya Mehra, Khushboo Sharma, Geetika Sharma, Mandeep Singh

First published in London, United Kingdom, 2021 by IntechOpen
IntechOpen is the global imprint of INTECHOPEN LIMITED, registered in England and Wales, registration number: 11086078, 5 Princes Gate Court, London, SW7 2QJ, United Kingdom
Printed in Croatia

British Library Cataloguing-in-Publication Data
A catalogue record for this book is available from the British Library

Additional hard and PDF copies can be obtained from orders@intechopen.com

Emerging Contaminants
Edited by Aurel Nuro
p. cm.
Print ISBN 978-1-83962-418-6
Online ISBN 978-1-83962-419-3
eBook (PDF) ISBN 978-1-83962-420-9

For EU product safety concerns:
IN TECH d.o.o., Prolaz Marije Krucifikse Kozulić 3, 51000 Rijeka, Croatia, info@intechopen.com or visit our website at intechopen.com.

Meet the editor

Aurel Nuro was born in Albania in 1977. He was graduated with a degree in Chemistry (2002) from the Faculty of Natural Sciences, Tirana University, Albania, in 2002. He obtained his Ph.D. from the same university in 2008 followed by "Docent" in 2010 and Associate Professor in 2012. Since 2002 he has worked as a lecturer and researcher in the working group of Organic Chemistry near the Department of Chemistry, FNS, UT. His main research areas include organic chemistry, instrumental analyses, gas chromatography, environmental pollution, and ecotoxicology, among others. Dr. Nuro has been a coordinator, expert, and participant in national and international projects. He has mentored more than 180 master's theses in Environmental Chemistry, Food Chemistry, Pharmacy, and other subjects. His publications include articles, books, and conference proceedings (altogether more than 150).

Contents

Preface

Emerging contaminants are synthetic or natural compounds and microorganisms produced and used by humans that tend to have adverse ecological and human health effects once they reach the environment. These contaminants are present in everyday life in personal care products, pharmaceuticals, detergents, surfactants, pesticides, and so on. Up until the 1990s, most of these contaminants were not identified by environmental monitoring programs because their production and use were not intended to have harmful consequences for the environment and living organisms. However, not only are these contaminants themselves problematic but so are their metabolic/degradation products. Today, there are many emerging contaminants, including domestic and industrial chemicals, hospital medicaments, and farming chemicals. All these compounds are persistent in the environment (water, soil, sediments), can spread far and wide, can bioaccumulate easily, and are toxic. This book, *Emerging Contaminants*, discusses a wide variety of toxins that have adverse effects on our lives.

The first section provides detailed information about the classification, sources, adverse effects, and toxicology of emerging contaminants as well as some of the most used analytical techniques for their quantification. People use many substances at home, in hospitals, and in agricultural and industrial processes. On one hand, these substances are necessary for our wellbeing, but on the other hand, they can damage the environment, organisms, and human health. Many of these substances end up in the environment through wastewater, farming activities, industrial wastes, water treatment processes, and so on. The classification of emerging contaminants is based on their use (such as pesticides, antibiotics, etc.). The chapters in the first section give a clear picture of the main classes of these pollutants as well as additional information concerning their origin in the environment and the problems that they cause. These contaminants are found in low levels (trace to ultra-trace) in the environment, in the ppt–ppm range. Even at these levels, emerging contaminants can have toxic effects. For the analyst, identifying contaminants at ultra-trace concentrations is a challenge, therefore special attention is paid in this section to analytical techniques for sample treatment and quantification, especially in water and biota samples.

One of the most important groups of emerging contaminants is pesticides. As such, the second section of this book presents some recent work on pesticides. Many chemical formulations are used for many years in agriculture as insecticides, fungicides, herbicides, rodenticides, and so on. The use of pesticides in agriculture is necessary, but it must be done carefully and in a controlled way because their use can cause soil pollution, food contamination, and broader environmental pollution that affects water ecosystems. Many pesticide formulations, due to their adverse environmental and health effects, have been banned, such as some organochlorine pesticides like DDT, aldrin, heptachlor, chlordane, etc.). Despite being banned, these compounds and their metabolites persist in the environment. Nowadays, pesticides continue to be in use in agriculture in many other formulations, mostly with N, S, and P; therefore chemical, biological, and toxicological monitoring should always be at the center of attention for institutions and scientists.

The third section of the book is dedicated to the impact of pharmaceutical active compounds (PhACs) in the environment. More than 3000 chemical substances are used as medicines for humans and animals. Human and veterinary drugs (antibiotics, antiseptics, antihistamines, endocrine disruptors, and many other classes) are vital, but their use can be problematic for the environment and living things. Hospital wastewaters are the main sources of pharmaceuticals in the environment. These molecules can promote drug tolerance or resistance to target organisms (e.g., antibiotic resistance in bacteria, or analgesic tolerance in humans) and unwanted effects in non-target organisms (e.g., alteration of sex ratio and decreased fertility) even in low concentrations. Their levels in hospital wastewaters and the environment are reported from ngL^{-1} to ugL^{-1}. Besides their parent molecules, PhACs are often excreted as metabolites that can be even more toxic than native molecules.

The fourth section of the book presents information on some other classes of emerging contaminants such as polybrominated biphenyl ethers, microplastics, industrial base contaminants, and others. Due to their persistence, these contaminants persist in the environment for a long time after their use. In addition, because of their chemical structures they can accumulate easily in organisms and thus cause various problems not only to those organisms but also to the food chain because of the biomagnification process.

Emerging Contaminants is an informative book for researchers in the field and we hope it will direct more attention to the numerous toxic compounds that we confront in our everyday lives.

Aurel Nuro
Department of Chemistry,
Faculty of Natural Sciences,
Tirana University,
Tirana, Albania

Section 1

What are Emerging Contaminants?

Chapter 1

Classification, Potential Routes and Risk of Emerging Pollutants/Contaminant

Yahaya Abdulrazaq, Abdulkareem Abdulsalam, A. Larayetan Rotimi, A. Aliyu Abdulbasit, Okpanachi Clifford, O. Abdulazeez Abdulsalam, O. Nayo Racheal, A. Akor Joy, F. Omale Victor, Z. Mbese Johannes, Muhammad Bilal and Salehdeen Umar M

Abstract

Emerging contaminants (ECs), encompass both natural and synthetic chemicals that are present or transformed to new chemical compounds in water bodies across the globe. They are presently not checked in the environment but poses a serious health threat to human and ecosystem as well as environmental damage. ECs are released into environment during the anthropogenic activities such as water treatments, fumigation, farming etc. More than 1036 ECs and their biotransformation have been identified by the NORMAN project, established in 2005 by the European Commission. They were further classified into different categorizes/classes including disinfection by-products, pesticides, pharmaceuticals and personal care products, nanomaterials, benzotriazoles, benzothiazoles among others. The potential sources, path route and their health implication on human were also discussed. The presence of ECs in our environments is global issue that requires urgent attention.

Keywords: emerging pollutants, contaminants, pharmaceuticals, personal care products cosmetics, disinfectant

1. Introduction

Emerging pollutants (EPs) are natural and synthetic chemicals as well as microbes that are of less concern to the researchers, national and international regulatory bodies [1, 2]. They have not been explicitly studied and there is limited information about their environmental effects, health implications and method of analysis. These pollutants are not new in our environments but they can stay for a longtime in the environment because of their biotransformation, formation of metabolites and by-products [2, 3].

EPs are recently recognized as re-emerging factory-made or naturally formed materials that are detrimental to human health after a long-term exposure and lacking regulatory health standard [4, 5]. They are classified as agricultural (pesticides), industrial and consumer waste products, pharmaceutical and illicit drug as well as personal care products [6–8]. More than 121 various types of unregulated chemicals and microbes are present in an untreated water and at least 25 were found in water

IntechOpen

treatment plants (WTP) [9]. Furthermore, unregulated chemicals such as nitrosamines (NAs), dioxane, nanomaterials, pharmaceuticals and personal care products (cosmetics, disinfectant, antiseptic, deodorant stick, soap, fragrances, insect repellent, sunscreen, surfactants and toothpaste) as well as perfluorinated alkyl acids (PFAA) are EPs [8, 10, 11]. For instance, N-Nitroso-dimethylamine (NDMA) an emerging contaminant produced as byproducts of chloramines in drinking water treatment plants [12]. Nanomaterials (NMs) are substances produced with a dimension in nanoscale range from 1 to 100 nanometer as to improve the chemical strength and reactivity [10]. However, some NMs introduced contaminants when used for water treatment. Polybrominated biphenyl ethers (PBDEs) PBDEs are brominated hydrocarbons used as flame retardants in the production of furniture, plastics, upholstery, electrical equipment, electronic devices and many other household products [10]. Perfluorooctane sulfonate (PFOS) and perfluorooctanoic acid (PFOA) fluorinated organic compounds with many industrial applications; such as surfactants in fluoropolymers and as grease, soil and water resistant in fluorinated polymer [13, 14].

2. Classification of emerging pollutants

Emerging pollutants are classified and categorized as follows [15–17]:

1. Pharmaceuticals (illicit and prescribed drugs).

2. Personal care products (cosmetic, surfactants, disinfectants, domestic biocides, food additives).

3. Industrial chemicals (food additives, pesticides, polychlorinated biphenyl, flame retardant, antimicrobial substances)

4. Disinfection by products (from water treatment plant: Nitrosamine, halonitromethanes, haloacetonitriles, trihalomethanes, halo acetic acids).

5. Algal toxins (toxic released from some algae: Cyanotoxins, microcystin)

6. Biocides and their metabolites (plants and agricultural preventive agents [pesticides])

7. Bioterrorism and disruption devices (Biological and Chemical weapons).

The personal care products and pharmaceuticals as ECs contain active substances in illicit, prescribed and non-prescribed drugs for human and animal consumption. Active ingredients are also found in disinfectants, biocides, preservatives and personal hygiene and beautification products [18]. PPCPs are classified (**Table 1**) base on the consumer product and their uses [19].

Personal care products	Pharmaceuticals
Fragrances, cosmetics, repellents, food supplements and their metabolites, and transformation products personal hygiene products, sunscreen agents, domestic insect,	Drugs such as cocaine and amphetamines, antiseptics, hormones, illicit analgesics, antibiotics, lipid regulators, steroids, anti-inflammatory drugs, diuretics, non-steroid stimulant drugs, antimicrobials, beta blockers,

Table 1.
Classification of pharmaceuticals and personal care products [16].

Category I: Personal care products compound	
ATII (traseolide)	Boisvelone/Iso-E super
Bayrepel	Butyl methoxydibenzoylmethane
Benzaldehyde, (phenylmethylene)	Cineole
hydrazone (Eusolex)	Damascone
Benzophenone	Decamethylcyclopentasiloxane (D5)
AHDI (phantolide)	Boisvelone/Iso-E super
alpha-Terpineol	Decamethyltetrasiloxane (MD2M)
2,4-Dihydroxybenzophenone	Dihydromethyljasmonate
4-Methylbenzylidene camphor	Dodecamethylcyclohexasiloxane(D6)
4-Oxoisophorone	
Acetylcedrene	Dodecamethylpentasiloxane(MD3M)
ADBI (celestolide)	Drometrizole
Ethylhexyl methoxycinnamate	Drometrizole trisiloxane (INCI)
Galaxolide	Methylsalicylate
g-Methylionone	Musk ambrette
Hexamethyldisiloxane (HM or HMDS)	Musk ketone
Hexylcinnamaldehyde	Musk xylene
Homosalate	Octamethylcyclotetrasiloxane (D4)
Isobornylacetate	Octamethyltrisiloxane (MDM)
Isobutyl paraben	Octocrylene
Methylparaben	Oxybenzone
Methyldihydrojasmonate (methyl 3-oxo-2-pentylcyclopentaneacetate)	Carvone
Methyl-iso-propylcyclohexenone	Propyl paraben
p-t-Bucinal (Lilial)	Tonalide
Category II: Personal care products/biocides compounds	Triclosan
N,N-Diethyltoluamide	D-Limonene
Category III: Personal care products/food additive compounds	
2,6-Di-*tert*-butylphenol	Triethylcitrate
Butylated hydroxyanisole	Butylated hydroxytoluene
2-Ethylthioacetic acid ethylester	Dipropyltrisulfide
2-Methylthioacetic acid ethylester	Ethylene brassylate
3-Methylthiopropionic acid	Habanolide
Category I: Pharmaceuticals compounds	
1-Hydroxy ibuprofen	Cyclophosphamide
2-Hydroxy ibuprofen	Desmethylnaproxen (metabolite of naproxene)
17-alpha-Estradiol	Danofloxacin
17-alpha-Ethinylestradiol	Dantrolene
17-beta-Estradiol	Dapsone
1,1,1-Trichloro-2,2-dihydroxyethane (chloral hydrate)	Daunorubicin

Acebutolol	Diethylstilbestrol
Acecarbromal	Difloxacin
Aceclofenac	Diphenhydramine
Acemetacin	Domperidone
Acetaminophen (Paracetamol)	Doxepine
Acetazolamide	Doxorubicin
Acetylsalicylic acid (Aspirin)	Doxycycline (anhydrous)
Acyclovir	Doxycycline (monohydrate)
Albuterol	Dexamethasone
Allobarbital	Diatrizoate
Alclofenac	Diazepam
Albuterol sulfate	Diclofenac
Alprazolam	Dicloxacillin
Amitriptyline	Enoxacin
Ampicillin	Epirubicin
Amoxicillin	Enrofloxacin
Amobarbital	Escitalopram
Anthracene-1,4-dione	Erythromycin
Apramycin	Esomeprazole
Aprobarbital	Estriol
Atenolol	Estrone
Azithromycin	Estrone sulfate
Bezafibrate	Ethosuximide
Baquiloprim	Etofibrate
Betamethasone	Fenofibric acid (metabolite of fenofibrate)
Beta-sitosterol	Fenfluramine
Betaxolol	Fenofibrate
Baclofen	Famotidine
Bisoprolol	Fenoprofen
Bromazepam	Flumequine
Butalbital	Fenoterol
Carbamazepine	Flucloxacillin
Carazolol	Fenoprofen calcium salt dihydrate
Caffeine	Fluorouracil
Cefacetrile	Fluoxetine
Cefapirin	Fluvoxamine
Cefalonium	Furosemide
Cefalexin	Gemfibrozil
Cefazoline	Gentamicin
Cefoperazone	Glibenclamide (glyburide)
Ciprofloxacin	Hexobarbital

Clofibric acid (metabolite of clofibrate)	Hydrocodone
Clarithromycin	Hydrochlorothiazide
Citalopram	Chlortetracycline
Clotrimazole	Chlorobutanol
Clenbuterol	Chloramphenicol
Crotamiton	Ibuprofen
Cloxacillin	Cholesterol
Codeine	Ifosfamide
Category II: Pharmaceutical Compounds	
Iopamidol	Penicillin G
Iminostilbene	Penicillin V
Indomethacin	Pentoxifylline
Iohexol	Pentobarbital
Iomeprol	Paroxetine
Imapramine	Phenazone
Iopromide	Phenylbutazone
Ivermectin	Phenobarbital
Josamycin	Phenytoin
Kanamycin sulfate	Pindolol
Lansoprazole	Prednisolone
Lamotrigine	Pravastatin
Ketoprofen	Pipamperon
Levetiracetam	Primidone
Lidocaine	Propranolol
Lincomycin	Salbutamol
Lorazepam	Ranitidine
Loratadine	Roxithromycin
Lithium carbonate	Propyphenazone
Lovastatin	Sarafloxacin
Marbofloxacin	Sotalol
Meprobamate	Secobarbital sodium
Meclofenamic acid	Sertraline
Medazepam	Streptomycin
Metformin	Secobarbital
Mebeverine	Spectinomycin
Mestranol	Spiramycin
Mefenamic acid	Simvastatin
Minocycline	Sulfamerazine
Methylphenobarbital	Sulfadimethoxin
Metoprolol	Sulfadoxin
Mevastatin	Sulfadiazine

Methicillin	Sulfamethazine
Nandrolone	Sulfamethoxazole
Nafcillin	Sulfapyridine
Nadolol	Tetracycline
Naproxen	Temazepam
N-Methylphenacetine	Terbutaline
Neomycin B	Taloxa
Norfloxacin	Tolfenamic acid
Nordiazepam	Tilmicosin
Novobiocin	Timolol
Ofloxacin	Tiamulin
Oxacillin	Tramadol
Omeprazole	Trimethoprim
Oleandomycin	Tylosin
Oxazepam	Verapamil
Oxytetracycline	Valnemulin
Oxprenolol	Zolpidem

Table 2.
Identified pharmaceuticals and personal care products (NORMAN [16, 20]).

More than 1036 EC and secondary metabolites as well as biotransformation products are identified on NORMAN List in **Table 2** (NORMAN [20]). The most common classes/categories of ECs include; industrial chemicals, disinfection by-products, pesticides, sweeteners endocrine disrupting compounds, nanoparticles, sunscreens, UV filters, pharmaceuticals and personal care products among others [21–23].

3. Potential routes emerging pollutants

Series of anthropogenic activities such as agricultural, domestic and industrial activities leads to the discharged of the pollutants into our environments (**Figure 1**). Water and sediments serve as a sink to these contaminants [25, 26]. Constant consumption of drugs due to less attention given to traditional medicine for the treatment aliments and frequent use of personal care products as vogue has led to the release of different by-products at low concentrations into our environments [27]. The concentration of these pollutants are based on the production processes of some products (drugs) applied in various countries. For example, the concentration of bisphenol A in European water and North American were 43 and 12 ppb correspondingly [28, 29]. Their presence in water has affected the water physicochemical parameter and required urgent attention to drinking issues. Essentially however, many wastewater treatment plant (WWTP) are not considered or designed for the removal of emerging contaminants as shown **Figure 1**. As a result of their hydrophobic nature PPCPs and other EC metabolites as well as associated particular mater settled below water surface when discharged in water. Treated and untreated urban water as well as WWTP (**Figure 1**) are the main sources and path route of EC.

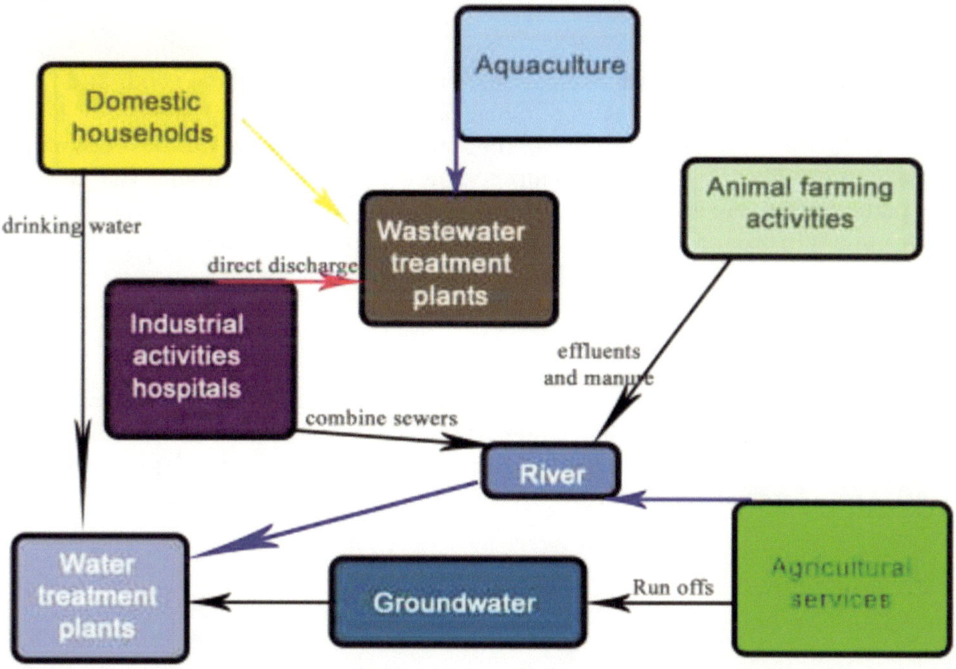

Figure 1.
Potential sources of PPCPs/EDCs in water cycle modified [24].

4. Risk of emerging pollutants

PPCPs are present in our environment at very low concentrations ranging from ng/L to μg/L but there is paucity of information about the various secondary metabolites generated during the biotransformation of parent compounds [16] The impact of the metabolites could be more toxic than the main compounds and their adverse effect on non-target organism is less understood [30]. The longtime exposure to PPCPs pose severe threat to aquatic biota and human life. Also, many studies on environmental toxicology on the exposure of non-target organs to PPCP are results from acute toxicity data [31–33]. Many PPCPs contained bioactive ingredient that could have a chronic effect non-target biota. Goldfish shows bio-centration factor of 113 when it is exposed to high concentration of PPCPs for 14 days [34]. Chronic effect could lead to gene mutation and decline in fish population. For instance, diclofenac may cause ill-effect on fish organs. Fish are susceptible to PPPCPs contaminants since they live in the environment where these contaminants are present [35]. The presence of carbamazepine and diclofenac in aquatic environs damaged algal chloroplasts [36]. Long time exposure to Sulfamethoxazole cause severe toxicity and inhibit photosynthesis process [37]. Also, ciprofloxacin is toxic to green algae [38]. Natural bacteria developed antibiotic tolerance in PPCP antibiotic contaminated environments [39]. Endocrine disrupting chemicals are also referred to as hormones in aquatic or aqueous environments. Some endocrine chemicals reduce the proper functioning of endocrine system (ES) that is physiological activities [40]. For example, retardation in reproductive processes such as sex variation, poor metabolic and embryonic developments [41]. These environmental hormones could cause serious effects on both aquatic and terrestrials animals through inhibition and modification of hormonal growth in ES and hormones in the cell correspondingly [42].

5. Conclusion

Nowadays there is increase in technologies for the production of goods and services as to meet up the demand of dense population, resulting to release of EC in our environs. EC is a serious environmental issue across globe that desires vital consideration. There is need to development new production technique that uses raw materials that are eco-friendly, less toxic metabolites and by-products in our environment in order to prevent the aquatic biota and ecosystem.

Acknowledgements

We extent our gratitude to Arch. Mohammed, Moshood Kayode at Federal University of Technology, Minna, Niger State, Nigeria, Dr. Babatunde, Olubayo and Engr. Adedoja, Samson Oluwaseye at Tshwane University of Technology, Pretoria, South Africa for their supports.

Author details

Yahaya Abdulrazaq[1,4*], Abdulkareem Abdulsalam[2], A. Larayetan Rotimi[1,4], A. Aliyu Abdulbasit[1], Okpanachi Clifford[1], O. Abdulazeez Abdulsalam[1], O. Nayo Racheal[1], A. Akor Joy[3], F. Omale Victor[1], Z. Mbese Johannes[4], Muhammad Bilal[5] and Salehdeen Umar M[6]

1 Department of Pure and Industrial Chemistry, Kogi State University, Anyigba, Kogi State, Nigeria

2 Department of Petroleum Engineering, Federal polytechnic of Oil and Gas, Bonny, Rivers State, Nigeria

3 Department of Integrated Science, Kogi State College of Education, Ankpa, Kogi State, Nigeria

4 Department of Pure Chemistry, Faculty of Science and Agriculture, University of Fort Hare, Alice, South Africa

5 School of Life Science and Food Engineering, Huaiyin Institute of Technology, Huaian, China

6 Federal University Lokoja, Department of Chemistry, Lokoja, Kogi State, Nigeria

*Address all correspondence to: yahayaabdulrazaq2012@gmail.com

References

[1] Dulio V, van Bavel B, Brorström-Lundén E, Harmsen J, Hollender J, Schlabach M, et al. Emerging pollutants in the EU: 10 years of NORMAN in support of environmental policies and regulations. Environmental Sciences Europe. 2018;*30*(1):1-13 https://doi.org/10.1186/s12302-018-0135-3

[2] Montes-Grajales D, Fennix-Agudelo M, Miranda-Castro W. Occurrence of personal care products as emerging chemicals of concern in water resources: A review. Science of the Total Environment. 2017;**595**:601-614

[3] NORMAN. NORMAN, NORMAN. Available from: https://www.norman-network.net/? In: q 5node/19. 2018

[4] https://www.whitehouse.gov/wp-content/uploads/2018/11/Plan-for-Addressing-Critical-Research-Gaps-Related-to-Emerging-Contaminants-in-Drinking-Water.pdf

[5] Villanueva CM, Kogevinas M, Cordier S, Templeton MR, Vermeulen R, Nuckols JR, et al. Assessing exposure and health consequences of chemicals in drinking water: Current state of knowledge and research needs. Environmental Health Perspectives. 2014;**122**(3):213-221

[6] Boxall ABA. New and emerging water pollutants arising from agriculture environment department, University of York. In: United Kingdom. Meeting the Policy Challenge, which is available at: OECD study Water Quality and Agriculture; 2012 www.oecd.org/agriculture/water

[7] EPA Method 1694, 2007. (Environmental Protection Agency). Pharmaceuticals and personal care products in water, soil, sediment, and biosolids by HPLC/MS/MS, December 2007, EPA-821-R-08-002

[8] Ferrer I, Zweigenbaum JA, Thurman EM. Analysis of 70 Environmental Protection Agency priority pharmaceuticals in water by EPA method 1694. Journal of Chromatography. A. 2010;**1217**(36):5674-5686. DOI: 10.1016/j.chroma.2010.07.002

[9] Glassmeyer ST, Furlong ET, Kolpin DW, Batt AL, Benson R, Boone JS, et al. Nationwide reconnaissance of contaminants of emerging concern in source and treated drinking waters of the United States. Science of the Total Environment. 2017;**581**:909-922

[10] US Department of Health and Human Services, Agency for Toxic Substances and Disease registry (ATSDR)(2015), Draft toxicological profile for Polybrominated diphenyl ethers (PBDEs). Public Health Service, Atlanta, GA.

[11] Yahaya A. Method development for the identification and quantitative analysis of seven nitrosamines using gas chromatography mass spectrometry. Chemical data collection. 2019;**21**:1-12 https://doi.org/10.1016/j.cdc.2019.100231

[12] Yahaya A, Babatunde DE, Olaniya LWB, Agboola O. Application of chromatographic techniques in the analysis of total nitrosamines in water. Heliyon Journal. 2020;**6**:1-8 https://doi.org/10.1016/j.heliyon.2020.e03447

[13] Houtz EF, Higgins CP, Field JA, Sedlak DL. Persistence of perfluoroalkyl acid precursors in AFFF-impacted groundwater and soil. Environmental Science & Technology. 2013;**47**(15):8187-8195

[14] US EPA, 2016. Drinking Water Health Advisory for Perfluorooctane Sulfonate (PFOS). U.S. Environmental

Protection Agency Office of Water (4304T) Health and Ecological Criteria Division Washington, DC. 20460. https://www.epa.gov/sites/production/files/201605/documents/pfos_health_advisory_final_508.pdf

[15] Carmichael WW. The cyanotoxins. In: Advances in Botanical Research. Vol. 27. Academic Press; 1997. pp. 211-256

[16] Dey, S., Bano, F., Malik, A., 2019. Pharmaceuticals and personal care product (PPCP) contamination—a global discharge inventory. In Pharmaceuticals and Personal Care Products: Waste Management and Treatment Technology (pp. 1-26). Butterworth-Heinemann.

[17] Yahaya, A., Sale, F. J., Salehdeen, M.U. 2020b. Analytical Methods for Determination of Regulated and Unregulated Disinfection by-Products in Drinking Water: A ReviewCaJoST, 2020, 1, 25-36

[18] Bu Q et al. Pharmaceuticals and personal care products in the aquatic environment in China: A review. Journal of Hazardous Materials. 2013;**262**:189 211. Available from: https://doi.org/. DOI: 10.1016/j.jhazmat.2013.08.040

[19] Yang, Y., OK, Y.S; Kim, K., Kwon, E.E., Tsang, Y.F. 2017. Occurrences and removal of pharmaceuticals and personal care products (PPCPs) in drinking water and water/sewage treatment plants: a review. Sci. Total Environ. 596 597, 303 320. Available from: https://doi.org/10.1016/j.scitotenv.2017.04.102.

[20] Network NORMAN. List of NORMAN emerging substances (mostly frequently discussed). In: Available from: ,https://www.norman-network.net/sites/default/files/files/ Emerging_substances_list_Feb_16/NORMAN list_2016_FINAL.XLSX. 2016

[21] Li, W.C., 2014. Occurrence, sources, and fate of pharmaceuticals in aquatic environment and soil. Environmental Pollution 187, 193 201. Available from: https://doi.org/10.1016/j.envpol.2014.01.015.

[22] Richardson SD, Ternes TA. Water analysis: Emerging contaminants and current issues. Analytical Chemistry. 2014 Availablc from: https://doi.org/10.1021/ac500508t

[23] Rodriguez-Narvaez, O.M., et al., 2017. Treatment technologies for emerging contaminants in water: A review. Chemical Engineering Journal 323, 361 380. Available from: https://doi.org/10.1016/j. cej.2017.04.106.

[24] Petrovic M, Gonalez S, Barcelo D. Analysis and removal of emerging contaminants in wastewater and drinking water. Trends in Analytical Chemistry. 2003;**22**(10):685-696

[25] Yahaya A, Okoh OO, Adeniji AO. Occurrences of organochlorine pesticides along the course of the Buffalo River in the eastern cape of South Africa and its health implications. International Journal of Environmental Research and Public Health. 2017;**1372**(14):1-16

[26] Yahaya, A., Adegbe, A. A. & Emorutu, J. E. 2012. Assessment of heavy metal content in the surface water of Oke – Afa canal, Isolo- Lagos Nigeria. Scholars Research Library: Archives of Applied Science Research, 4, 6, pp.2322-2326.

[27] Yu C-P, Chu K-H. Occurrence of pharmaceuticals and personal care products along the west prong little Pigeon River in East Tennessee, USA. Chemosphere. 2009;**75**:1281-1286

[28] Klecka GM, Staples CA, Clark KE, Van der Hoeven N, Thomas DE, Hentges SG. Exposure analysis of bisphenol-a in surface water

systems North America and Europe. Environmental Science and Technology. 2009:6145-6150

[29] Rogers JA, Metz L, Wee Yong V. Review: Endocrine disrupting chemicals and immune responses: A focus on bisphenol-a and its potential mechanisms. Molecular Immunology. 2013;**53**:421-430

[30] World Health Organization. WHO Multi-Country Survey Reveals Widespread Public Misunderstanding about Antibiotic Resistance. World Health Organization; 2015

[31] Harrison EZ, Oakes SR, Hysell M, Hay A. Organic chemicals in sewage sludges. Science of the Total Environment. 2006;**367**:481-497

[32] Kimberly, G. J. (2007). Applications of Radiation Chemistry to Understand the fate and Transport of Emerging Pollutants of Concern in Coastal Waters. A dissertation submitted to the Graduate Faculty of North Carolina State University in partial fulfillment of the requirements for the Degree of Doctor of Philosophy Marine, Earth, and Atmospheric Sciences, USA.

[33] Tijani JO, Fatoba OO, Petrik LF. A review of pharmaceuticals and endocrine-disrupting compounds: Sources, effects, removal, and detections. Water. Air, & Soil Pollution. 2013;**224**(11):1770

[34] Mimeault C, Woodhouse AJ, Miao XS, Metcalfe CD, Moon TW, Trudeau VL. The human lipid regulator, gemfibrozil bioconcentrates and reduces testosterone in the goldfish, Carassius auratus. Aquatic Toxicology. 2005;*73*(1):44-54 https://doi.org/10.1016/j.aquatox.2005.01.009

[35] Mehinto AC, Hill EM, Tyler CR. Uptake and biological effects of environmen-tally relevant concentrations of the nonsteroidal anti-inflammatory pharmaceutical diclo-fenac in rainbow trout (Oncorhynchus mykiss). Environmental Science & Technology. 2010 Available from: https://doi.org/10.1021/es903702m

[36] Vannini C, Domingo G, Marsoni M, De Mattia F, Labra M, Castiglioni S, et al. Effects of a complex mixture of therapeutic drugs on unicellular algae Pseudokirchneriella subcapitata. Aquatic Toxicology. 2011;**101**(2):459-465 https://doi.org/10.1016/j.aquatox.2010.10.011

[37] Liu BY, Nie XP, Liu WQ, Snoeijs P, Guan C, Tsui MT. Toxic effects of erythromycin, ciprofloxacin and sulfamethoxazole on photosynthetic apparatus in Selenastrum capricornutum. Ecotoxicology and Environmental Safety. 2011;**74**(4):1027-1035 https://doi.org/10.1016/j.ecoenv.2011.01.022

[38] Halling-Sorensen B. Environmental risk assessment of antibiotics: Comparison of mecillinam, trimethoprim and ciprofloxacin. The Journal of Antimicrobial Chemotherapy. 2000 Available from: https://doi.org/10.1093/jac/46.suppl_1.53

[39] Novo, A., André, S., Viana, P., Nunes, O.C. and Manaia, C.M., 2013. Antibiotic resistance, antimicrobial residues and bacterial community composition in urban wastewater. Water Research, 47(5), pp.1875-1887. https://doi.org/10.1016/j.watres.2013.01.010.

[40] Schug TT, Janesick A, Blumberg B, Heindel JJ. Endocrine disrupting chemicals and disease susceptibility. The Journal of Steroid Biochemistry and Molecular Biology. 2011;**127**:204-215

[41] Flint S, Markle T, Thompson S, Wallace E. Bisphenol a exposure, effects, and policy: A wildlife

perspective. Journal of Environmental Management. 2012;**104**:19-34

[42] Olujimi OO, Fatoki OS, Odendaal JP, Okonkwo JO. Endocrine disrupting chemicals (phenol and phthalates) in the south African environment: A need for more monitoring. Review. Water SA. 2010;**36**(5):671-682

Chapter 2

Emerging Compounds in Mexico: Challenges for Their Identification and Elimination in Wastewater

José Gustavo Ronderos-Lara, Hugo Saldarriaga-Noreña, Pedro Guillermo Reyes-Romero, Luis Alberto Chávez-Almazán, Josefina Vergara-Sánchez, Mario Alfonso Murillo-Tovar and César Torres-Segundo

Abstract

In recent years, the presence of organic pollutants has received great attention due to their effects on public health and biota. Within this set of compounds, a new range of compounds that are characterized by their high persistence and low degradation have been identified, called Emerging Compounds. Emerging pollutants include a wide variety of products for daily use of different structures, domestic and industrial applications, such as: pesticides, industrial and personal hygiene products, hormones, and drugs, most of which are toxic, persistent and bioaccumulative. A characteristic of these types of pollutants is that current wastewater treatment plants are unable to remove them; they are designed to remove organic matter and nutrients in higher concentrations. In Mexico there is little information on the concentration levels of these compounds, due to the lack of public policies aimed at providing resources to institutions and researchers trained to carry out this type of study. On the other hand, the technological infrastructure of the wastewater treatment plants is insufficient for the country's demand. This situation represents one of the greatest challenges for the authorities responsible for the management of water resources, in the immediate time if it is intended to preserve said resource and therefore take care of the health of the population.

Keywords: emerging compounds, monitoring, wastewater, removal

1. Introduction

Currently one of the greatest challenges worldwide is the conservation of the quality of water resources. On a daily basis, a large amount of waste from different industrial, urban and livestock activities is discharged into water bodies, mainly through wastewater. According to UNESCO, 59% of total water consumption in developed countries is destined for industrial use, 30% for agricultural consumption and 11% for domestic activities [1]. It has been reported that more than 80% of hazardous waste in the world is produced in industrialized countries; it is also known that in developing countries 70% of the waste generated in industry is dumped to bodies of water without any type of previous treatment [1]. Specifically in Mexico, 54% of wastewater is not treated, which has become one of the biggest public health problems, since this type of water is reused for agricultural activities and in some cases for human consumption [2]. **Table 1** shows the percentage proportions of water uses according to their origin in Mexico.

In recent decades, the use of new chemical products has intensified in different anthropic activities, which has caused the degradation of water resources throughout the planet [3]. Within this set of compounds, a new range of compounds that are characterized by their high persistence and low degradation have been identified, called emerging compounds (EC). The term EC is used to refer to compounds of different origin and chemical nature, whose presence in the environment is not considered significant in terms of distribution and/or concentration, so they go unnoticed. What constitutes a high risk for the environment and the health of the population [4]. Emerging pollutants include a wide variety of products for daily use of different structures, domestic and industrial applications, such as: pesticides, industrial and personal hygiene products, hormones, and drugs, most of which are toxic, persistent and bioaccumulative. **Figure 1** briefly describes the classification of this type of compound by families.

It has been established that these compounds enter the environment through different sources, such as domestic and industrial wastewater [6], from waste, treatment plants [7], hospital effluents [8], agricultural and livestock activities [9] and septic tanks, among others [10], which are produced at different concentrations in surface waters, whose environmental quality criteria have not yet been specified [11].

One of the main problems of this type of pollutant is that the current wastewater treatment plants are unable to eliminate them. They are designed to remove organic matter and nutrients in higher concentrations (g L^{-1}). Therefore, emerging pollutants are present in surface water, groundwater and in purified water. In addition, the primary degradation of some of these compounds in wastewater treatment plants or in the environment itself, generate more persistent and more dangerous products, and synergistic effects may even occur if the compounds share the mechanisms of action [12].

This situation has been a matter of concern for the scientific community and for regulatory environmental entities, given the multiple impacts that they can cause on the environment and human health [11].

	Source			
	Superficial	**Groundwater**	**Total volume**	**Percentage of extraction**
Use	**(thousands of hm^3)**	**(thousands of hm^3)**	**(thousands of hm^3)**	**(%)**
Agricultural	42.0	23.2	65.2	76.7
Public supply	4.8	7.3	12.1	14.2
Industry	1.6	2.0	3.6	4.2
Electric power excluding hydroelectricity	3.7	0.5	4.2	4.9
Total	52.0	32.9	84.9	100

Table 1.
Water uses according to the source in Mexico [2].

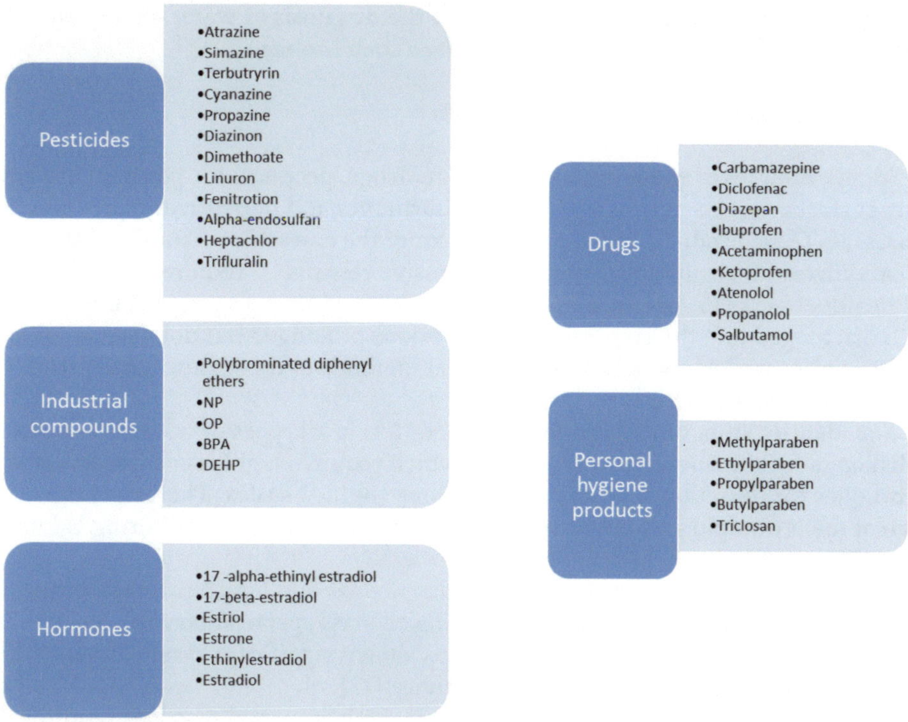

Figure 1.
Classification of emerging compounds by family. (Adapted from [5]).

Since 1989, the World Health Organization (WHO) has developed guidelines for the safe use of wastewater, and on this basis, each country has established its own regulatory framework. In this sense, Mexico has implemented a decentralized policy framework for managing water resources. In particular, the National Water Commission (CONAGUA, by its acronym in Spanish) was created, whose main function is the management of water resources, likewise is responsible for the formulation of public policies for water management. Water management legally incorporates Integrated Water Resources Management (IWRM), whose purpose is to promote stakeholder participation in coordinating the development and management of water, land and related resources [13]. Within the functions of the IWRM is the management of wastewater treatment and its reuse. However, the treatment and reuse of wastewater has not yet been adequately implemented within the sanitation services in terms of comprehensive water management, this is partly due to the fact that sanitation is not defined within the water legislation, in addition to institutional fragmentation, making it difficult to carry out such activity [14].

2. Challenges in the management of water resources in Mexico

Until now, all strategies and policies for the administration of water resources in Mexico have been ineffective, mainly because of the economic and political interests of some groups in society, which has not allowed the application of the principles established in IWRM, considering the participation of interest groups. This situation has not allowed the investment of resources to address the environmental problem generated by the presence of EC in water bodies. Currently only a few very specific studies have been carried out in a few states of the republic which indicates that there are many pending tasks on the part of the entities responsible for the management of water resources, in terms of the diagnosis of water quality and the development of advanced technologies to face such problem.

2.1 Identification of emerging compounds

Some examples of these contaminants are drugs, products for personal use and care, surfactants, fire retardants, steroids, hormones and derivatives of disinfection processes. These products correspond in most of the cases to contaminants that may be candidates for regulation; however, extensive research is required on its potential health effects [15, 16]. In some cases it is assumed that several of the EC have been discharged into the environment for periods prolonged but not detected due to the little information and a lack of analytical methods to detect low concentrations in different matrices [16].

The identification of this family of compounds in all types of waters has become a challenge for the scientific community, which requires highly sensitive analytical techniques for detection at nanograms per liter (ngL^{-1}) scales. Therefore, the development of rapid and sensitive analytical methods for EC monitoring is important [17].

The analytical techniques most used today are gas and liquid chromatography, both coupled to mass spectrometry. Coupling to mass spectrometry for the identification of EC in environmental matrices has shown significant results, mainly due to its high sensitivity, specificity and selectivity [17].

The detection of this type of compounds in environmental matrices requires efficient sample treatment procedures to concentrate analytes of interest and eliminate interferences [17].

Sample preparation techniques include solid phase extraction, solid phase microextraction, liquid-liquid extraction, microwave assisted extraction, liquid phase microextraction techniques, stir bar sorption extraction, and pressurized liquid extraction, among others [18].

However, access to these techniques requires large investments of money and highly specialized personnel for the development and validation of adequate methodologies. This situation has not made it possible to carry out diagnoses of the real situation of the presence of these compounds and in the main water bodies of the republic, since the states do not have the necessary resources.

In Mexico, there are few studies that have determined the concentration levels of this type of compounds in wastewater, groundwater, and surface water, almost all made in the center of the country (Guanajuato, Hidalgo, Jalisco, Morelos states, and Mexico city). Among the reported compounds are estradiol, ethinylestradiol, 4-nonylphenol, bisphenol A, 4-tert-octylphenol, naproxen, acetaminophen, diclofenac, bezafibrate, atenolol and carbamazepine, among others. This situation is worrying if we consider the great industrial and agricultural activity that takes place in a large part of the republic (**Table 2**).

2.2 Removal of EC from wastewater

The pollution of water bodies is a technical, social, and environmental challenge, attributable to continuous population increase and limited waste elimination strategies coupled with poor public management of water contaminants [24, 25]. The treatment of wastewater has been carried out for a long time, with the intention of reducing adverse effects on the environment and human health.

Although wastewater treatment plants are designed to remove solid materials, dissolved organic matter, nutrients and reduce the levels of metals, bacteria, and other pathogens. Most are not designed to efficiently remove organic pollutants,

Site		Compound	Concentration (ng/L)	Reference
Hidalgo state	**Residual water**	4-nitrophenol	16.7	[19]
		Bisphenol A	2.5	
		Estradiol	0.022	
Xochimilco channel	**Farming and livestock**	Bisphenol A	15200-22370	[20]
		Estradiol	980-1680	
Morelos state	**Surface water**	4-nitrophenol	85.5	[21]
		Bisphenol A	88.8	
		Estradiol	103.6	
		ethinylestradiol	91.5	
Morelos state	**Surface water**	Acetaminophen	2400-4460	[22]
		Diclofenac	1100-1276	
		Ibuprofen	502-1106	
		Indomethacin	112-164	
		Naproxen	3000-4820	
		Salicylic acid	200-664	
		Sulfamethoxazole	76-222	
		Atenolol	12-16	
		Carbamazepine	52-276	
Guanajuato state	**WWTP (influent, dry season)**	Atenolol	277	[23]
		Atorvastatin	18.7	
		Enalapril	149	
		Cotinine	1580	
		Metformin	94,600	
		Ranitidine	2720	
		Ibuprofen	1800	
		Naproxen	12,800	
		Triclosan	926	
		Paracetamol	66,000	
		Caffeine	31,100	
		Carbamazepine	167	
		Sulfamethoxazole	1100	
		Valsartan	1620	
		Androstenedione	390	
		Androsterone	750	
		Mestranol	741	
		Estrone	39.9	
		17 b estradiol	20	

Table 2.
Emergent compounds concentrations detected in surface and wastewater in different states of the Mexican Republic.

since the presence of different ECs has been detected in the wastewater and in sludge at high concentrations of up to thousands of µg/L or µg/kg [26].

In this context, contamination of water with EC represents a technical problem for its treatment and purification, since conventional treatments: aerobic biological, anaerobic, coagulation-flocculation, inverse osmosis filtration and disinfection with chlorine are not enough to completely eliminate or degrade this type of compounds [27, 28].

For this reason, the latest technological developments have focused on advanced oxidation processes (AOP), which focus on the generation of hydroxyl radicals (° OH), which have a greater oxidation potential than ozone or chlorine. The interactions with the compounds of interest are controlled mainly by diffusion and eventually result in the fragmentation of organic compounds and mineralization to CO2 [29].

In order to provide these radicals, several processes have been implemented that are based on the application of electrical energy (electrochemical oxidation), radiation (UV), ultrasound (US), chemical additives (O_3, H_2O_2) photo-fenton (Fe^{2+}/UV/H_2O_2) or a combination of these methods (**Table 3**). A consequence of the high reactivity of the oxidizing agent (° OH) is its low selectivity; which is a desirable feature in the case of wastewater pollutant removal.

Homogeneous processes	
(a) No external energy input	
	* Ozonolysis in alkaline medium (O_3/^-OH)
	* Ozonolysis with hydrogen peroxide (O_3/H_2O_2) and (O_3/H_2O_2/^-OH)
	* Hydrogen peroxide and catalyst
(b) With external energy input	
(b1) Energy from UV radiation	
	* Ozonolysis and UV radiation (O_3/UV)
	* Hydrogen peroxide and UV radiation (H_2O_2/UV)
	* Ozone, hydrogen peroxide and UV radiation (O_3/H_2O_2/UV)
	* Photo-fenton (Fe^{2+}/H_2O_2/UV)
(b2) Energy from ultrasound (US)	
	* Ozonolysis and US (O_3/US)
	* Hydrogen peroxide and US (H_2O_2/US)
(b3) Electrochemistry	
	* Electrochemical oxidation
	* Anodic oxidation
	* Electro-fenton
Heterogeneous processes	
	* Catalytic ozonolysis (O_3/TiO_2)
	* Photocatalytic ozonolysis (O_3/TiO_2/UV)
	* Heterogeneous photocatalysis (H_2O_2/TiO_2/UV)

Table 3.
Summary of the main AOPs used for the degradation of organic compounds.

On the other hand, these are processes that use expensive reagents such as hydrogen peroxide or ozone, so its use should be restricted to situations in which other processes cheaper, such as biological, are not possible. Their full potential is exploited when they are achieved integrate with other treatments, such as adsorption or biological treatments, in order to achieve the maximum oxidant economy.

Faced with this panorama, the challenge for Mexico is great if we consider that in the country there are 2540 wastewater treatment plants, of which 3.2% apply primary treatment, 96% secondary treatment and only 0.12% apply tertiary treatment (**Table 4**).

In recent years, various government agencies responsible for carrying out research in the management processes and development of water remediation strategies have led to some studies aimed at the application of advanced technologies for the removal and/or degradation of organic compounds in wastewater.

Type of treatment	Process	Number of plants	Treated flow (m^3/s)
Primary	Primary	13	0.035
	Advanced primary	10	4.431
	Imhoff tank	58	0.326
Secondary	Aerobic	20	1.849
	Anaerobe	100	0.625
	Biodiscs	30	0.872
	Biological	30	0.737
	Dual	24	27.402
	Biological filters	39	5.13
	Septic tank	100	0.142
	Septic tank + biological filter	40	0.044
	Septic tank + wetland	115	0.207
	Wetlands	74	1.249
	Aerated lagoons	29	7.024
	Stabilization lagoons	774	13.739
	Activated sludge	725	70.239
	UAR + biological filter	62	0.577
	UAR + wetland	34	0.331
	Upflow Anaerobic Reactor (UAR)	133	1.175
	Enzymatic reactor	44	0.097
	Sedimentation + wetland	21	0.04
	Imhoff tank + biological filter	26	0.181
	Imhoff tank + wetland	6	0.017
	Oxidation trenches	13	0.985
Tertiary	Tertiary	3	0.044
Not specified	Others	17	0.203
	Total	**2540**	**137.701**

Table 4.
Main municipal wastewater treatment processes (source CONAGUA [2]).

For example, since 2014 the Mexican Institute of Water Technology (IMTA, for its acronym in Spanish), has been developing different technologies for the removal of EC. Within these developments, they used biofiltration systems for biodegradation of two drugs, metformin and ciprofloxacin. Obtaining biodegradation efficiencies of 83 and 71% respectively, during 103 days of operation [30]. Likewise, in another study carried out in two wastewater treatment plants located in the states of Guanajuato and Mexico, they used a system integrated by oxidation ditches and UV light lamps, obtaining EC elimination efficiencies between 20% and 22% % (Guanajuato). Likewise, while in the other treatment that consisted of anaerobic / anoxic/aerobic tanks together with two disinfection processes; chlorine dioxide and ultraviolet lamps, the removal of EC was significant (up to 80%) (Mexico) [23].

Also used a submerged membrane bioreactor for the degradation of compounds estrone, estradiol and 17α-ethinylestradiol, obtaining removals close to 96% for all compounds [31]. Meanwhile Flores and Mijaylova 2017, evaluated the removal of three pharmaceutical micropollutants (fluoxetine, mefenamic acid and metoprolol) from municipal wastewater, by using four aerated submerged attached growth bioreactors, with removal efficiencies of 95, 82 and 73% for fluoxetine, mefenamic acid and metoprolol, respectively [32]. In another study conducted by García-Espinosa et al. 2018, obtained degradation percentages of Carbamazepine in wastewater of 88.7%, using an electrochemical oxidation process [33].

3. Conclusions

The main challenge facing Mexico for the comprehensive management of water resources, to do with current legislation has some structural deficiencies, for example, the sanitation process is not defined within the water legislation, as well as to institutional fragmentation. On the other hand, it must be considered that decision-making is strongly influenced by political interests and social pressure, which makes it difficult to align common goals in public health and environmental protection between local authorities and different sectors of society. It is also important to note that many official guidelines for water management are generally prepared by new presidential administrations every six years, which prevents the continuity of plans and programs, which causes waste of economic resources, which accelerates the deterioration of water and sanitation services.

As can be seen in scientific reports and publications, in Mexico there is little information on the real level of concentration levels of emerging compounds, which is worrisome considering that there are currently no laws that regulate said compounds in bodies of Water. Some of the studies carried out reveal alarming concentrations of some compounds. The foregoing suggests the implementation of intensive programs in the areas with the highest population, and regions with high industrial and agricultural activity; however, access to this type of methodologies requires highly qualified personnel, as well as high investments in the acquisition of supplies and equipment.

Finally, the number of waste treatment plants is insufficient; in addition, the vast majority are concentrated in primary and secondary treatments, and only 0.12% apply tertiary treatments. Although some advanced methods have been implemented for the removal of organic compounds, some of them with high efficiencies, which is encouraging, however these technologies continue to be expensive, which suggests the participation of government and private companies to support projects, that yields mutual benefits for both parties; that is to say, environmental, social and economic.

Acknowledgements

The authors wish to thank PRODEP (Program for the Development of Teachers), for the support in the financing of this publication. Likewise, they also wish to National Council for Science and Technology (CONACYT) for the grant awarded to José Gustavo Ronderos Lara.

Conflict of interest

The authors declare no conflict of interest.

Author details

José Gustavo Ronderos-Lara[1], Hugo Saldarriaga-Noreña[1]*, Pedro Guillermo Reyes-Romero[2], Luis Alberto Chávez-Almazán[3], Josefina Vergara-Sánchez[4], Mario Alfonso Murillo-Tovar[5] and César Torres-Segundo[4]

1 Chemical Research Center, Autonomous University of Morelos State, Cuernavaca, Mexico

2 Advanced Physics Laboratory, Faculty of Sciences, Autonomous University of Mexico State, Toluca, Mexico

3 Guerrero Health Department, Central Zone Blood Bank, Guerrero, Mexico

4 Laboratory of Analysis and Environmental Sustainability, School of Higher Studies of Xalostoc, Autonomous University of Morelos State, Ayala, Mexico

5 National Council of Science and Technology, Cuernavaca, Mexico

*Address all correspondence to: hsaldarriaga@uaem.mx

References

[1] UNESCO. Agua para todos, agua para la vida, s.l.: Ediciones UNESCO y Mundi-Prensa. 2003.

[2] CONAGUA. Estadísticas del Agua en México, s.l.: s.n. 2015.

[3] Cherfi A, Achour M, Cherfic M, Otmani S. and Morsli A. Health risk assessment of heavy metals through consumption of vegetables irrigated with reclaimed urban wastewater in Algeria. Process Saf. Environ. Protec. 2015; 98: 245-252.

[4] Barceló D, and. L. M. Contaminación y calidad química del agua: el problema de los contaminantes emergentes. Panel científico-técnico de seguimiento de la política de aguas. 2007.

[5] Moyá-Llamas M. Tratamientos combinados para la degradación y eliminación de compuestos emergentes en aguas. 2ª Reunión de la red Traguanet. 2016b.

[6] Daughton C G. "PPCPs in the Environment: Future Research — Beginning with the End Always in Mind," In Pharmaceuticals in the Environment, Kümmerer K (Ed.), 2nd edition, Springer, 2004, Chapter 33, pp. 463-495.

[7] Dana W, Kolpin Edward T, Furlong Michael T, Meyer E, Michael Thurman Steven D, Zaugg Larry B, Barber and Herbert T. Buxton Pharmaceuticals, Hormones, and Other Organic Wastewater Contaminants in U.S. Streams, 1999–2000: A National Reconnaissance. Environmental Science & Technology 2002; 36 (6): 1202-1211.

[8] Kümmerer K. Emerging contaminants. Treatise on Water Science, 2011; 3: 69-87.

[9] Watanabe N, Bergamaschi B.A, Loftin K.A, Meyer M.T, and Harter T. Use and environmental occurrence of antibiotics in freestall dairy farms with manured forage fields. Environ. Sci. Technol. 2010; 44: 6591– 6600.

[10] Christopher H Swartz, Sharanya Reddy, Mark J. Benotti, Haifei Yin, Larry B. Barber, Bruce J. Brownawell, and Ruthann A. Rudel. Environmental Science & Technology. 2006; 40 (16): 4894-4902.

[11] Trine Eggen, Monika Moeder, Augustine Arukwe. Municipal landfill leachates: A significant source for new and emerging pollutants, Science of The Total Environment. 2010; 408 (21): 5147-5157.

[12] Giger W, Brunner P.H. & Schaffner, C. 4-Nonylphenol in sewage sludge: accumulation of toxic metabolites from non-ionic surfactants. Science. 1984; 225: 623–625.

[13] Agarwal A, de los Angeles M.S., Bhatia R., Chéret I., Davila-Poblete S., Falkenmark M., González-Villarreal, F., Jønch-Clausen T., Kadi M.A., Kindler J. et al. Integrated Water Resources Management. Global Water Partnership (GWP); Tack Background Papers No. 4; Global Water Partnership Technical Advisory Committee (TAC): Stockholm, Sweden. 2000; p. 71. ISBN 91-630-9229-8.

[14] OECD. Getting It Right Una Agenda Estratégica Para las Reformas en México (An Strategic Agenda for the Mexican Reforms); OECD Publishing: Mexico City, Mexico, 2013; p. 312. ISBN 978-92-64-19036-8.

[15] USEPA 2014. Contaminants of Emerging Concern including Pharmaceuticals and Personal Care Products.https://www.epa.gov/wqc/contaminants-emerging-concern-including-pharmaceuticals-and-personal-care-products. (Accessed Aug 2020).

[16] Erickson M.L., Langer S.K., Roth J. L., and Kroening S.E. Contaminants of emerging concern in ambient groundwater in urbanized areas of Minnesota. 2009–12 U.S. Geological Survey Scientific Investigations Report 2014–5096, 38 p., with appendix, http://dx.doi.org/10.3133/sir20145096.

[17] Agüera Ana,AU - Martínez Bueno, María Jesús AU - Fernández-Alba, Amadeo R. PY. New trends in the analytical determination of emerging contaminants and their transformation products in environmental waters. Environmental Science and Pollution Research. 2013; 20 (6): 3496-3515.

[18] K. Mogolodi Dimpe, Philiswa N. Nomngongo. Current sample preparation methodologies for analysis of emerging pollutants in different environmental matrices, TrAC Trends in Analytical Chemistry. 2016; 82: 199-207.

[19] Gibson R., Becerril-Bravo E., Silva-Castro V., and Jiménez B. Determination of acidic pharmaceuticals and potential endocrine disrupting compounds in wastewaters and spring waters by selective elution and analysis by gas chromatography-mass spectrometry. Journal of Chromatography A. 2007; 1169 (1-2): 31-39.

[20] Díaz-Torres E., Gibson R., González-Farías F. et al. Endocrine Disruptors in the Xochimilco Wetland, Mexico City. Water Air Soil Pollut. 2013: 224, 1586.

[21] Ronderos-Lara J.G., Saldarriaga-Noreña H. Murillo-Tovar M.A., Vergara-Sánchez J. Optimization and Application of a GC-MS Method for the Determination of Endocrine Disruptor Compounds in Natural Water. *Separations.* 2018; (*5*), 33.

[22] José Abraham Rivera-Jaimes, Cristina Postigo, Rosa María Melgoza-Alemán, Jaume Aceña, Damia Barceló, Miren López de Alda, Study of pharmaceuticals in surface and wastewater from Cuernavaca, Morelos, Mexico: Occurrence and environmental risk assessment, Science of The Total Environment. 2018; (613–614): 1263-1274.

[23] Edson Baltazar Estrada-Arriaga, Juana Enriqueta Cortés-Muñoz, Arturo González-Herrera, César Guillermo Calderón-Mólgora, Ma. de Lourdes Rivera-Huerta, Esperanza Ramírez-Camperos, Leticia Montellano-Palacios, Silvia Lucila Gelover-Santiago, Sara Pérez-Castrejón, Lina Cardoso-Vigueros, Alejandra Martín-Domínguez, Liliana García-Sánchez, Assessment of full-scale biological nutrient removal systems upgraded with physico-chemical processes for the removal of emerging pollutants present in wastewaters from Mexico, Science of The Total Environment. 2016; (571): 1172-1182.

[24] Corcoran E. (Editor); Nellemann C. (Editor); Baker E. (Editor); Bos R. (Editor); Osborn D. (Editor); Savelli H. (Editor) (2010): Sick Water? The centra l role of wastewater management in sustainable development. A Rapid Response Assessment. United Nations Environment Programme (UNEP), UN-HABITAT, GRID-Arendal. URL [Accessed August 2020].

[25] M. Bilal, M. Asgher, R. ParraSaldivar, H. Hu, W. Wang, X. Zhang, H.M.N. IqbalImmobilized ligninolytic enzymes: an innovative and environmental responsive technology to tackle dye-based industrial pollutants–a review Sci. Total Environ. 2017; (576): 646-659.

[26] Klaus Kümmerer. Antibiotics in the aquatic environment – A review – Part I, Chemosphere. 2009; (75): 417-434.

[27] Teodora Basile, Andrea Petrella, Mario Petrella, Giancarlo Boghetich,

Valentina Petruzzelli, Sara Colasuonno, and Domenico Petruzzelli. Review of Endocrine-Disrupting-Compound Removal Technologies in Water and Wastewater Treatment Plants: An EU Perspective. Industrial & Engineering Chemistry Research. 2011; 50 (14): 8389-8401.

[28] Mark J. Benotti, Rebecca A. Trenholm, Brett J. Vanderford, Janie C. Holady, Benjamin D. Stanford, and Shane A. Snyder. Environmental Science & Technology. 2009; 43 (3): 597-603.

[29] Giri R.R., Ozaki H., Ota S. et al. Degradation of common pharmaceuticals and personal care products in mixed solutions by advanced oxidation techniques. Int. J. Environ. Sci. Technol. 2010; (7): 251–260.

[30] Estrada-Arriaga, E. Removal of emerging compounds through the biofiltration. (2014). Mexican Institute of Water Technology (IMTA). (Consulted August 2020).

[31] Estrada-Arriaga Edson Baltazar, Mijaylova-Nacheva Petia, Moeller-Chavez Gabriela, Mantilla-Morales Gabriela, Ramírez-Salinas Norma y Sánchez-Zarza Manuel. Presencia y tratamiento de compuestos disruptores endócrinos en aguas residuales. Ingeniería Investigación y Tecnología. 2013; (2) , abril-junio: 275-284.

[32] Y. Flores Velázquez, P. Mijaylova Nacheva. Removal of pharmaceuticals from municipal wastewater by aerated submerged attached growth reactors, Journal of Environmental Management. 2017; (192): 243-253.

[33] Josué Daniel García-Espinoza, Petia Mijaylova-Nacheva, Martha Avilés-Flores. Electrochemical carbamazepine degradation: Effect of the generated active chlorine, transformation pathways and toxicity, Chemosphere. 2018; (192): 142-151.

Chapter 3

Various Methods for Removal, Treatment, and Detection of Emerging Water Contaminants

Wan Norfazilah Wan Ismail and Siti Umairah Mokhtar

Abstract

This review covers various methods to remove, treat, and detect emerging contaminants (ECs) in water and wastewater. ECs have drawn the attention of many countries due to their potential threat to human health as well as the environment. They are found in many human everyday products that are continuously released into the environment and will accumulate over time. In order to remove ECs, a number of methods have been developed, which include adsorption, membrane technology, biological treatment, and advanced oxidation process. In addition, advances in detection techniques and instrumentation are now able to detect ECs in which they occur at low concentrations. All the removal, treatment, and detection methods will be covered in this review. The removal, treatment, and detection of ECs and their transformation products in water and wastewater are challenging tasks due to their complexity in water samples. Therefore, such information should be emphasized in order to improve the current methods and develop new advanced methods.

Keywords: emerging contaminants, adsorption, membrane technology, biological treatment, advanced oxidation process, chromatography, spectroscopy

1. Introduction

In recent years, emerging contaminants (ECs) have been of great concern to environmentalists and governmental agencies due to their potential threat to human health as well as the environment. Three main sources of ECs have been identified, which include daily products used by humans, hormones, or medicines used by livestock and pesticides or nanomaterials (to improve the uptake of nutrient) used on plants [1]. They enter the environment in various ways. For example, the daily use products produced by humans will enter the wastewater, which will then be treated in conventional wastewater treatment plant (WWTP) without any ECs removal facility. Treated wastewater and wastewater sludge that still contain ECs will be released back into water sources and used as fertilizer on soil, respectively. ECs from livestock will go into their manure, while ECs on plants will be directly applied on soil from which they are eventually leached out by rain water thereby going all the way into nearby water sources. **Figure 1** summarizes the release of ECs into water from three main sources.

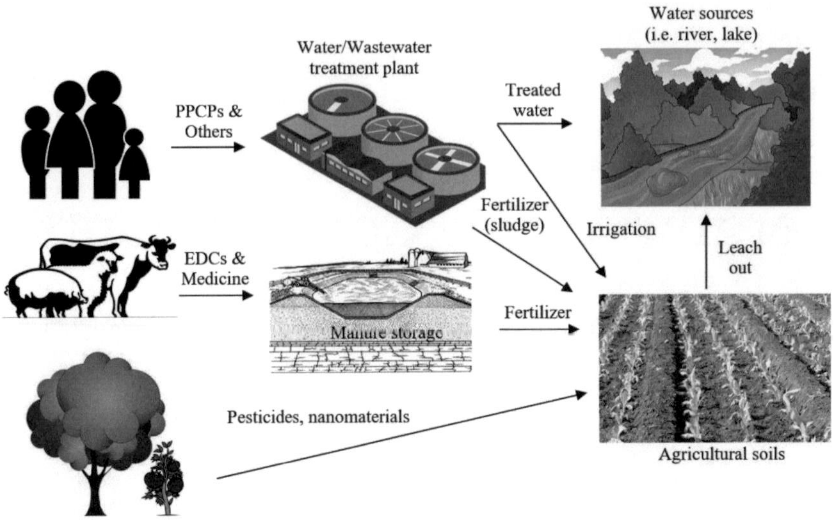

Figure 1.
Sources of ECs and their release to the water sources.

Water consumption in particular potable water increases every year due to population growth, urbanization, industrial development as well as changes in agricultural and land use practices [2]. One of the most serious problems faced by billions of people today is the availability of fresh water. Governments and organizations all over the world have realized that sustainable water and wastewater management are necessary components of functioning communities. However, current situation has shown that different classes of ECs have been identified in waters with low elimination rates [3]. Furthermore, for most of the ECs, it is difficult to predict their circumstances in the aquatic environment due to insufficient data on their occurrence, risk assessment, and ecotoxicological [4]. The reason behind this is partially due to the limitation of analytical methods for low concentrations measurement of ECs (usually part per billion or part per trillion levels), diversity of their chemical properties, and the complexity of matrices [5].

In order to improve the current analytical methods by removing or treating ECs in water and wastewater, a number of methods have been developed and reviewed in this paper, which included adsorption, membrane technology, biological treatment, and advanced oxidation process. Such information should be taken into consideration in order to improve the current methods or develop new advanced methods.

2. Emerging contaminants (ECs)

According to US Geological Survey (2015), ECs are defined as "any synthetic or naturally occurring chemicals or microorganisms that are not commonly monitored in the environment and cause known or suspected adverse ecological and/or human health effects" [6]. Three kinds of target can be generally defined as "ECs" in scientific research, namely (i) new compounds and molecules that recently appeared, (ii) existed contaminants with undiscovered environmental issues or emerging interests, and (iii) new information on their environmental risks that challenges the understanding of some legacy contaminants [7].

Obviously, ECs are found in many of our everyday products, such as human and veterinary pharmaceuticals, personal care products, surfactants and residues of surfactants, plasticizers, and different industrial additives that are continuously released into the environment and will accumulate over time. Although they bring great convenience and benefits, humans face the risks resulting from their growing applications. Potential carcinogenicity, reproductive dysfunction, and endocrine disrupting influences are some of the major adverse effects on human health associated with ECs. In the 1990s, research started on the hormonal disruption of fish in surface waters. Jobling et al. were the first to document widespread sexual endocrine disruption in wild fish—consistent with exposure to hormonally active substances in the surface waters of Great Britain [8]. Hormones found in surface water mostly came from sewage water and were not removed by treatment [9].

Beyond the health concerns, these substances that are entering the environment inevitably are not only sourced from daily use but also from the field of healthcare, industry, transport, agriculture, and so on [10]. Most of these emerging contaminants are released into the environmental waters mainly due to their incomplete removal in the sewage effluent via the WWTP. The determined concentrations of most emerging contaminants in surface water generally range from under ng/L to thousands of μg/L level [11]. Such low concentrations may sometimes lead to underestimation of their environmental impacts. However, it has been reported that some of these organic compounds showed strong persistence after being released into the environmental waters. Due to the lack of risk assessment for ECs, this is not included in policy and monitoring programs [12]. However, the absence of standard does not imply that a substance is safe.

There are many ECs that act as endocrine disrupting chemicals (EDCs) [13–15] which can disrupt animal reproduction and development at low level exposure by mimicking the naturally occurring hormones that bind to a receptor and block the endogenous hormone, or interfere with the normal hormonal function [16]. EDCs are defined by the World Health Organization (WHO) as chemicals that may alter the reproductive function in males and females, increase incidence of breast cancer, abnormal growth patterns, and neurodevelopmental delays in children as well as changes in immune function [17]. EDCs include natural hormones created in the body or synthetic hormones as well as industrial/commercial compounds which can have some hormonal functions, such as pharmaceuticals and personal care products (PPCPs) [18–20], pesticides [21–23], plasticizers [24, 25], nanoparticles [26, 27], and so on.

Natural estrogen hormones such as estrone, estradiol, and estriol are widely found in wastewater streams that mainly derive from human urines [28]. Human excretes estrogens from the body, even without taking hormonal drugs. As a result, natural hormones are believed to be present in a wide range of concentrations in wastewater from households, which are conveyed to the WWTP through the sewer system. Consequently, the aquatic species in downstream waters will pose elevated dosages of estrogenic compounds. These compounds can cause feminism at certain dosage to the exposed fish [29]. Länge et al. reported that a low concentration of 4 ng/L of ethinylestradiol can block the development of secondary sexual characteristics for male fathead minnows [30].

Pharmaceutically active compounds, namely prescription drugs, over-the-counter therapeutics, and veterinary drugs as well as personal care products, such as microbial disinfectants, represent the major portion of the domestically produced EDCs [31]. The health risks of these contaminants are a real concern in preserving a healthy ecosystem for aquatic life creatures and for water reuse purposes [32]. Even though the concentrations of pharmaceuticals in the aquatic environment are generally reported to be low, these compounds possess a high biological activity so often associated with high stability and their potential impact on the aquatic wildlife even at trace levels [33]. Bisphenol-A (BPA) is one of the most common EDCs found in every house, a monomer for the production of polycarbonate and epoxy resins found in some plastics that has been linked to heart disease, infertility, and behavioral and developmental problems in children exposed in utero [34].

Furthermore, pesticides are a class of compounds that may produce a wide range of potentially hazardous toxic side effects to the environment despite their benefits. Their extensive past or present use contributes to their prevalence as environmental contaminants in groundwater and surface water [35]. Reemtsma et al. studied the occurrence of 150 pesticides and their metabolites in 58 groundwater and surface water samples. The results revealed that 17 of the 27 highly ranked metabolites were denoted as emerging metabolites. Therefore, they concluded that metabolites could be found more frequently in groundwater than their parent compounds [36]. Besides the use of these compounds as herbicides or pesticides, some synthetic steroid hormones are used as growth promoters in beef cattle. It had shown that the soil and runoff from large feedlots contain a large amount of bioactive steroids that may affect wildlife and the environment around these cattle feeding operations [37]. Consequently, although they are in low concentrations, the elimination of these trace contaminants from entering the water resources is required.

3. Removal or treatment method of ECs

The traditional wastewater treatment process such as activated sludge is designed to remove solids, organic loadings, and pathogens. Although several treatment steps in a wastewater treatment plant can contribute to the partial removal of ECs, complete removal has been proven to be a challenge due to high variety, extreme low concentration, and unique characteristics of ECs [38]. Due to the increased number of trace ECs detected in water and wastewater streams and the fact that EDCs have adverse effects on the human endocrine system, investigation on the removal or treatment of these micro contaminants grabs considerable attention of many researchers worldwide [39].

Adsorption technology, membrane technology, biological treatment, and advanced oxidation method are among various remediation options considered to be effective removal or treatment methods for ECs from wastewater or water sources. The removal and treatment of ECs and their transformation products in water and wastewater are challenging tasks due to their complexity in water samples. Therefore, information on the mentioned treatment methods should be taken into consideration in order to improve the current methods and develop new advanced methods.

3.1 Adsorption technology

Adsorption is a process which involves the mass transfer of substances between two phases, namely liquid-liquid, liquid-solid, gas-liquid, or gas-solid interface [40]. Adsorbents are used to adsorb any particular pollutant (adsorbate) from wastewater with the help of intermolecular forces [41]. There are two types of interaction between the solid surface and adsorbates, namely physisorption and chemisorption. The process is called physisorption if the interaction has a weak physical nature such as van der Waals forces and the process results are reversible [42]. Additionally, it occurs at lower or close to the critical temperature of the adsorbate. Contrary to physisorption, chemisorption involves the chemical bonding between solid surface and adsorbates. In contrast, it occurs only as a monolayer, and adsorbates are hardly removed because of the strong interaction. Both processes can occur simultaneously or alternatively, depending on the circumstances [43]. To study on adsorption method, one should consider the factors that affect the adsorption process, such as (i) surface area, (ii) nature and initial concentration of adsorbate, (iii) solution pH, (iv) temperature, (v) interfering substances, and (vi) nature and dose of adsorbent [44].

Adsorption is simple and effective, but it requires high operating costs associated with the generation of commercial adsorbents (i.e., activated carbons, minerals, and natural clays) [45]. Various researchers have attempted to prepare alternative adsorbents from agricultural and industrial wastes to replace the commercial adsorbents in order to reduce the costs as well as the environmental effects [46]. Therefore, this review presents the use of unconventional adsorbents, especially agriculture wastes instead of commercial adsorbents. Hemicellulose, starch, lipids, proteins, water, lignin, hydrocarbons, and simple sugars are the basic components of agricultural waste materials that contain a variety of functional groups [47]. Many previous studies have shown that agricultural materials containing cellulose have a high sorption capacity of various pollutants. Their applications as adsorbents can be extended through chemical treatment with numerous chemicals, such as to increase their chelating functional groups [41]. Generally, the adsorption takes place by complexation, ion exchange, and hydrogen bonding [48, 49]. Cheap and readily available resources of agricultural solid wastes such as rice husk and straw [50, 51], coconut husk and coir dust [52, 53], wood sawdust and chips [54, 55], and fruit peels and stones [56–59] have been investigated for the removal of pollutants from aqueous solutions.

Mandal et al. [50] studied on bamboo chips, corn cob, eucalyptus bark, rice husk, and rice straw as low-cost adsorbent for the removal of pesticides, namely atrazine and imidacloprid. The experimental data obtained were modeled with Freundlich and Elovich isotherm equations. The authors tested the effect of contact time, pH, pesticide concentration, and adsorption-desorption behavior of the adsorbents. The authors concluded that the kinetics study for both pesticides on rice straw adsorbent was well represented by the modified Elovich model, with a maximum adsorption capacity of 70.7% (atrazine) and 77.8% (imidacloprid). They also carried out further studies to evaluate the adsorption efficiency of rice straw adsorbent by treating it with phosphoric acid and found that it was able to further enhance the sorption of both pesticides. From the results obtained, the authors suggested that among the five unconventional adsorbents used, rice straw adsorbent has great potential for pesticide industrial waste water treatment.

The removal of heavy metals from wastewater using rice husk as adsorbent was carried out by Hegazi [51] who found that the low-cost adsorbent was able to effectively remove Fe, Pb, and Ni simultaneously within a concentration range of 20–60 mg/L. The author evaluated the effect of adsorbent amount on the removal of the selected heavy metals, and the results showed that the removal percentage increased with the increase of adsorbent amount. The maximum adsorption (76–96%) of heavy metals was achieved at room temperature with the optimum contact time of 2 h and pH range of 6.0–7.0. Another analyte that is removed by rice husk absorbent is tetracycline, a commonly used personal care and veterinary drug [60]. Moreover, Jing and co-authors have modified the rice husk by methanol to improve the adsorption capacity of tetracycline and reduce the inherent organic compound content in the rice husk. Comparison between the modified rice husks with untreated rice husk adsorbents showed approximately 45.6% enhancement of adsorption capacity in 12 h and 17.2% in equilibrium time. They reported that the main factor responsible for the enhancement of tetracycline adsorption was the change in O-containing groups in the modified adsorbent, which affects π-π electron-donor-acceptor interactions between the adsorbent and tetracycline.

The ability to remove dye using coconut husk and its coir dust as adsorbents was investigated by Bello et al. [53] and Etim et al. [52], respectively. The coconut coir dust is a remaining dust after the extraction of fiber from the coconut husk, which is about 70% of the coconut husk weight. Bello and colleagues used acid-activated coconut husk for the removal of rhodamine-B (Rh-B) dye, and the effects of different operational parameters were examined, namely initial concentration, contact time, and solution temperatures. Adsorption data were fitted to Langmuir, Freundlich, Dubinin-Radushkevich, and Temkin isotherm models. Langmuir isotherm was found to be the most fitted model with the maximum adsorption capacity of 1666.67 mg/g. The authors indicated that the coconut husk adsorbent has greater affinity for Rh-B dye adsorption due to the increase of pore development via acid activation. Meanwhile, Etim et al. used coconut coir dust to remove methylene blue dye, and the effects of adsorbent amount, pH, and concentration with time were studied. Three isotherm models were tested, namely Langmuir, Freundlich, and Temkin, and adsorption was found to fit well into these models with $R^2 \geq 0.90$. The authors suggested that the increase in adsorption capacity of the dye was due to the chemical interaction of the functional groups on the surface of the adsorbent according to the Fourier transform infrared (FTIR) results obtained.

Studies outlined by Thue et al. on wood chips and wood sawdust have revealed their potential as low-cost adsorbents for phenol [54] and metals [55] removal from aqueous solutions, respectively. Former study has modified the wood chips by mixing them with inorganic components before they were pyrolyzed and treated with hydrochloric acid to produce inorganic:organic ratios of 1.0 and 1.5 (ZnCW-1.0; ZnCW-1.5; FeZnCW-1.0; FeZnCW-1.5). The maximum amounts of phenol adsorbed onto activated carbons at 25°C were 434.2, 667.9, 256.5, and 233.5 mg/g for ZnCW-1.0, ZnCW-1.5, FeZnCW-1.0, and FeZnCW-1.5, respectively. The results indicated that ZnCW-1.0 and ZnCW-1.5 exhibited excellent performance in the treatment of simulated effluents contaminated with mixtures of phenols in a complex medium. On the other hand, the latter study focused on the use of first-row transition metals (Co, Ni, Cu, and Zn) in the preparation of activated carbons from wood biomass via microwave-assisted irradiation. Results showed that the metals were bound successfully in different amounts with surface functional groups of the wood biomass through ion exchange and surface complexation interaction during the impregnation step. An adsorption experiment revealed that samples prepared using $ZnCl_2$ showed the highest sorption capacities for the tested adsorbates, followed by $CuCl_2$, $CoCl_2$, and $NiCl_2$.

Three different fruit peels, such as orange, pomelo, and passion fruit peels have been used by Nhung et al. to develop unconventional adsorbents to remove Rh-B from aqueous solution [56]. Equilibrium isotherms were determined and analyzed using the Freundlich equation. Capacities of adsorbent were found to be in the order of passion fruit peels > orange peels > pomelo peels for Rh-B. Other than that, orange peel as an adsorbent has also been studied by Foo and Hameed [57] for the removal of methylene blue. The authors investigated the effects of initial dye concentration (50, 100, 200, 300, 400, and 500 mg/L) and pH (2–12) at room temperature with the contact time of 48 h and adsorbent dosage of 0.20 g. The adsorption capacity of methylene blue was found to be 382.75 mg/g at the initial pH 6.33. In addition, Ben-Ali et al. [58] studied the removal of copper ions onto untreated pomegranate peel in order to explore its potential use as low-cost adsorbents from an agricultural waste. Adsorption isotherms were determined at 313 K, and the experimental data obtained were modeled with the Langmuir, Freundlich, Dubinin-Radushkevich, and Temkin isotherm models. The authors concluded that the equilibrium data were well represented by the Langmuir isotherm equation with the maximum adsorption capacity of 30.12 mg/g.

Torrellas et al. [59] used activated carbons from peach stones as adsorbent for the removal of caffeine (stimulant), diclofenac (anti-inflammatory drug), and carbamazepine (psychiatric drug) from aqueous solution and reported that the adsorption capacity of carbamazepine was higher than caffeine and diclofenac, reaching to 335 mg/g. The modified adsorbents exhibited an increase in adsorption capacity for carbamazepine due to the existence of hydrophobic character and water solubility properties. Oxidation of the activated carbon had greatly enhanced the hydrophilic character of the material, thereby decreased the adsorption capacity and highly affected the breakthrough times and adsorption capacity values in the fixed-bed adsorption process. The same adsorbent was used to remove dye (methylene blue) [61] and metal (platinum) [62].

The economical and readily available adsorbent would certainly make an adsorption-based process, a viable alternative for the treatment of wastewater that contains pollutants. Depending upon the adsorbent and adsorbate characteristics, the selection of an appropriate adsorbent is the most crucial part to ensure a maximum removal of different types of pollutant. Various environmental conditions and variables are used to ensure the effectiveness of adsorption process, such as initial adsorbate and adsorbent concentration, particle size of adsorbent, temperature, pH, selectivity, ionic strength, contact time, and the rate of rotation. A summary of various agricultural wastes as sources of adsorbents for the removal of various types of ECs is summarized in **Table 1**.

3.2 Membrane technology

Membrane technology is one of the promising technologies for the excellent removal of micropollutants in water. This technology uses both biological (membrane bioreactors) and nonbiological processes (reversed osmosis, ultrafiltration, and nanofiltration). Membrane bioreactors (MBRs) are the combination of membrane-based filtration processes, such as microfiltration (MF) or ultrafiltration (UF) system with suspended growth biological reactors. MBRs are the most prominent and proven processes these days to achieve a relatively clean water from wastewater through the combination of membrane and biological treatments [63, 64].

Agricultural waste	Emerging contaminants	References
Bamboo chips, corn cob, eucalyptus bark, rice husk, and rice straw	Pesticides (atrazine and imidacloprid)	[50]
Rice husk	Metal ions (Fe, Pb and Ni)	[51]
Coconut coir dust	Dye (methylene blue)	[52]
Coconut husk	Dye (Rhodamine-B)	[53]
Wood chips	Phenol	[54]
Wood sawdust	Metal ions (Co, Ni, Cu, and Zn)	[55]
Orange, pomelo, and passion fruit peels	Dye (Rhodamine-B)	[56]
Orange peel	Dye (methylene blue)	[57]
Pomegranate peel	Metal ion (Copper)	[58]
Peach stones	Stimulant (caffeine), anti-inflammatory drug (diclofenac) and psychiatric drug (carbamazepine)	[59]
Rice husk	Personal care and veterinary drug (tetracycline)	[60]
Peach stones	Dye (methylene blue)	[61]
Peach stones	Metal ion (Platinum)	[62]

Table 1.
Summary of various agricultural wastes as sources of adsorbents for the removal of diverse types of ECs.

Nonbiological processes or pressure-driven membrane technology known as reverse osmosis (RO), nanofiltration (NF), MF, or UF utilizes high pressures across the membranes to accomplish the filtration of contaminants from the produced water [65]. These technologies are the most common membrane techniques of water purification. The membranes are also continuously upgraded or modified for further improvement in performance and usage. Consequently, membrane processes are ideal for removing turbidity and microbiological contaminants. However, high-operation costs still limit its full-scale usage. Membranes easily suffer from fouling problems that could result in unexpected interruptions during the treatment of aqueous contaminants [66].

In Keucken et al. report published in 2017, it proved that using UF can achieve a stable operation even in difficult circumstances and excellent effluent water quality can be achieved as well [67]. Cui and Choo concluded that although fouling cannot be prevented since there must be natural organic matter (NOM) existed in water, it can still be minimized [68]. Potential of hydrogen (pH), ionic strength, and calcium ion concentration are parameters that need to be considered to affect membrane filtration process. Sun et al. conducted experiments to measure the fouling potential when the membrane system is maintained at different values of pH, ionic strength, and calcium ion concentration. They discovered that the fouling potential increased with the increase of the feed solution acidity, ionic strength, and calcium ion concentration [69]. Martin et al. and Bu et al. conducted UF experiments with powdered activated carbon (PAC) adsorption and coagulation as a pretreatment to improve removal efficiency, which provided good examples [70, 71]. The advantage of combining UF with PAC is mainly in terms of the adsorption capacity of PAC and retaining ability of membrane.

In a pilot scale membrane treatment system, Wang et al. compared the removal efficiency of MBR system with integrated membrane systems (MBR/RO or MBR-NF) toward 27 PPCPs. The results suggested that the integrated membrane systems can achieve even higher removal rates of above 95% for most of them. Further studies showed that MBR-RO achieved even better results with the removal of 20 compounds to below detection limits as compared to 13 compounds by MBR/NF [72]. A study conducted by Baransi-Karkaby et al. reported the improved rejection of multiple EDCs and pharmaceutically active compounds (PhACs) by RO membrane elements using concentration polarization- and surfactant-enhanced surface polymerization [73]. The researchers performed experiments using RO membranes modified by grafting poly(glycidyl methacrylate) and as compared to commercial brackish water RO membranes. One of the conclusions was that the modified membrane generally exhibited enhanced rejection comparable with the commercial membrane. However, both membranes did not exhibit a complete rejection for the tested compounds.

3.3 Biological treatment

Many studies have shown that the major removal of ECs is by biodegradation and adsorption during biological treatment (secondary treatment). Therefore, this section will focus on the circumstances and behavior of ECs during the biological treatment processes, which usually consist of two main kinds of treatment, namely activated sludge process (ASP) and trickling filter (TF). The presence of aerobic bacteria and other microorganisms is one of the features of this process in order to oxidize or incorporate into cells of organic matter. The bacteria need a sufficient amount of oxygen in order for the treatment to occur [74]. TF contains fixed surface with a large population of microorganisms, while ASP is where large population of microorganisms combine with wastewater. Normally, after going through TF or ASP, the dense microbial biomass is separated from water by secondary sedimentation [75].

The most common and widely used ASP is to treat both industrial and domestic wastewater. The general process of activated sludge system involves the injection of air into the reactor and continuous re-circulation of biomass into the aeration tank. There are three main divisions in an activated sludge system [76], namely (i) reactor (to keep the microorganisms in suspension and aerators that are used for treatment purposes), (ii) separator system or (allow solids to sediment in sedimentation tank), and (iii) recycle system (solids from sedimentation tank are moved to the aeration tank) (**Figure 2**).

Depending on the type of ASP, there are many benefits as well as drawbacks from using it. Reduced level of ammonia, small space usage, and odor free are a few benefits of some types of ASP. However, a high energy is required to operate aeration tank, and the changes of the effluent characteristics are very rigid [77]. Some examples of factors that affect ASP are the availability of oxygen, temperature, features of the treated wastewater, detergents that produce foam, and return rate [78]. The performance of ASP to remove steroid estrogens in WWTPs was better (up to 100%) than that of the trickling filter (up to 75%) [79]. Han et al. investigated the total removal of organic pollutants in eight full-scale municipal WWTPs by three modified ASP, namely anoxic/oxic (A/O), anaerobic/anoxic/oxic (A2/O), and oxidation ditch process [80]. From the results obtained, it was concluded that the oxidation ditch process appeared to have the advantage of removing tyrosine-like substances and presented a slightly better removal efficiency of tryptophan-like fluorescent substances than the A/O and A2/O processes.

On the other hand, microbial biomass grows on an inert or solid medium in film form with continuous wastewater being sprinkled over them, known as ultrafiltration (fixed film or attached growth systems) [81]. Nowadays, the treatment capacity of the conventional TF has been enhanced with the use of plastic packing as the packing material to replace circular rock. A lot of benefits have been highlighted on this type of biological treatment, including better sludge thickening, low maintenance cost, less energy consumption, and ease of operation. However, the TF system has poor effluent quality as compared to suspended growth systems [82]. The effluent quality of TF is based on the biochemical oxygen demand (BOD) and suspended solids (SS). Temperature, retention time, type of media and its depth, and hydraulic and organic matter are some parameters that might affect the TF performance [83] which lead to a relative removal efficiency difference of some

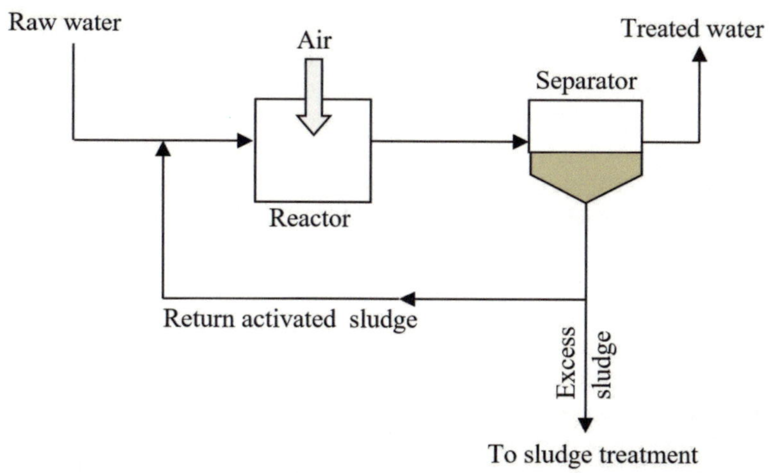

Figure 2.
Diagram of activated sludge process.

ECs, like estrogens in TF [28]. The removal efficiency of 18 different ECs from 42 WWTPs in five different regions of India was studied using different treatment solutions, including TF by Visanji et al. [84]. The results of using TF showed that this treatment is able to remove ofloxacin, dichlofenac, and ibuprofen but unable to remove ampicillin, ciprofloxacin, and naproxen. The authors concluded that although ECs classed under the same category have similar physical, chemical, and biological parameters, these contaminants are not always removed in the same way.

3.4 Advanced oxidation method

Advanced oxidation method (i.e., ozone, O_3/H_2O_2, H_2O_2/UV, Fenton, ultrasound, photocatalytic, and electrochemical oxidation) can achieve effective degradation of EDCs. Highly reactive oxidant species, such as hydroxyl radical, ion superoxide, hydroperoxyl radical, and organic peroxide radical can highly react with a wide range of organic compounds in a nonselective oxidation way [85]. The target compounds could be quickly and efficiently converted into small inorganic molecules, such as CO_2 and H_2O. Among the advanced oxidation methods used to remove or reduce the concentrations of some hormones and PPCPs from wastewater, ozone oxidation is the most promising technique [86–92]. In secondary effluents, ozonation was found to be very useful in decreasing the concentrations of bisphenol-A, estriol, and 17-β estradiol and their estrogenicity [92, 93]. In addition, the high removal efficiency of estrogens has been observed at low pH by Cesaro and Belgiorno [93]; however, it was impractical to change the sewage pH.

A study by Sun et al. [94] reported that ozone could completely oxidize three estrogens, namely estrone, 17-β estradiol, and 17-α ethinylestradiol in wastewater samples at the dose of 0.3 mg/L. Similar results were achieved by Si et al. [95], showing that combined process of 5.5 mg/L ozone and UF could completely remove the estrogens, with ultraviolet absorbance reduction enhanced from 11–18% or 24–26% as compared to the UF or ozonation alone. Another study by Li et al. [96] found that the application of 1.0–5.0 mg/L ozone was able to decrease the concentration of antibiotics and EDCs to below the limit of detection (LOD). However, other compounds like nitrite and carbonyl compounds as by-products were still detected with a similar level of ozone dosage. In order to remove the by-products, the authors have combined the ozone oxidation process with biological aerated filtration.

The ozone dose applied was found to affect the removal efficiency of some PPCPs. Apart from ozone dose, the reaction time was also found to be one of the main factors to remove ECs, as can be seen in the study conducted by Paucar et al. to remove 38 PPCPs [97]. Among the 38 PPCPs detected in the secondary effluent, 31 PPCPs were degraded to or below their LOD at the ozone dosage of 6 mg/L in 10 min. Of these, eight PPCPs, including dipyridamole, diclofenac, mefenamic acid, diltiazem, ciprofloxacin, chloramphenicol, naproxen, and isopropylantipyrine were readily degraded (below LOD) at the ozone dosage of 1 mg/L in 5 min, and seven PPCPs, namely clarithromycin, sulpiride, N,N-diethyl-3-methylbenzamide, ketoprofen, theophylline, azithromycin, and roxithromycin, were undegraded (below LOD) at the ozone dosage of 6 mg/L with the reaction time of 10 min. As a result, it was found that ozone dose, pH, temperature, compounds of interest and wastewater characteristics, and contact time were among the many factors that could affect the performance of ozonation [98].

Advanced oxidation method with low concentration of ECs would result in low reaction rates and efficiencies, while the short lifetime of the oxidants requires the use of large amounts of oxidants to oxidize the trace levels of ECs and thus leading to the increase in operation costs. In addition, this method also generates toxic intermediate or by-products under suboptimum conditions. Unfortunately, the common disadvantage shared between all advanced oxidation methods is the high operating cost, which has a somewhat limited large-scale application of this otherwise very powerful technology. However, with the advent of higher efficiency UV lamps, visible light catalysts, and improved reactor design, with the help of computational fluid dynamics and energy modeling, both UV and solar-based photocatalysis have great potential for large-scale application [99]. However, further research is still needed in developing new immobilized photocatalysis reactors and improving the performance, immobilization, and illumination in the reactor of the photocatalyst.

4. Detection methods of emerging contaminants

Previous studies reported that the release of emerging chemical or microbial contaminants to the environment had likely occurred for a long time but may not have been recognized until new detection methods were developed [100]. The detection methods of ECs can be performed by using chromatographic and spectroscopic techniques as well as metal analysis. Chromatographic methods are the most common analytical methods for the identification and detection of various compounds in any kind of samples. Nonpolar, thermostable, and volatile ECs, such as flame retardants, filters, and some pesticides are determined by gas chromatography (GC), while non-volatile, polar, and thermolabile ECs are analyzed by liquid chromatography (LC).

4.1 Liquid chromatography

High-performance liquid chromatography (HPLC) or LC is a gold analytical method used to analyze a wide number of ECs that are relatively polar and nonvolatile in various samples, as can be seen in **Table 2**. In many cases, ultra-high-performance liquid chromatography (UHPLC) has been selected instead of conventional HPLC since these LC techniques provide a more efficient stationary phase through the reduction in particle size, which leads to a better resolution and shorter time [129]. UHPLC applies columns with particle size <1.7 μm and almost always carried out in the reversed phase mode mainly using C18 stationary phases [130]. Generally, for the reversed phase separation of ECs acidified water (with small amounts of formic or acetic acid), methanol, or acetonitrile as organic solvents (in some cases also acidified with formic acid or acetic acid) are used as mobile phases. Furthermore, as stated by Díaz-Cruz et al., among the LC modes, reversed phase (RP) with octadecyl C18-bonded or octyl C8-bonded, silica packing is the most commonly used stationary phase for pharmaceutical analysis [131].

LC coupled with mass spectrometry (MS) allows a successful detection with high sensibility and selectivity of extremely polar, thermally unstable, and large molecular weight compounds [100]. Although MS has been used for quantifying compounds or residues in the environment, MS/MS detection is preferred due to high analytical sensitivity and selectivity in complex matrices. Generally, MS/MS method is based on monitoring selected MS/MS transitions (precursor-product) of the compounds of interest [132]. This approach is known as the most sensitive and rugged for the quantitation purposes of environmental samples at trace levels. In fact, MS/MS showed excellent specificity, but the instrument eliminates the matrix background, which could not be visualized [133]. However, this method is blind to other untargeted species, and thus, the untargeted analysis of unexpected compounds is not feasible.

As an alternative to MS/MS, the use of high-resolution mass spectrometers (LC-HRMS) enables the acquisition of a theoretically unlimited number of species by means of accurate mass measurements in full-scan mode. This allows obtaining the elemental composition of acquired ions which is useful for the identification

Methods	Extraction methods	Instrumental analysis	Emerging contaminants	Samples	References
LC	HFLPME	LC-MS/MS	Pesticides, pharmaceutical product, PCPs, Industrial products, lifestyle products	Water samples	[101]
	SPE	LC-QTOF/MS	Endocrine active compounds	Lakewater	[102]
	SPE	LC-TOFMS	Pharmaceuticals, lifestyle products, drug of abuse, pesticides, nitrosamines, flame retardants, plasticizers, perfluorinated compounds	Surface water and wastewater	[103]
	SPE	LC-Orbitrap-MS	Pharmaceuticals	Wastewater	[104]
	MAE	UHPLC-MS/MS	PCPs, pharmaceuticals and illicit drugs	Crude wastewater, final effluent and river water	[105]
	SPE	UHPLC-Q-Exactive Orbitrap MS	Pesticides, drug residues	Tap water	[106]
	SPE	UHPLC-HRMS	Pharmaceutical, herbicides, stimulant, illicit drugs, preservative agent	Surface water, wastewater, suspended particulate matter, sediments	[107]
	SPE	UHPLC-Q-Orbitrap-MS	Pharmaceutical and personal care products	Wastewater samples	[108]
	MASE	UHPLC-Q-orbitrap	Antibiotics	Aquifer sediments	[109]
	SPE	LC-MS	Drug of abuse, pharmaceuticals	Drinking water	[110]
	Direct injection	UHPLC-QqLIT-MS/MS	Pharmaceticals, transformation products, pesticides	Treated wastewater	[111]
	SPE	LC-HRMS/MS	Artificial sweeteners, flame retardants, fungicides, herbicides, industrial chemicals, insecticides, pharmaceuticals, plasticizers.	Surface waters	[112]
	SPE	LC-HR MS/MS	Pesticides, veterinary drugs	European rivers and canals	[113]
		LC-QTOFMS	Pesticides, pharmaceuticals	Groundwater and stormwater runoff	[114]
	Soxhlet extraction	UHPLC-QTOFMS	Pharmaceuticals	Freshwater sediments	[115]
	SD-DLLME	LC-MS/MS	Pharmaceuticals, personal care products, pesticides	Water samples	[116]
	SPE	LC–MS/MS	Pharmaceuticals, personal care products	Wastewater samples	[117]

Methods	Extraction methods	Instrumental analysis	Emerging contaminants	Samples	References
GC	QuEChERS	GCxGC-TOFMS	Pesticides, organohalogens	Wastewater samples	[118]
	Rotating-disk sorptive extraction (RSDE)	GC-MS with derivatization	Parabens, hormones, anti-inflammatory drugs, triclosan, bisphenol A	Water samples	[119]
	Disk-based SPE and ultrasound-assisted extraction (UAE)	GC-MS/MS	Polycyclic and nitro-aromatic musks, brominated, chlorinated flame retardants, methyl triclosan, chlorobenzenes, organochlorine pesticides, polychlorinated biphenyls	Seawater	[120]
Metal Analysis	—	SP-ICP-MS	Metal contaminants (Cd)	Wastewater	[121]
	LLE	GC-ICP-MS	Flame retardants	Water samples	[122]
	Online-SPE	HPLC-ICP-MS	Mercury	River and seawater	[123]
	Microwave assisted digestion	ICP-MS	Trace metals	Atmospheric and lake water samples	[124]
	—	ICP-OES	Heavy metals	Soil samples	[125]
	Cross-flow ultrafiltration and LLE	ICP-MS	Colloidal trace metals	Seawater	[126]
	Microwave assisted enzymatic extraction	HPLC/ICP-MS	Heavy metals and organic contaminants	Ground water samples	[127]
Spectroscopic	—	FT-NIR	pharmaceuticals	wastewater	[128]

Table 2.
Several studies on the determination of emerging contaminants in water by different techniques for the last 5 years.

of targeted and untargeted compounds, metabolites, or transformation products. In HRMS, mass accuracy is critical for identification purposes. Normally, mass errors of below 5 ppm are observed in routine analysis [129]. Therefore, the use of LC-HRMS for screening contaminants in environmental samples has been extensively reported in literature, such as time-of-flight (TOF), Orbitrap, quadrupole-time-of-flight (Q-TOF), or Q-Orbitrap, as tabulated in **Table 2**.

With the advantage of high resolution, Orbitrap, an ion trap mass analyzer in MS has been used widely for a large number of compounds screening. Referring to hybrid quadrupole Orbitrap mass spectrometry, the qualification and quantification of complicated compounds can be obtained in one analysis [109]. This method is a promising analytical technique, especially as the full scan mode enables the post-target and nontarget analyses of chemical fingerprints which are proven by Cotton et al. [106]. In the study, the screening of a large number of ECs was developed by the combination of SPE and UHPLC-Orbitrap-MS in water samples. As reported by Robles-Molina et al., over 400 ECs have been detected in the analysis of surface water and wastewater, such as pharmaceuticals, lifestyle products, drugs of abuse, pesticides, flame retardants, and plasticizers compounds. The combination of LC with time-of-flight mass spectrometry (TOFMS) has provided a high resolution data from accurate masses of target ions along with retention time data and characteristic insource fragment ions [103].

4.2 Gas chromatography

GC is one of the analytical methods that can be used to separate, analyze, and identify the chemical compounds of any sample. GC coupled with MS is the most common technique and able to produce accurate result. However, GC is less often used than LC to separate ECs because the majority of the studies focused on the determination of very polar contaminants, such as pharmaceuticals and personal care products [100]. Due to the high polarity and low volatility of analytes, such as hydroxyl, phenolic EDCs, amines, and amides, GC method demands the inclusion of a derivatization reaction to improve the chromatographic behavior of the analytes [134]. In fact, derivatizing the ECs into more volatile derivatives is compulsory to enhance detection sensitivity and selectivity as well as improve separation.

Different strategies for derivatization usually involve the use of acylation, alkylation, or silylation reagents. Silylation is the principal derivatization technique in which active hydrogen in the functional groups of ECs is replaced by trimethylsilyl (TMS) which increased the volatility and stability of the compounds. The most common silylation agents are N-methyl-N-(trimethylsilyl) trifluoroacetamide (MSTFA), N,O-bis(trimethylsilyl)-trifluoroacetamide (BSTFA) and N-(t-butyldimethylsilyl)-N-methyltrifluoroacetamide (MTBSTFA). Based on the study by Arismendi et al., for derivatization method, water sample extract was evaporated under a N_2 stream. A derivatizing agent of N-methyl-N-(trimethylsilyl) trifluoroacetamide (MSTFA) with pyridine was then added and heated at 80°C for 35 min [119] before the derivatized solution was analyzed by using GC-MS.

However, derivatization requires a complicated, tedious procedure, resulting in a number of disadvantages, such as losses and unexpected oxidation of some compounds, less stability of derivatives, and incomplete derivatization and extraction. Although derivatization procedures may be time-consuming and have some limitations, they are still widely used and well established for routine work. In fact, in situ derivatization has been developed as an alternative to avoid preparative steps, accelerate reaction rates, and reduce evaporative losses. This method offers higher chromatographic response, reduction of reagent usage, and less sample preparation time as compared to the manual derivatization. However, the limitation of in situ derivatization is that a high amount of reagent needs to be injected into the instrument and thus affects the performance of the chromatographic column over time [135].

GC-MS/MS provides a high selectivity and sensitivity as compared to other conventional GC-MS. MS/MS is able to improve the detection by reducing co-extractive compound interferences and sample matrix effects, especially for complex samples [136]. A study by Zhang et al. showed the application of GC-MS/MS in determining the organic pollutants and emerging organic contaminants in seawater samples [120]. It has proven the sensitivity of this method of which the detection limits typically went down to parts per quadrillion (ppq) ranges, while in another study, GC-MS/MS method was developed to determine the ECs, such as personal care, hormones, pharmaceuticals, and organophosphorus flame retardants and plasticizers in various types of vegetables (salads, carrots, lettuce, onions, and tomatoes) [135]. In this study, in situ derivatization was carried out in the injection port of the GC system as an alternative to manual procedures because it simplifies the sample preparation and avoids the manipulation of hazardous reagents, and higher chromatographic response can be achieved. The high selectivity of MS/MS detection allowed the identification and quantification of various selected analytes without performing an additional clean-up step [135].

A recent advanced technique namely comprehensive two-dimensional gas chromatography (GC × GC) is a technique that is capable of enhancing the separation of 150–250 relevant compounds with high sensitivity within a complex matrix. It means that many compounds are now completely separated, so there is less need of GC-MS identification/quantitation for routine samples. The two orthogonally aligned columns comprised two different stationary phase chemistries which allow for the increase of peak capacity [118]. Based on Jover et al., three main benefits of GC × GC are (i) increase in chromatographic resolution; (ii) improved analyte detectability due to the cryofocusing that occurs in the thermal modulator, and (iii) chemical ordering in the contour plots [137]. Due to the mentioned benefits, various studies have revealed the application of this technique in various types of samples. In the previous years, GC × GC coupled with TOFMS was applied to determine the numbers of organic contaminants, such as pharmaceuticals, plasticizers personal care products, herbicides, triazines, organophosphorus compounds, pheylureas, organochlorine biocides, polycyclic aromatic hydrocarbons, benzothiazoles, and benzotriazoles in various water samples [118, 137–139].

Prebihalo et al. have reported the identification of some ECs in wastewater and soil samples by using GC × GC-TOFMS. Coupling fast-scanning TOFMS with GC × GC provided the capabilities necessary for the identification of compounds within a complex matrix. In this work, the results showed the presence of halogenated benzotriazoles in wastewater and soil samples. It has also proven that this method is able to decrease the data interpretation time by 50% as compared to traditional methods [118].

4.3 Metal analysis

One of the ECs is heavy metal ions that are gradually being introduced into the environment from various resources. Inductively coupled plasma mass spectrometry (ICP-MS) is the technique of choice for the determination of toxic heavy metal contaminants and nanoparticles in numerous types of samples. The environmental pollution caused by trace metals is a long-term, irreversible process and can be toxic even at low concentrations. ICP-MS is considered as one of the most sensitive techniques for the multi-elemental analysis of trace metals in various elemental samples [124]. The main advantages of this technique lie in the high precision, low economic cost, multi-element, and simultaneous analysis of most elements and isotopes present in the periodic table in a few minutes [100].

ICP-MS also offers extremely low detection limits, ranging from subpart per billion (ppb) to trillion (ppt) for most elements. It has a rapid multi-element scanning capability over a wide range of masses with lower detection limits as compared to graphite furnace-atomic absorption spectroscopy (GF-AAS) and inductively coupled plasma-atomic emission spectroscopy (ICP-AES) [140]. In most cases, the detection limits were 100–1000 times superior than those achieved by ICP-AES. These detection limits were broadly achieved for almost all the elements across the periodic table.

Lu et al. reported an analysis of trace metals, namely Cu, Cd, and Pb in seawater by ICP-MS. The developed method demonstrated a simple, efficient, and convenient analysis with low detection limit (0.13–1.18 pmol/L). The method was successfully applied for the analysis of seawater samples collected from the sites off the south coast North Yellow Sea in Shandong Peninsula, China. In the study, prior to the analysis of the water samples by ICP-MS, the seawater samples were first treated with cross-flow ultrafiltration (CFUF) for the separation of colloidal matter with different molecular weights and was then continued with the liquid-liquid extraction (LLE) procedure. This method was performed to eliminate the possible interference of salt matrix and chloroform during instrumental analysis and concentrate the metals in each sample [126].

In addition, ICP-MS combined with chromatographic techniques is of great importance in the characterization and identification of impurities, degradation products, and speciation studies in pharmaceuticals [141]. It provides valuable information on impurity profiling of drugs and pharmaceuticals. It is becoming the method of choice for quality control and assurance within the pharmaceutical industry. A number of chromatographic techniques, viz., HPLC, capillary electrophoresis (CE), gel permeation chromatography (GPC), ion chromatography (IC), supercritical fluid chromatography (SFC), and GC have been coupled with ICP-MS for the purpose of speciation. A study of GC-ICP-MS was performed for the detection of the six polybrominated diphenyl ether (PBDEs) congeners in environmental water samples [122]. These PBDEs belong to group of brominated flame retardants which are considered as one of the ECs that are added into different industrial products. Apart from that, Jia et al. reported on a simultaneous analysis of mercury speciation in environmental water samples by using HPLC-ICP-MS [123]. HPLC was preferred for the mercury speciation, while ICP-MS is the most used detection technique for elemental analysis. The study revealed a good repeatability and wide linear range achievements of this coupled technique.

4.4 Other techniques

Recently, another technique has been developed to overcome the drawbacks from the traditional techniques, which is Fourier transform near-infrared (FT-NIR) spectroscopy. This technique offers no reagents use, nondestructive character, allowing the reuse of sample after measurement, fast technique, and there is a possibility of online monitoring. Based on Quientelas et al. (2018), the combination between FT-NIR and chemometrics was developed to determine pharmaceutical compounds, such as ibuprofen, carbamazepine, β-estradiol, ethinylestradiol, and sulfamethoxazole in wastewaters [128]. A chemometric approach was used by employing a Kolmogorov-Smirnov test to check the normality of the data, a boxplot analysis for outliers' identification and a principal component analysis (PCA) aiming to identify samples interrelationships and define the data sets. Next, a partial least squares (PLS) regression analysis was performed in order to obtain a prediction model that is suitable for pharmaceuticals quantification purposes.

5. Conclusion

Although EC compounds occur in trace concentrations of waters, their adverse effects to aquatic organisms, animals, and humans cannot be underestimated due to their continuous release into the water systems. The removal and detection of ECs and their transformation products in natural and drinking waters are challenging tasks due to the complexity of contaminants in water samples. However, tremendous progress has been made on the assessment of many ECs due to the great efforts and times committed by many scientists working in different research fields. The future trends in the removal of emerging water contaminants will be on the advanced oxidation method because it can effectively degrade the contaminants and has not yet been fully explored. However, it should also be noted that the presence of degradation intermediates can be more toxic than their parent compounds, which remains a great challenge to overcome in advanced oxidation method. In short, feasible techniques need to be in place to remove ECs or at least reduce them below the regulated levels and more research is necessary to completely describe the mechanisms and optimize the main parameters related to the removal of emerging water contaminants in real applications.

Acknowledgements

The authors gratefully acknowledge the grant provided by the Ministry of Education, Malaysia (MOE) under the Fundamental Research Grant Scheme (FRGS) (vote number: RDU1901101) and Universiti Malaysia Pahang (UMP) under the RDU Grant (vote number: RDU1903109).

Declaration of interest statement

The authors certify that they have NO affiliations with or involvement in any organization or entity with any financial or non-financial interests in the subject matter or materials discussed in this manuscript.

Author details

Wan Norfazilah Wan Ismail[1*] and Siti Umairah Mokhtar[2]

1 Faculty of Industrial Sciences and Technology, Universiti Malaysia Pahang, Kuantan, Pahang, Malaysia

2 Faculty of Chemical and Process Engineering Technology, Universiti Malaysia Pahang, Kuantan, Pahang, Malaysia

*Address all correspondence to: norfazilah@ump.edu.my

References

[1] Fawell J, Ong CN. Emerging contaminants and the implications for drinking water. International Journal of Water Resources Development. 2012;**28**:247-263

[2] Falconer IR, Chapman HF, Moore MR, Ranmuthugala G. Endocrine-disrupting compounds: A review of their challenge to sustainable and safe water supply and water reuse. Environmental Toxicology. 2006;**21**:181-191

[3] Petrovic M. Analysis and removal of emerging contaminants in wastewater and drinking water. TrAC Trends in Analytical Chemistry. 2003;**22**:685-696

[4] Huang Q, Bu Q, Zhong W, Shi K, Cao Z, Yu G. Derivation of aquatic predicted no-effect concentration (PNEC) for ibuprofen and sulfamethoxazole based on various toxicity endpoints and the associated risks. Chemosphere. 2018;**193**:223-229

[5] Oetjen K, Giddings CGS, McLaughlin M, Nell M, Blotevogel J, Helbling DE, et al. Emerging analytical methods for the characterization and quantification of organic contaminants in flowback and produced water. Trends in Environmental Analytical Chemistry. 2017;**15**:12-23

[6] Philip JM, Aravind UK, Aravindakumar CT. Emerging contaminants in Indian environmental matrices—A review. Chemosphere. 2018;**190**:307-326

[7] Sauvé S, Desrosiers M. A review of what is an emerging contaminant. Chemistry Central Journal. 2014;**8**:15

[8] Jobling S, Nolan M, Tyler CR, Brighty G, Sumpter JP. Widespread sexual disruption in wild fish. Environmental Science & Technology. 1998;**32**:2498-2506

[9] Mompelat S, Le Bot B, Thomas O. Occurrence and fate of pharmaceutical products and by-products, from resource to drinking water. Environment International. 2009;**35**:803-814

[10] Lei M, Zhang L, Lei J, Zong L, Li J, Wu Z, et al. Overview of emerging contaminants and associated human health effects. BioMed Research International. 2015;**2015**:1-12

[11] Petrie B, Barden R, Kasprzyk-Hordern B. A review on emerging contaminants in wastewaters and the environment: Current knowledge, understudied areas and recommendations for future monitoring. Water Research. 2015;**72**:3-27

[12] Lapworth DJ, Baran N, Stuart ME, Ward RS. Emerging organic contaminants in groundwater: A review of sources, fate and occurrence. Environmental Pollution. 2012;**163**:287-303

[13] Tri TM, Anh DH, Hoai PM, Minh NH, Nam VD, Viet PH, et al. Emerging Endocrine Disrupting Chemicals and Pharmaceuticals in Vietnam: A Review of Environmental Occurrence and Fate in Aquatic and Indoor Environments. Washington, DC: American Chemical Society; 2016

[14] Petrovic M, Eljarrat E, Lopez de Alda MJ, Barceló D. Endocrine disrupting compounds and other emerging contaminants in the environment: A survey on new monitoring strategies and occurrence data. Analytical and Bioanalytical Chemistry. 2004;**378**:549-562

[15] Barceló D, Rodriguez-Mozaz S, Marco M-P, Lopez de Alda M. Biosensors for environmental monitoring of endocrine disruptors: A review article. Analytical and Bioanalytical Chemistry. 2004;**378**:588-598

[16] Schug TT, Janesick A, Blumberg B, Heindel JJ. Endocrine disrupting chemicals and disease susceptibility. The Journal of Steroid Biochemistry and Molecular Biology. 2011;**127**:204-215

[17] Bergman Å, Heindel JJ, Jobling S, Kidd KA, Zoeller RT. State of the Science of Endocrine Disrupting Chemicals - 2012. United Nations Environment Programme (UNEP) and WHO; 2013

[18] Ebele AJ, Abou-Elwafa Abdallah M, Harrad S. Pharmaceuticals and personal care products (PPCPs) in the freshwater aquatic environment. Emerging Contaminants. 2017;**3**:1-16

[19] Liu J-L, Wong M-H. Pharmaceuticals and personal care products (PPCPs): A review on environmental contamination in China. Environment International. 2013;**59**:208-224

[20] Mottaleb MA, Meziani MJ, Matin MA, Arafat MM, Wahab MA. Emerging Micro-Pollutants Pharmaceuticals and Personal Care Products (PPCPs) Contamination Concerns in Aquatic Organisms - LC/MS and GC/MS Analysis. Washington: American Chemical Society; 2015

[21] Kairo G, Biron DG, Ben Abdelkader F, Bonnet M, Tchamitchian S, Cousin M, et al. Nosema ceranae, Fipronil and their combination compromise honey bee reproduction via changes in male physiology. Scientific Reports. 2017;7:8556

[22] De Jesús Andino F, Lawrence BP, Robert J. Long term effects of carbaryl exposure on antiviral immune responses in *Xenopus laevis*. Chemosphere. 2017;**170**:169-175

[23] Ewence A, Brescia S, Johnson I, Rumsby PC. An approach to the identification and regulation of endocrine disrupting pesticides. Food and Chemical Toxicology. 2015;**78**:214-220

[24] Nagel SC, Bromfield JJ. Bisphenol A: A model endocrine disrupting chemical with a new potential mechanism of action. Endocrinology. 2013;**154**:1962-1964

[25] Rocha BA, Azevedo LF, Gallimberti M, Campiglia AD, Barbosa F. High levels of bisphenol A and bisphenol S in Brazilian thermal paper receipts and estimation of daily exposure. Journal of Toxicology and Environmental Health. Part A. 2015;**78**:1181-1188

[26] Ema M, Kobayashi N, Naya M, Hanai S, Nakanishi J. Reproductive and developmental toxicity studies of manufactured nanomaterials. Reproductive Toxicology. 2010;**30**:343-352

[27] Schilling K, Bradford B, Castelli D, Dufour E, Nash JF, Pape W, et al. Human safety review of 'nano' titanium dioxide and zinc oxide. Photochemical & Photobiological Sciences. 2010;**9**:495-509

[28] Koh YKK, Chiu TY, Boobis A, Cartmell E, Scrimshaw MD, Lester JN. Treatment and removal strategies for estrogens from wastewater. Environmental Technology. 2008;**29**:245-267

[29] Vajda AM, Barber LB, Gray JL, Lopez EM, Woodling JD, Norris DO. Reproductive disruption in fish downstream from an estrogenic wastewater effluent. Environmental Science & Technology. 2008;**42**:3407-3414

[30] Länge R, Hutchinson TH, Croudace CP, Siegmund F, Schweinfurth H, Hampe P, et al. Effects of the synthetic estrogen 17α-ethinylestradiol on the life-cycle

of the fathead minnow (*Pimephales promelas*). Environmental Toxicology and Chemistry. 2001;**20**:1216-1227

[31] Maeng SK, Sharma SK, Lekkerkerker-Teunissen K, Amy GL. Occurrence and fate of bulk organic matter and pharmaceutically active compounds in managed aquifer recharge: A review. Water Research. 2011;**45**:3015-3033

[32] Kashiwada S, Ishikawa H, Miyamoto N, Ohnishi Y, Magara Y. Fish test for endocrine-disruption and estimation of water quality of Japanese rivers. Water Research. 2002;**36**:2161-2166

[33] Collier AC. Pharmaceutical contaminants in potable water: Potential concerns for pregnant women and children. EcoHealth. 2007;**4**:164-171

[34] vom Saal FS, Hughes C. An extensive new literature concerning low-dose effects of bisphenol A shows the need for a new risk assessment. Environmental Health Perspectives. 2005;**113**:926-933

[35] Hayes TB, Collins A, Lee M, Mendoza M, Noriega N, Stuart AA, et al. Hermaphroditic, demasculinized frogs after exposure to the herbicide atrazine at low ecologically relevant doses. Proceedings of the National Academy of Sciences. 2002;**99**:5476-5480

[36] Reemtsma T, Alder L, Banasiak U. Emerging pesticide metabolites in groundwater and surface water as determined by the application of a multimethod for 150 pesticide metabolites. Water Research. 2013;**47**:5535-5545

[37] Bartelt-Hunt SL, Snow DD, Kranz WL, Mader TL, Shapiro CA, Donk SJ, et al. Effect of growth Promotants on the occurrence of endogenous and synthetic steroid hormones on feedlot soils and in runoff from beef cattle feeding operations. Environmental Science & Technology. 2012;**46**:1352-1360

[38] Ross I, McDonough J, Miles J, Storch P, Kochunarayanan PT, Kalve E, et al. A review of emerging technologies for remediation of PFASs. Remediation Journal. 2018;**28**:101-126

[39] Archer E, Petrie B, Kasprzyk-Hordern B, Wolfaardt GM. The fate of pharmaceuticals and personal care products (PPCPs), endocrine disrupting contaminants (EDCs), metabolites and illicit drugs in a WWTW and environmental waters. Chemosphere. 2017;**174**:437-446

[40] Sharma SK. Heavy Metals in Water: Presence, Removal and Safety. Cambridge: Royal Society of Chemistry; 2014

[41] De Gisi S, Lofrano G, Grassi M, Notarnicola M. Characteristics and adsorption capacities of low-cost sorbents for wastewater treatment: A review. Sustainable Materials and Technologies. 2016;**9**:10-40

[42] Deng F, Luo X-B, Ding L, Luo S-L. Application of Nanomaterials and Nanotechnology in the Reutilization of Metal Ion from Wastewater. Amsterdam: Elsevier; 2019

[43] Tareq R, Akter N, Azam MS. Biochars and Biochar Composites. Amsterdam: Elsevier; 2019

[44] Patel H. Fixed-bed column adsorption study: A comprehensive review. Applied Water Science. 2019;**9**:45

[45] Uddin MT, Rahman MA, Rukanuzzaman M, Islam MA. A potential low cost adsorbent for the removal of cationic dyes from aqueous solutions. Applied Water Science. 2017;7:2831-2842

[46] Kyzas G, Kostoglou M. Green adsorbents for wastewaters: A critical review. Materials. 2014;7:333-364

[47] Sen TK. Air, Gas, and Water Pollution Control Using Industrial and Agricultural Solid Wastes Adsorbents. Boca Raton: CRC Press; 2017

[48] Fang Y, Zhou A, Yang W, Araya T, Huang Y, Zhao P, et al. Complex formation via hydrogen bonding between rhodamine B and montmorillonite in aqueous solution. Scientific Reports. 2018;**8**:229

[49] Crini G, Lichtfouse E, Wilson LD, Morin-Crini N. Adsorption-Oriented Processes Using Conventional and Non-conventional Adsorbents for Wastewater Treatment. Cham: Springer International Publishing; 2018

[50] Mandal A, Singh N, Purakayastha TJ. Characterization of pesticide sorption behaviour of slow pyrolysis biochars as low cost adsorbent for atrazine and imidacloprid removal. Science of the Total Environment. 2017;**577**:376-385

[51] Hegazi HA. Removal of heavy metals from wastewater using agricultural and industrial wastes as adsorbents. HBRC Journal. 2013;**9**:276-282

[52] Etim UJ, Umoren SA, Eduok UM. Coconut coir dust as a low cost adsorbent for the removal of cationic dye from aqueous solution. Journal of Saudi Chemical Society. 2016;**20**:S67-S76

[53] Bello OS, Adegoke KA, Fagbenro SO, Lameed OS. Functionalized coconut husks for rhodamine-B dye sequestration. Applied Water Science. 2019;**9**:8

[54] Thue PS, Adebayo MA, Lima EC, Sieliechi JM, Machado FM, Dotto GL, et al. Preparation, characterization and application of microwave-assisted activated carbons from wood chips for removal of phenol from aqueous solution. Journal of Molecular Liquids. 2016;**223**:1067-1080

[55] Thue PS, Lima EC, Sieliechi JM, Saucier C, Dias SLP, Vaghetti JCP, et al. Effects of first-row transition metals and impregnation ratios on the physicochemical properties of microwave-assisted activated carbons from wood biomass. Journal of Colloid and Interface Science. 2017;**486**:163-175

[56] Nhung NTH, Quynh BTP, Thao PTT, Bich HN, Giang BL. Pretreated fruit peels as adsorbents for removal of dyes from water. IOP Conference Series Earth and Environmental Sciences. 2018;**159**:012015

[57] Foo KY, Hameed BH. Preparation, characterization and evaluation of adsorptive properties of orange peel based activated carbon via microwave induced K2CO3 activation. Bioresource Technology. 2012;**104**:679-686

[58] Ben-Ali S, Jaouali I, Souissi-Najar S, Ouederni A. Characterization and adsorption capacity of raw pomegranate peel biosorbent for copper removal. Journal of Cleaner Production. 2017;**142**:3809-3821

[59] Torrellas SÁ, García Lovera R, Escalona N, Sepúlveda C, Sotelo JL, García J. Chemical-activated carbons from peach stones for the adsorption of emerging contaminants in aqueous solutions. Chemical Engineering Journal. 2015;**279**:788-798

[60] Jing X-R, Wang Y-Y, Liu W-J, Wang Y-K, Jiang H. Enhanced adsorption performance of tetracycline in aqueous solutions by methanol-modified biochar. Chemical Engineering Journal. 2014;**248**:168-174

[61] Alvarez-Torrellas S, Garcia-Lovera R, Rodriguez A, Garcia J. Removal of methylene blue by adsorption on mesoporous carbon from peach stones. Chemical Engineering Transactions. 2015;**43**:1963-1968

[62] Martin-Martinez M, Álvarez-Torrellas S, García J, Silva AMT, Faria JL, Gomes HT. Exploring the activity of chemical-activated carbons synthesized from peach stones as metal-free catalysts for wet peroxide oxidation. Catalysis Today. 2018;**313**:20-25

[63] Iorhemen OT, Hamza RA, Tay JH. Membrane bioreactor (MBR) technology for wastewater treatment and reclamation: Membrane fouling. Membranes. 2016;**6**:33

[64] Nofiana R, Istirokhatun T, Susanto H. Reuse of domestic wastewater by membrane technologies towards sustainable city development. IOP Confernce Series Earth and Environmental Sciences. 2019;**248**:012053

[65] Abdel-Fatah MA. Nanofiltration systems and applications in wastewater treatment: Review article. Ain Shams Engineering Journal. 2018;**9**:3077-3092

[66] Maddah HA, Alzhrani AS, Bassyouni M, Abdel-Aziz MH, Zoromba M, Almalki AM. Evaluation of various membrane filtration modules for the treatment of seawater. Applied Water Science. 2018;**8**:6

[67] Keucken A, Heinicke G, Persson K, Köhler S. Combined coagulation and ultrafiltration process to counteract increasing NOM in brown surface water. Watermark. 2017;**9**:697

[68] Cui X, Choo K-H. Natural organic matter removal and fouling control in low-pressure membrane filtration for water treatment. Environmental Engineering Research. 2014;**19**:1-8

[69] Sun C, Zhang N, Li F, Ke G, Song L, Liu X, et al. Quantitative analysis of membrane fouling mechanisms involved in microfiltration of humic acid–protein mixtures at different solution conditions. Watermark. 2018;**10**:1306

[70] Schulz M, Bünting S, Ernst M. Impact of powdered activated carbon structural properties on removal of organic foulants in combined adsorption-ultrafiltration. Watermark. 2017;**9**:580

[71] Bu F, Gao B, Yue Q, Liu C, Wang W, Shen X. The combination of coagulation and adsorption for controlling ultra-filtration membrane fouling in water treatment. Watermark. 2019;**11**:90

[72] Wang Y, Wang X, Li M, Dong J, Sun C, Chen G. Removal of pharmaceutical and personal care products (PPCPs) from municipal waste water with integrated membrane systems, MBR-RO/NF. International Journal of Environmental Research and Public Health. 2018;**15**:269

[73] Baransi-Karkaby K, Bass M, Freger V. In situ modification of reverse osmosis membrane elements for enhanced removal of multiple micropollutants. Membranes. 2019;**9**:28

[74] Akhbari A, Kutty PK, Chuen OC, Ibrahim S. A study of palm oil mill processing and environmental assessment of palm oil mill effluent treatment. Environmental Engineering Research. 2019;**25**:212-221

[75] Conserva S, Tatti F, Torretta V, Ferronato N, Viotti P. An integrated approach to the biological reactor–sedimentation tank system. Resources. 2019;**8**:94

[76] Samer M. Biological and Chemical Wastewater Treatment Processes. London: IntechOpen; 2015

[77] Drewnowski J, Remiszewska-Skwarek A, Duda S, Łagód G. Aeration process in bioreactors as the Main energy consumer in a wastewater treatment plant. Review of solutions and methods of process optimization. Processes. 2019;**7**:311

[78] Kolev S, A. Dairy wastewaters—General characteristics and treatment possibilities—A review. Food Technology and Biotechnology. 2017;**55**:14-28

[79] Ting YF, Praveena SM. Sources, mechanisms, and fate of steroid estrogens in wastewater treatment plants: A mini review. Environmental Monitoring and Assessment. 2017;**189**:178

[80] Han X, Zuo YT, Hu Y, Zhang J, Zhou MX, Chen M, et al. Investigating the performance of three modified activated sludge processes treating municipal wastewater in organic pollutants removal and toxicity reduction. Ecotoxicology and Environmental Safety. 2018;**148**:729-737

[81] Sehar S, Naz I. Role of the Biofilms in Wastewater Treatment. London: IntechOpen; 2016

[82] Naz I, Saroj DP, Mumtaz S, Ali N, Ahmed S. Assessment of biological trickling filter systems with various packing materials for improved wastewater treatment. Environmental Technology. 2015;**36**:424-434

[83] Oakley S, von Sperling M. Media Filters: Trickling Filters and Anaerobic Filters. California: Michigan State University; 2017

[84] Visanji Z, Sadr SMK, Johns MB, Savic D, Memon FA. Optimising wastewater treatment solutions for the removal of contaminants of emerging concern (CECs): A case study for application in India. Journal of Hydroinformatics. 2020;**22**:93-110

[85] Yang L, Yang L, Ding L, Deng F, Luo X-B, Luo S-L. Principles for the Application of Nanomaterials in Environmental Pollution Control and Resource Reutilization. Amsterdam: Elsevier; 2019

[86] Jasim SY, Irabelli A, Yang P, Ahmed S, Schweitzer L. Presence of pharmaceuticals and pesticides in Detroit river water and the effect of ozone on removal. Ozone Science and Engineering. 2006;**28**:415-423

[87] Esplugas S, Bila DM, Krause LGT, Dezotti M. Ozonation and advanced oxidation technologies to remove endocrine disrupting chemicals (EDCs) and pharmaceuticals and personal care products (PPCPs) in water effluents. Journal of Hazardous Materials. 2007;**149**:631-642

[88] Gagnon C, Lajeunesse A, Cejka P, Gagné F, Hausler R. Degradation of selected acidic and neutral pharmaceutical products in a primary-treated wastewater by disinfection processes. Ozone Science and Engineering. 2008;**30**:387-392

[89] Zhang H, Yamada H, Tsuno H. Removal of endocrine-disrupting chemicals during ozonation of municipal sewage with brominated byproducts control. Environmental Science & Technology. 2008;**42**:3375-3380

[90] Segura P, Kaplan P, Yargeau V. Identification and structural elucidation of ozonation transformation products of estrone. Chemistry Central Journal. 2013;**7**:74

[91] Tizaoui C. Novel Ozone Oxidation Process for Emerging Contaminants Removal in Water. Cham: Springer International Publishing; 2018

[92] Tayo LL, Caparanga AR, Doma BT, Liao C-H. A review on the removal of pharmaceutical and personal care

products (PPCPs) using advanced oxidation processes. Journal of Advanced Oxidation Technologies. 2018;**21**:196-214

[93] Cesaro A, Belgiorno V. Removal of endocrine disruptors from urban wastewater by advanced oxidation processes (AOPs): A review. Open Biotechnology Journal. 2016;**10**:151-172

[94] Sun Q, Zhu G, Wang C, Yang Z, Xue Q. Removal characteristics of steroid estrogen in the mixed system through an ozone-based advanced oxidation process. Water, Air, and Soil Pollution. 2019;**230**:9

[95] Si X, Hu Z, Huang S. Combined process of ozone oxidation and ultrafiltration as an effective treatment technology for the removal of endocrine-disrupting chemicals. Applied Sciences. 2018;**8**:1240

[96] Li X, Shi H, Li K, Zhang L. Combined process of biofiltration and ozone oxidation as an advanced treatment process for wastewater reuse. Frontiers of Environmental Science & Engineering. 2015;**9**:1076-1083

[97] Paucar NE, Kim I, Tanaka H, Sato C. Effect of O3 dose on the O3/UV treatment process for the removal of pharmaceuticals and personal care products in secondary effluent. ChemEngineering. 2019;**3**:53

[98] Yargeau V, Leclair C. Impact of operating conditions on decomposition of antibiotics during ozonation: A review. Ozone Science and Engineering. 2008;**30**:175-188

[99] Chong MN, Jin B, Chow CWK, Saint C. Recent developments in photocatalytic water treatment technology: A review. Water Research. 2010;**44**:2997-3027

[100] Martín-Pozo L, de Alarcón-Gómez B, Rodríguez-Gómez R, García-Córcoles MT, Çipa M, Zafra-Gómez A. Analytical methods for the determination of emerging contaminants in sewage sludge samples. A review. Talanta. 2019;**192**:508-533

[101] Salvatierra-stamp V, Muñiz-Valencia R, Jurado JM, Ceballos-Magaña SG. Hollow fiber liquid phase microextraction combined with liquid chromatography-tandem mass spectrometry for the analysis of emerging contaminants in water samples. Microchemical Journal. 2018;**140**:87-95

[102] Guyader ME, Warren LD, Green E, Butt C, Ivosev G, Kiesling RL, et al. Prioritizing potential endocrine active high resolution mass spectrometry (HRMS) features in Minnesota lakewater. Science of the Total Environment. 2019;**670**:814-825

[103] Robles-Molina J, Lara-Ortega FJ, Gilbert-López B, García-Reyes JF, Molina-Díaz A. Multi-residue method for the determination of over 400 priority and emerging pollutants in water and wastewater by solid-phase extraction and liquid chromatography-time-of-flight mass spectrometry. Journal of Chromatography. A. 2014;**1350**:30-43

[104] Llorca M, Lucas D, Ferrando-Climent L, Badia-Fabregat M, Cruz-Morató C, Barceló D, et al. Suspect screening of emerging pollutants and their major transformation products in wastewaters treated with fungi by liquid chromatography coupled to a high resolution mass spectrometry. Journal of Chromatography. A. 2016;**1439**:124-136

[105] Petrie B, Youdan J, Barden R, Kasprzyk-Hordern B. Multi-residue analysis of 90 emerging contaminants in liquid and solid environmental matrices by ultra-high-performance liquid chromatography tandem mass spectrometry. Journal of Chromatography. A. 2016;**1431**:64-78

[106] Cotton J, Leroux F, Broudin S, Poirel M, Corman B, Junot C, et al. Development and validation of a multiresidue method for the analysis of more than 500 pesticides and drugs in water based on on-line and liquid chromatography coupled to high resolution mass spectrometry. Water Research. 2016;**104**:20-27

[107] Comtois-Marotte S, Chappuis T, Vo Duy S, Gilbert N, Lajeunesse A, Taktek S, et al. Analysis of emerging contaminants in water and solid samples using high resolution mass spectrometry with a Q Exactive orbital ion trap and estrogenic activity with YES-assay. Chemosphere. 2017;**166**:400-411

[108] Althakafy JT, Kulsing C, Grace MR, Marriott PJ. Liquid chromatography—Quadrupole Orbitrap mass spectrometry method for selected pharmaceuticals in water samples. Journal of Chromatography. A. 2017;**1515**:164-171

[109] Tong L, Liu H, Xie C, Li M. Quantitative analysis of antibiotics in aquifer sediments by liquid chromatography coupled to high resolution mass spectrometry. Journal of Chromatography. A. 2016;**1452**:58-66

[110] Peng Y, Gautam L, Hall SW. The detection of drugs of abuse and pharmaceuticals in drinking water using solid-phase extraction and liquid chromatography-mass spectrometry. Chemosphere. 2019;**223**:438-447

[111] Campos-Mañas MC, Plaza-Bolaños P, Sánchez-Pérez JA, Malato S, Agüera A. Fast determination of pesticides and other contaminants of emerging concern in treated wastewater using direct injection coupled to highly sensitive ultra-high performance liquid chromatography-tandem mass spectrometry. Journal of Chromatography. A. 2017;**1507**:84-94

[112] Osorio V, Schriks M, Vughs D, de Voogt P, Kolkman A. A novel sample preparation procedure for effect-directed analysis of micro-contaminants of emerging concern in surface waters. Talanta. 2018;**186**:527-537

[113] Casado J, Brigden K, Santillo D, Johnston P. Screening of pesticides and veterinary drugs in small streams in the European Union by liquid chromatography high resolution mass spectrometry. Science of the Total Environment. 2019;**670**:1204-1225

[114] Pinasseau L, Wiest L, Fildier A, Volatier L, Fones GR, Mills GA, et al. Use of passive sampling and high resolution mass spectrometry using a suspect screening approach to characterise emerging pollutants in contaminated groundwater and runoff. Science of the Total Environment. 2019;**672**:253-263

[115] Terzic S, Ahel M. Nontarget analysis of polar contaminants in freshwater sediments influenced by pharmaceutical industry using ultra-high-pressure liquid chromatography–quadrupole time-of-flight mass spectrometry. Environmental Pollution. 2011;**159**:557-566

[116] Caldas SS, Rombaldi C, de Oliveira Arias JL, Marube LC, Primel EG. Multi-residue method for determination of 58 pesticides, pharmaceuticals and personal care products in water using solvent demulsification dispersive liquid–liquid microextraction combined with liquid chromatography-tandem mass spectrometry. Talanta. 2016;**146**:676-688

[117] Andrés-Costa MJ, Carmona E, Picó Y. Universal method to determine acidic licit and illicit drugs and personal care products in water by liquid chromatography quadrupole time-of-flight. MethodsX. 2016;**3**:307-314

[118] Prebihalo S, Brockman A, Cochran J, Dorman FL. Determination of emerging

contaminants in wastewater utilizing comprehensive two-dimensional gas-chromatography coupled with time-of-flight mass spectrometry. Journal of Chromatography. A. 2015;**1419**:109-115

[119] Arismendi D, Becerra-Herrera M, Cerrato I, Richter P. Simultaneous determination of multiresidue and multiclass emerging contaminants in waters by rotating-disk sorptive extraction–derivatization-gas chromatography/mass spectrometry. Talanta. 2019;**201**:480-489

[120] Zhang H, Bayen S, Kelly BC. Multi-residue analysis of legacy POPs and emerging organic contaminants in Singapore's coastal waters using gas chromatography–triple quadrupole tandem mass spectrometry. Science of the Total Environment. 2015;**523**:219-232

[121] Vidmar J, Oprčkal P, Milačič R, Mladenovič A, Ščančar J. Investigation of the behaviour of zero-valent iron nanoparticles and their interactions with Cd2+ in wastewater by single particle ICP-MS. Science of the Total Environment. 2018;**634**:1259-1268

[122] Novak P, Zuliani T, Milačič R, Ščančar J. Development of an analytical procedure for the determination of polybrominated diphenyl ethers in environmental water samples by GC–ICP-MS. Analytica Chimica Acta. 2014;**827**:64-73

[123] Jia X, Zhao J, Ren H, Wang J, Hong Z, Zhang X. Zwitterion-functionalized polymer microspheres-based solid phase extraction method on-line combined with HPLC–ICP-MS for mercury speciation. Talanta. 2019;**196**:592-599

[124] Ahmed M, Chin YH, Guo X, Zhao X-M. Microwave assisted digestion followed by ICP-MS for determination of trace metals in atmospheric and lake ecosystem. Journal of Environmental Sciences. 2017;**55**:1-10

[125] Gutiérrez C, Fernández C, Escuer M, Campos-Herrera R, Rodríguez MEB, Carbonell G, et al. Effect of soil properties, heavy metals and emerging contaminants in the soil nematodes diversity. Environmental Pollution. 2016;**213**:184-194

[126] Lu Y, Gao X, Chen C-TA. Separation and determination of colloidal trace metals in seawater by cross-flow ultrafiltration, liquid-liquid extraction and ICP-MS. Marine Chemistry. 2019;**215**:103685

[127] Marcinkowska M, Barałkiewicz D. Multielemental speciation analysis by advanced hyphenated technique—HPLC/ICP-MS: A review. Talanta. 2016;**161**:177-204

[128] Quintelas C, Mesquita DP, Ferreira EC, Amaral AL. Quantification of pharmaceutical compounds in wastewater samples by near infrared spectroscopy (NIR). Talanta. 2019;**194**:507-513

[129] Hernández F, Ibáñez M, Bade R, Bijlsma L, Sancho JV. Investigation of pharmaceuticals and illicit drugs in waters by liquid chromatography-high-resolution mass spectrometry. TrAC Trends in Analytical Chemistry. 2014;**63**:140-157

[130] Pico Y, Belenguer V, Corcellas C, Diaz-Cruz MS, Eljarrat E, Farré M, et al. Contaminants of emerging concern in freshwater fish from four Spanish Rivers. Science of the Total Environment. 2019;**659**:1186-1198

[131] Díaz-Cruz MS, García-Galán MJ, Guerra P, Jelic A, Postigo C, Eljarrat E, et al. Analysis of selected emerging contaminants in sewage sludge. TrAC Trends in Analytical Chemistry. 2009;**28**:1263-1275

[132] Hird SJ, Lau BP-Y, Schuhmacher R, Krska R. Liquid chromatography-mass spectrometry for the determination of chemical contaminants in food. TrAC Trends in Analytical Chemistry. 2014;**59**:59-72

[133] Farré M, Barceló D, Barceló D. Analysis of emerging contaminants in food. TrAC Trends in Analytical Chemistry. 2013;**43**:240-253

[134] Poole CF. Alkylsilyl derivatives for gas chromatography. Journal of Chromatography. A. 2013;**1296**:2-14

[135] Albero B, Sánchez-Brunete C, Miguel E, Tadeo JL. Application of matrix solid-phase dispersion followed by GC–MS/MS to the analysis of emerging contaminants in vegetables. Food Chemistry. 2017;**217**:660-667

[136] Rutkowska E, Łozowicka B, Kaczyński P. Modification of multiresidue QuEChERS protocol to minimize matrix effect and improve recoveries for determination of pesticide residues in dried herbs followed by GC-MS/MS. Food Analytical Methods. 2018;**11**:709-724

[137] Jover E, Matamoros V, Bayona JM. Characterization of benzothiazoles, benzotriazoles and benzosulfonamides in aqueous matrixes by solid-phase extraction followed by comprehensive two-dimensional gas chromatography coupled to time-of-flight mass spectrometry. Journal of Chromatography. A. 2009;**1216**:4013-4019

[138] Matamoros V, Jover E, Bayona JM. Part-per-trillion determination of pharmaceuticals, pesticides, and related organic contaminants in river water by solid-phase extraction followed by comprehensive two-dimensional gas chromatography time-of-flight mass spectrometry. Analytical Chemistry. 2010;**82**:699-706

[139] Lima Gomes PCF, Barnes BB, Santos-Neto ÁJ, Lancas FM, Snow NH. Determination of steroids, caffeine and methylparaben in water using solid phase microextraction-comprehensive two dimensional gas chromatography–time of flight mass spectrometry. Journal of Chromatography. A. 2013;**1299**:126-130

[140] Wilschefski SC, Baxter MR. Inductively coupled plasma mass spectrometry: Introduction to analytical aspects. Clinical Biochemist Reviews. 2019;**40**:115-133

[141] Pröfrock D, Prange A. Inductively coupled plasma–mass spectrometry (ICP-MS) for quantitative analysis in environmental and life sciences: A review of challenges, solutions, and trends. Applied Spectroscopy. 2012;**66**:843-868

Section 2

Emerging Agricultural Contaminants

Chapter 4

Pesticides and Human Health

Riaz Shah

Abstract

Pesticides are used in managing pests and their use will continue in future because of food security and vector control. Most pesticides are potentially toxic to human beings resulting in severe health consequences. There is also evidence that parental exposure, as well as, exposure in early life or adolescence could increase the longer-term risks. Pesticide exposures have been linked to many human diseases such as Alzheimer, Parkinson, amyotrophic lateral sclerosis, asthma, bronchitis, infertility, birth defects, attention deficit hyperactivity disorder, autism, diabetes, and obesity, respiratory diseases, organ diseases and system failures. People who are exposed to pesticides are at a greater risk to develop various cancers including non-Hodgkin lymphoma (NHL), leukemia, brain tumors, and cancers of the breast, prostate, lung, stomach, colorectal, liver, and the urinary bladder. The cell culture is an excellent experimental model reflecting human exposure to pesticides at a molecular level which is necessary to understand the hazards. Pesticide users should be aware of their risks and proper handling, as well as must use personal protective equipment which is effective in reducing damage to human health. Carcinogenic pesticides must be eliminated and sustainable and new approaches in pest management should be encouraged.

Keywords: pesticides, cancer, endocrine disruption, pesticide residues, toxicity

1. Introduction

A pesticide is any substance which is used to prevent, destroy or repel any pest from causing any damage. The term pest represents any living organism that may cause harm to human in respect to food competition, destruction of property and spread of disease. Pests include insects, rodents, microbes, fungi and weeds (unwanted plants), etc. of agricultural, medical and veterinary importance, and therefore, a pesticide can be an insecticide, an insect and plant growth regulator, a fungicide, an herbicide, a molluscicide, and an algaecide, etc. based on the target pest organism.

The major site of action for most pesticides are the nervous and endocrine systems and, therefore, are also potentially toxic to human with serious direct or indirect adverse health effects. Human beings are exposed to pesticides directly or indirectly. Direct exposure occurs during pesticide application process in agriculture, public health and livestock, and fumigation while indirect exposure involves ingestion of contaminated food and water, and inhalation of pesticides droplets from the drift. Children are more susceptible to pesticides than adults due to their physical makeup, behavior and physiology, and exposure to very low levels at early developmental stages can cause adverse health effects. Codex Alimentarius committee and the Pesticide Data Program of the United States Department of Agriculture have established pesticide maximum residue limits in edible food which must be followed to avoid any health risks.

Pesticide exposures have been linked to the elevated incidence of human diseases such as cancers, Alzheimer, Parkinson, amyotrophic lateral sclerosis, asthma, bronchitis, infertility, birth defects, attention deficit hyperactivity disorder, autism, diabetes, and obesity, respiratory diseases, organ diseases and system failures. People who are exposed to pesticides are at a greater risk to develop various cancers including non-Hodgkin lymphoma (NHL), leukemia, brain tumors, and cancers of the breast, prostate, lung, stomach, colorectal, liver, and the urinary bladder.

Pesticides cause genetic and epigenetic changes by involving various processes at cellular levels. Pesticides may be involved in endocrine disruption and induction of inflammatory signals which result in production of reactive oxygen species (ROS) causing oxidative stress. ROS disrupt the cellular functions of mitochondria and endoplasmic reticulum.

This chapter covers different types, importance and modes of action of pesticides. Human exposure to pesticides and pesticide residues in food are also discussed. Finally, the impacts of pesticide exposure on human health with focus on the major chronic health effects (neurotoxic, genotoxic and carcinogenic, and reproductive effects) and recent findings regarding health effects associated with exposure to common types of pesticides, i.e., organochlorines, organophosphates, carbamates, pyrethroids and neonicotinoids insecticides, fungicides and herbicides are discussed.

2. Types of pesticides and pesticide formulations

2.1 Types of pesticides

Pesticides can be classified based on chemical classes, functional groups, mode of action, and toxicity. The active ingredients of most pesticides are either organic (contain carbon) or inorganic (minerals e.g. copper sulfate, ferrous sulfate, copper, lime, sulfur, etc.). Organic pesticides are hydrophobic and more complex than those of inorganic pesticides. Organic pesticides can be natural (produced from naturally available sources) or synthetic (artificially produced by chemical synthesis in factories). The major types of pesticides used in agriculture, forestry, landscape, medical and veterinary sectors are listed in **Table 1**.

2.2 Pesticide formulations

Pesticides are sold as formulated products. Pesticide formulations are a combination of one or more active ingredients (a.i.) and several inert ingredients. Active ingredients control the pests. The inert ingredients help in solubility and stability of the product. A ULV (Ultra Low Volume) formulation need specialized spray equipment and the Ready-to-Use formulations are already diluted and are appropriate for indoor or small areas, for example, aerosols (A), granules (G), and most baits (B) [1].

Most liquid formulations are diluted with water according to the label directions. The three main types of liquid formulations are solutions, suspensions, and emulsions. A true solution is a mixture that cannot be separated by a filter or other mechanical means while a suspension is an even mixture of very small solid particles throughout a liquid and an emulsion is a mixture of droplets of one liquid

Type of pesticide	Active ingredient	Target pests
Insecticides	Natural and synthetic	Insect (6-legged) pests of agricultural, forestry, landscape, medical and veterinary importance
Miticides/acaricides	Natural and synthetic	Mites (8-legged) pests of agricultural, forest, landscape, medical and veterinary importance
Fungicides	Natural and synthetic	Fungal diseases (molds, mildews, rust) of agricultural, forestry and landscape importance
Herbicides	Natural and synthetic	Unwanted plants (weeds) of agricultural and landscape importance
Insect growth regulators	Synthetic	Disrupt the growth and reproduction of insect pests. IGR are species or genus specific.
Pheromones	Natural and synthetic	Attract and trap male insects and are often species-specific.
Plant growth regulators	Synthetic	Alter plants growth, e.g., induce or delay flowering
Algaecides	Natural and synthetic	Algae growing on different surfaces, e.g., patios
Molluscicides	Natural and synthetic	Slugs and snails of agricultural, forestry and landscape importance
Biopesticides	Natural	Can be insecticides, fungicides or herbicides
Antimicrobials	Synthetic	Microbes (mostly bacteria) of medical and veterinary importance
Rodenticides	Natural and synthetic	Rodents (mice, rats) in agriculture, landscape, building, storages and hospitals
Treated seeds	Synthetic	Seeds coated with an insecticide or fungicide or both to prevent damage from soil insect pests and fungus diseases
Wood preservatives	Synthetic	Pesticides to protect wood from insect pests, fungus and other diseases
Minimum risk pesticides	Natural and synthetic	Any pesticides which have been proven safe for human and are exempt from registration by any regulatory authorities

Table 1.
Major types of pesticides used in agriculture, forestry, landscape, medical and veterinary sectors. (adopted from: National Pesticides Information Center at http://npic.orst.edu/ingred/ptype/index.html).

in another liquid. Common Liquid Formulations are Emulsifiable Concentrate (E or EC), Solutions (S, CS), Emulsions in Water (EW), Flowables (F, L, or SC), Microencapsulated Pesticides (M or ME) and Aerosol (A).

In dry formulations the active ingredient is on the surface of a solid carrier, such as talc, clay, or ground corncobs. Common solid formulations include Granules (G), Wettable Powders (WP or W), Soluble Powder (SP or S), Water-Dispersible Granules (WDG) or Dry Flowables (DF Water-Soluble Bags/Packages (WSB) and Baits (B).

3. Importance of pesticides

The United Nations population division estimates 9.7 billion people by the year 2050 and to feed them, the Food and Agriculture Organization (FAO) of the United Nations estimates that an 80% increase in food production is necessary.

This increase in production will come from an increase in yields of crops as well as a decrease of damage to crops due to pests. There are approximately 9000 species of insects/mites (14% loss), 50,000 species of plant pathogens (13% loss) and 8000 weeds species (13% loss) worldwide [2]. Without pesticide application the pest losses to fruits, vegetables and cereals would reach 78%, 54% and 32%, respectively. Pesticides are, therefore, indispensable in agricultural production and there will be a need for pesticide based pest control and food security in the future. Pesticides are also used to control vector-born infectious diseases such as Zika virus, Lyme disease, and rabies, household pests like cockroaches, bed bugs, and as repellents etc. More than 1000 active ingredients are used in pesticides around the world to ensure food safety and prevention from pests and the highest amount (~45%) is spent on herbicides followed by insecticides, fungicides, and other types of pesticides.

4. Human exposure to pesticides and exposure risks

4.1 Human exposure to pesticides

Human beings get exposed to pesticides either actively through occupational exposure or passively through non-occupational exposure. Pesticides occupational exposure may occur during manufacturing, transportation, sale, and application process including exterminators. For example, in an incident of occupational exposure, 2800 workers were poisoned during malathion spray for malaria vector control in Pakistan [3]. Parents working in agriculture industry usually take pesticide contaminated clothing, equipment home, which has been associated with the development of cancers in their children.

Non-occupational exposure may include pesticides residues ingestion with contaminated food and water and inhalation of pesticides droplets from the air through drift from point of release or fumigation. Human beings are also exposed to residual indoor sprays and outdoor fogging of insecticides applied against insect pests of public health importance and homeowners exposed to structural pest control pesticides. Additionally, treatment of ectoparasites in pets, e.g. fleas, is also a source of exposure, especially for children.

Exposure through the intact skin (dermal exposure) is the most common route and may occur as a result of a splash, spill, or spray drift, during mixing, loading, disposing, and/or cleaning of application equipment especially when proper protective equipment are not used. Dermal absorption can be influenced by the amount/concentration, duration of exposure and temperature/humidity. Absorption is high through groin areas, the eyes and ear canal. Liquid formulations (e.g., emulsifiable concentrates) are readily absorbed through the skin compared to the solid formulations (e.g., powders, dusts, and granules).

Accidental ingestion of pesticides (oral exposure) occurs by drinking from unlabeled containers when pesticides are stored in food/drink container, water stored in pesticide-contaminated bottles, eating or smoking while, or after handling pesticides or through application equipment or pesticide residues in food and water. Inhalation of pesticides (respiratory exposure) may occur due to application of fumigants (which change into toxic gas after coming in contact with moisture in air) or presence of fine droplets in air (particle or vapor drift) after application of pesticides. Pesticides can enter blood stream after absorption through lungs.

Pesticides are distributed throughout the human body through the bloodstream and are excreted through urine, skin, and exhaled into air after metabolism. These pathways also determine the toxicity of any pesticide. Pesticides recognized as persistent organic pollutant (POP) are fat soluble and are easily accumulated within the human fat-tissues, breast milk, and maternal blood placenta.

4.2 Pesticides exposure risks

The amount of risk from pesticide exposure depends on the toxicity and the exposure to the pesticide. Toxicity is a measure of how harmful or poisonous a pesticide is (causing sickness or other unwanted effects), while exposure is a measure of the contact (duration) with a pesticide. Toxicity of a pesticide is measured as lethal dose (LD_{50}). The LD_{50} value is the statistical estimate of a pesticide (mg/kg of body weight) which will kill 50% of the test animals within a stated period of time (24 hours to 7 days). The LD_{50} value also depends on the route of entry of a pesticide; oral LD_{50} for oral ingestion, dermal LD_{50} for skin contact exposure and Lethal Concentration (LC_{50}) for inhalation of fumigants and pesticide vapors.

A short term exposure or exposure to a single dose will cause acute toxicity with its health effects. Chronic toxicity results from repeated exposure to a pesticide over a longer period of time from several months to years. Hazard symbols, signal words and color on the primary display panel of a pesticide label are based on their dermal toxicity.

5. Pesticides modes of action

5.1 Insecticides

Insecticides Resistance Action Committee (IRAC) has classified insecticides into 32 groups based on their mode/site of action, in addition, there are 5 other types of insecticides with unknown modes of action. Most commonly used insecticides work at different sites in the nervous system of insects. Insecticides target the same sites of action in human nervous system and cause toxicity with adverse health effects. Carbamate (group 1A) and Organophosphate (OP) (group 1B) insecticides inhibit the enzyme Acetyl Choline Esterase (AChE) and cause hyper-excitation. AChE terminates the action of the excitatory neurotransmitter acetylcholine at the nerve synapses. Examples of pesticides inhibiting AChE include dichlorvos, malathion, phorate, carbaryl, carbofuran, etc. Cyclodiene organochlorine insecticides (OC) (group 2A) and phenylpyrazoles (group 2B) block the gamma amino butyric acid (GABA)-activated chloride channel causing hyper-excitation and convulsions. GABA is the major inhibitory neurotransmitter in insects. Examples of insecticides inhibiting GABA include endosulfan and fipronil. Synthetic pyrethroids and natural pyrethrins (group 3A) and DDT (group 3B) keep sodium channels open causing hyper-excitation and, in some cases, nerve blockage. Sodium channels are involved in the propagation of action potentials along nerve axons. Examples include deltamethrin and permethrin. Neonicotinoid insecticides (group 4A) bind to the acetylcholine site on nicotinic acetylcholine receptor (nAChRs) causing a range of symptoms from hyper-excitation to lethargy and paralysis. Examples include acetamiprid, clothianidin, imidacloprid, thiacloprid and thiamethoxam. Other groups of insecticides that work on nervous system includes those which allosterically activate nAChRs (e.g. spinetoram, spinosad) or glutamate-gated chloride channels (GluCls) (e.g. abamectin, emamectin benzoate), or allosterically inhibit the GABA-activated chloride channel and cause paralysis (e.g. broflanilide and fluxametamide). Glutamate is an important inhibitory neurotransmitter in insects. Other insecticides will block the nAChR ion channel or sodium channels, e.g. indoxacarb, cause nervous system shutdown and paralysis.

5.2 Fungicides

Fungicides inhibit fungal growth by interfering with critical cellular processes. Fungicide resistance action committee (FRAC) classify fungicides and bactericides into 50 groups based on the site of action. Within each group, there are target sites, which are the specific enzymes to which the fungicides bind. The different known target sites include nucleic acids metabolism, cytoskeleton and motor protein, respiration, amino acids and protein synthesis, signal transduction, lipid synthesis or transport/membrane integrity or function, sterol biosynthesis in membranes, cell wall biosynthesis, melanin synthesis in cell wall and host plant defense induction. Some fungicides and herbicides are considered endocrine disrupting pesticides.

5.3 Herbicides

Herbicides are pesticides that inhibit or interrupt normal plant growth and development. Herbicides are widely used in agriculture, landscape industry, and non-crop areas for weed management. Herbicides resistance action committee (HRAC) has classified herbicides into 27 groups. These include: growth regulators (synthetic auxins; auxin transport inhibitors), seedling growth inhibitors, photosynthetic inhibitors, amino acid synthesis inhibitors, lipid synthesis inhibitors, cell membrane disrupters, pigment inhibitors.

Growth regulator herbicides consist of the synthetic auxin and auxin transport inhibitory compounds and the most commonly used synthetic auxins include 2,4-*Dichlorophenoxyacetic acid* (2,4-D), fluroxypyr, dicamba, quinclorac, dichlorprop, MCPA (2-methyl-4-chlorophenoxyacetic acid), mecoprop and picloram. These are commonly used systemic herbicides which mimic the plant growth hormone auxin (indole acetic acid) [4]. Some of these synthetic auxin herbicides disrupt human hormonal system. Atrazine is also a commonly used photosynthetic inhibitor herbicide. Glyphosate (Roundup) is an amino acid derivative and inhibits synthase of EPSPS enzyme, which is involved in the synthesis of the aromatic amino acids (tyrosine, tryptophan, and phenylalanine). Paraquat (gramoxone) is an electron diverter, and as a respiratory inhibitor can be a significant risk to humans if inhaled or ingested.

6. Pesticide residues in food, water and air

6.1 Pesticide residues

'Pesticide residue' means any specified substance in food, agricultural commodities, or animal feed resulting from the use of pesticides. The term also includes any derivatives of a pesticide, such as conversion products, metabolites, reaction products, and impurities considered to be toxic. Application of pesticides during the production or storage of agricultural commodities result in pesticide residues in food (fruits, vegetables, grain, meat, etc). Pesticide residues are also found in the drinking water. Pesticide residues can build up to harmful levels through bio-accumulation and bio-magnification within the food chain.

WHO, in collaboration with FAO performs pesticide risk assessment to humans, both through direct exposure and through residues in food. The WHO core assessment group on pesticide residues review toxicological data and establish the acceptable daily intakes (ADIs) and acute reference doses (ARfDs) of pesticide residues for different commodities through a lifetime of food consumption. The ADIs are amount of pesticide residues which will not result in adverse health effects. Codex

Alimentarius Commission (the intergovernmental standards-setting body for food) establishes maximum residue limits (MRLs) for pesticides in food based on ADIs.

The MRL depends on the crop it is used on, and the same pesticide active ingredient may have different MRL values when used on different crops. Extraneous maximum residue limit (EMRL) refers to the maximum permitted limit of residues of mostly POP pesticides, which were previously used as pesticides but not registered any more, and residues arising from environmental contamination (including previous agricultural use) or residues from uses of these pesticides other than for agricultural purpose, e.g. DDT, Aldrin, etc.

6.2 Pesticide residues in food, water and air

There are several reports of pesticide residues detected on food exceeding the MRL values. For example, in India, vegetable samples were tested for the presence of OC, OP and pyrethroid insecticides, and 15.3% samples exceeded the MRL. In two Brazilian pesticide residue monitoring programs less than 3% of the samples had residue levels above the MRL. Pesticide residues were detected in 34% of samples of cereal grains collected throughout Poland and 3% samples contained residues over the maximum limit. A study from Maule Region (Talca, Chile) found pesticide residues on the fruits and vegetables schoolchildren brought as snack [5].

The pesticide residues detected in fruits and vegetables from Lithuania had multiple pesticides; 9 residues in grapes and tea, 5-9 residues in orange, mandarins, lemons, peaches, pears and 3-5 residues in pomegranates, plums, cucumbers, tomatoes and strawberries, and found that 2.6% samples exceeded the MRL values [6]. In a European Union study 14–23% of the samples had detectable residues of more than one active ingredient where 3.0–5.5% samples had residues levels above the MRL [7]. Exposure to multiple pesticide residues could be due to intake from a single food item containing multiple residues or from several food items each containing one or more residues. The combined toxic effects of two or more compounds can be independent, additive or synergistic.

Both recreational and medicinal cannabis samples contained high levels of residual pesticides and pesticides not legally allowed to be used on cannabis products in Oregon. Medicinal cannabis products were found to have mean levels of residual pesticides that were 3-12 times higher than recreational products, and 9 of the 50 pesticides identified were classified highly or extremely hazardous by the WHO [8].

Pesticide residues have been found in surface, groundwater and potable water samples from India [9]. Pesticide residues levels in river water and in drinking water samples in Turkey were significantly high compared with guideline values set by Turkey, EU and WHO as hazardous to human health [10]. Higher concentrations of pesticides in ambient air were recorded from potato farm sites in Prince Edward Island, Canada, Taihu Lake region of China and Kaweah Reservoir, CA, USA. A total of 87 pesticides were identified in the household dust samples from the rural Yakima Valley of Washington state, 47 of these have evidence of neurotoxicity included in the EPA list [11].

7. Impacts of pesticide use on human health

7.1 Acute health effects of pesticide exposure

The short-term acute adverse effects pesticide exposure on human health are stinging eyes, rashes, blisters, skin irritations, blindness, nausea, dizziness, diarrhea and death. Exposure to pesticides in agricultural work can cause serious risks to the respiratory system causing chronic cough, dyspnea, wheezing and expectoration, decreased lung capacity, asthma, and bronchitis. These respiratory problems were found in workers in flower crops in Ethiopia, coffee plantations in Brazil and banana plantations in Costa Rica. In banana farming in Rio Grande do Norte (Brazil), the use of pesticides was related to the symptoms of burning in the throat and lungs, airway congestion, cramps, skin peeling, diarrhea, headache, chest pain, weakness, cough and skin irritation.

In banana production region of the Ribeira Valley (Brazil), workers (majority males, low schooling, mean age 39.6 years and 13.8 years of working time) had moderate obstructive disorder (10.0%) and mild obstructive disorder (13.3%) with decreased FEV1 (forced expiratory volume in 1 second) and FEV1/FVC (the ratio between forced expiratory volume in the first second and forced vital capacity and is very important for the detection of obstructive disorders). Similarly, exposures to mixtures (pollutants and pesticides) in children with asthma in California were also associated with reduced lung function measures FEV1 and FVC [12].

Many studies have found positive associations with pesticide exposure and children's respiratory and allergic effects such as asthma, wheezing, coughs, acute respiratory infections, hay fever, rhinitis, eczema, chronic phlegm, and lung function impairments. A study of school-age children with asthma in the agricultural community of Yakima Valley (Washington State) found that increase in exposures to OP insecticides was related with increase in LTE4 levels which was associated with a higher risk of asthma morbidity [13]. The neonicotinoid insecticides (e.g. imidacloprid, nitenpyram) are nicotinic receptors agonists and their exposure cause nausea, vomiting, muscle weakness, respiratory effects, headache, lethargy, and tachycardia.

7.2 Chronic effects of pesticide exposure

The long-term chronic adverse effects of pesticides exposure are cancers, birth defects, reproductive harm, neurological and developmental toxicity, immunotoxicity, and disruption of the endocrine system. The chronic effects of pesticides on human can be categorized into three major groups; neurotoxic effects, genotoxic and carcinogenic effects, and reproductive effects.

7.2.1 Neurotoxic effects

Neurotoxicity can be defined as any adverse effect on the central or peripheral nervous system caused by chemical, biological or physical agents. A developing nervous system in children (during replication, migration, differentiation, myelination of neurons, and synapse formation) is more susceptible to neurotoxic chemicals including pesticides. Chemicals (pesticides) can cause neuronal cell death by disruption of the cytoskeleton, induction of oxidative stress, calcium overload, or by damaging mitochondria. Most of the synthetic insecticides, some fungicides and herbicides, currently in use are neurotoxicants.

Pesticide molecules are small and lipophilic in nature, and can enter from blood to brain and then in neurons, glial cells and brain micro vessels. Pesticides can disrupt blood-brain barrier receptors in the central nervous system which enhance chronic toxicity and affect the receptor-mediated transcytosis. Neuronal cells are more susceptible to oxidative stress due to their high polyunsaturated fat content in the myelin sheaths, low anti-oxidative capabilities, enzymatic systems with transient metals that aid in the production of free radicals, and demand for high oxygen and glucose metabolism rate.

OPs and carbamates bind to and phosphorylate/carbamalate the AChE which causes accumulation of acetylcholine at cholinergic synapses causing overstimulation of muscarinic and nicotinic cholinergic receptors. Neuropsychiatric disorders, such as anxiety and depression, are observed in patients with acute and long-term poisoning from OPs. OPs may also cause an intermediate syndrome and OP-induced delayed polyneuropathy (OPIDP) 1-3 weeks after a single exposure. In carbamates, the AChE inhibition is reversible and acute intoxication is generally resolved within a few hours.

The OP insecticides can disturb the function of mitochondria by inducing oxidative stress in central nervous system through critical depletion of mitochondrial energy, the activation of proteolytic enzymes, and DNA fragmentation leading to apoptosis. The dysfunction of mitochondria and oxidative stress is responsible for several neurological diseases, including Parkinson's disease, seizure, cognitive dysfunction, attention and memory deficits, dementia, depression, and Alzheimer's disease. OP triggered induction of a xanthine oxidase may play a role in cognitive impairment.

In a study, increased inhibition of cholinesterase enzyme with increased exposure to OP insecticides was confirmed in both occupationally exposed (OE) and environmentally exposed (EE) groups of people. The OP exposure, mainly in the EE group, was associated with a diminished neuropsychological performance; general mental status, language, memory, attention, executive function, praxis and psychomotricity.

Acute poisoning due to exposure to OP (particularly chlorpyrifos) was reported with higher prevalence of peripheral polyneuropathy, and deterioration of cognitive functions (verbal fluency, and visual and auditory memory) was observed in agricultural workers and in inhabitants of rural agricultural areas. Exposure to OP insecticides in rural schoolchildren was associated with a lower processing speed in children and an IQ lower than expected for their age.

Exposure to type I pyrethroids cause tremor syndrome (behavioral arousal, aggressive sparring, increased startle response, and fine body tremor progressing to whole-body tremor, and prostration) while type II pyrethroids exposure cause salivation syndrome (profuse salivation, coarse tremor progressing to choreoatetosis, and clonic seizure). The poisoned cerebral cortex affect learning, memory, emotions, and movement. Pyrethroids exposure has been positively associated with hearing loss in U.S. adolescents. Pyrethroids exposure induced Tau protein malfunction which may be the mechanism underlying cognitive impairment. Paraquat, triazine and pyrazole (herbicides) through oxidative stress, raised influx of calcium and the stimulation of nitrogen oxide species, and aggravated Aβ amyloidogenesis cause cognitive impairment.

Exposure to endocrine disrupting chemicals (EDCs) including many pesticides can disrupt maternal thyroid imbalance which can result in permanent and lifelong neurodevelopmental consequences for their children, including attention-deficit disorder, autism spectrum disorder, and cognitive and behavioral dysfunction. Workers of fruit and seed export companies in a rural area of Santiago exposed to methyl bromide (CH3Br, a fumigant) had increased concentration of CH3Br in blood after application which resulted in a higher frequency of insomnia, headaches, paresthesias, mood swings, memory loss, and decreased concentration [14].

Parkinson's disease (PD) is characterized by progressive degeneration of dopaminergic neurons of the nigrostriatal pathway and the formation of alpha-synuclein (α-syn)-containing Lewy bodies. Dieldrin (OC) is selectively toxic to dopaminergic cells, disrupts striatal dopamine activity, and may promote α-syn aggregation while ziram (dithiocarbamate fungicide) increases the probability of synaptic vesicle release by dysregulation of the ubiquitin signaling system and increases excitability in both aminergic and glutamatergic neurons leading to PD.

7.2.2 Genotoxic and carcinogenic effects

A genotoxic agent can be a physical, chemical or biological agent that can interact with the genetic material (DNA) causing alterations, damage or ruptures, and those that interfere with enzymatic processes of repair, genesis or polymerization of proteins involved in chromosome segregation. These alterations could lead to impaired embryonic development or be the initial steps in the development of cancer. Pesticides exposure can cause genomic damage. Genetic damage caused by pesticides is broadly classified into three classes; (i) Pre-mutagenic damage like DNA strand breaks and DNA adducts (ii) gene mutations like insertion, deletion, inversion and translocation (iii) chromosomal aberrations, including loss or gain of whole chromosome (aneuploidy), deletion or breaks (clastogenicity), and chromosomal rearrangements.

Farmers exposed to pesticide mixtures in Greece had possible clastogenic (chromosome breakage cause mutation) and aneugenic (abnormal number of chromosomes) effect of pesticides on the genetic material. DNA methylation changes in the placenta were significantly associated with the maternal plasma concentrations of OCs in early pregnancy causing prenatal toxicity. OPs affect DNA methylation, induce the AChE gene expression and activate the NMDA glutamate receptors resulting in calcium influx in the post-synaptic neurons leading to degeneration.

Genetic damage has been reported from exposure to malathion (OP), carbofuran (carbamate), triflumuron (Insect growth regulator), imidacloprid, acetamiprid and thiamethoxam (neonicotinoid insecticides), pentachlorophenol (OC), Emamectin benzoate (used in agriculture, household, and veterinary medicine), and tembotrione (novel post-emergence herbicide) (**Table 2**).

Cancer is characterized by an uncontrolled cell growth with limitless replication, resistance to apoptosis, alteration of growth factors (GFs), resistance to chemotherapy, metastasis and angiogenesis. Cancer develops as a result of multifactorial complex interactions of genetic and lifestyle factors including, diet, stress, physical and biological agents, infections, and exposure to the hazardous chemical substances. Pesticides exposure acts as a stimulant to cancer and chronic low-dose is considered one of the important risk factors for the increasing cancer incidence. **Table 3** presents a list of pesticides suggesting carcinogenicity in different types of studies.

WHO Hazard Class	Band color	Signal word	Dermal LD_{50} (mg/Kg)	
			Solid formulation	Liquid formulation
Class Ia Extremely Hazardous	Red	VERY TOXIC	<10	<40
Class Ib Highly Hazardous	Red	TOXIC	10–100	40–400
Class II Moderately Hazardous	Yellow	HARMFUL	100-1000	400-4000
Class III Slightly Hazardous	Blue	CAUTION	>1000	>4000
Class U Products unlikely to present a hazard	Green			

Table 2.
Pesticides hazard classification by FAO.

Type of cancer	ToP	Name of pesticide	Type of study	Reference
Non-Hodgkin lymphoma (NHL) and Hodgkin lymphoma (HL)	OC	P,p'-DDT	Case control	[15]
		P,p'-DDE	Agricultural health	[16]
		HCH	Case control	
	MoC	Nonachlor/ trans-nonachlor hexachlorobenzene	Blood Agricultural health	
	OC	Mirex	Case control	
		Chlordane	Case control	
		Lindane	Case control	
	OP	Malathion	Case control	[17]
		Diazinon		
		Terbufos	Case control	[18]
		Dimethoate chlorpyrifos	Agricultural health	[15]
	PYR	Permethrin	Case control	[16]
	NPYR	Pyrethrum	Agricultural health	[17]
	PHE	2,4-D	Case control	[19]
		Mecoprop	Epidemiological	[20]
	CHL	Dichlorprop	Case control	[21]
	BNZ	Dicamba	Case control	[20]
	GLY	Glyphosate	Case control	[16]
Breast	OC	Pp'-DDT	Histopathology	[22]
		Pp'-DDD	Histopathology	[23]
		P,p′-DDE		
		β-HCH	Histopathology	[24]
		Heptachlor		
		Hexachlorobenzene		
	OP	Chlorpyrifos	MCF-7 breast cancer cells	[25]
		Malathion		
		Terbufos	Case control/MCF-7/ MCF-10F	[26]
		Diazinon		
		Dimethoate		
	PYR	Flucythrinate	AutoDock Vina 1.1.1	[27]
		Fluvalinate		
		Bifenthrin		
		Cyhalothrin		
		Cypermethrin		
	NEO	Thiacloprid imidacloprid	Hs578t cells	[28]
	PTH	Captan	Agricultural health	[29]
	GLY	Glyphosate	Case control	[30]

Prostate	OC	Pp'-DDT Lindane	Case-control	[31]
		Endosulfan	Human prostate cancer PC3 and DU145 cell	[32]
	OB	Methyl bromide	Agricultural health	[33]
	OP	Chlorpyrifos	Prostate epithelial lines	[34]
		Dimethoate	Agricultural health	[35]
		Malathion Carbaryl	Case-control	[31]
	PYR	λ-Cyhalothrin	Prostate epithelial lines	[34]
		Bifenthrin	PC3 human	[36]
		Deltamethrin	Prostate cancer cell	[37]
	QUI	Dichlone	Case control	[31]
	IMI	Prochloraz	PC-3 prostate cancer cells	[38]
	DIC	Vinclozolin		
	MoV	M2		
	CHL	2,4-D	Case control	[31]
		2,4-DB	Histopathology	[39]
		2,4,5-T		
	CHP	Picloram	Histopathology	[39]
	ORG	Cacodylic acid	Case control	[31]
	TRI	Simazine Atrazine	RM1 cells	[40]
	Mo2	2, 4-dichlorophenol (DCP)	Case control	[31]
	MoD	Dinoseb amine		
	GLY	Glyphosate	Prostate epithelial lines	[34]
Lung cancer	OP	Diazinon	Epidemiological	[41]
	PYT	Cypermethrin	Lewis lung cancer cells	[42]
	αCH	Acetochlor	Agricultural health	[43]
	TRI	Atrazine		
Bladder	IMZ	Imazethapyr imazaquin	Agricultural health	[44]
Hepatocellular carcinoma	OC	Pp'-DDT	Serum levels	[45]
		Pp'-DDE	Toxicological	[46]
	OC	Endosulfan	Human liver carcinoma cells (HepG2)	[47]
	CAR	Carbaryl	Toxicological	[46]
	BEZ	Fluopyram	Female rat	[48]
	BED	Carbendazim	Toxicological	[46]
	BEN	Dicamba	Agricultural health	[49]
	αCH	Acetochlor	Human liver carcinoma cells (HepG2)	[47]
Stomach	TRI	Atrazine	Agricultural health	[50]

Thyroid	OP	Malathion	Agricultural health	[51]
	TRZ	Penconazole	Agricultural health	[52]
	TRI	Atrazine	Agricultural health	[53]
		Amitrole	Nthy-ori-3-1 cell	[54]
Ovarian	OC	Pp'-DDT	Blood	[55]
		Pp'-DDE		
		β-HCH		
		Endosulfan		
	OP	Diazinon	Agricultural health	[51]
	PYR	λ-Cyhalothrin	BG-1 ovarian cancer cells	[56]
		Cypermethrin		
		Cyprodinil		
	HYD	Fenhexamid	Mouse model with transplanted BG-1 cells	[56]
Colorectal	OC	Pp'-DDE		[57]
		Endosulfan		
	OP	Chlorpyrifos	Human colorectal adenocarcinoma H508 cells	[58]
	CAR	Aldicarb		
	αCH	Acetochlor	Agricultural health	[51]
Brain	OP	Dichlorvos	Male albino Wistar rats	[59]

ToP, type of pesticide; OC, organochlorine insecticide; MoC, metabolites of chlordane; OP, organophosphate insecticide; PYT, pyrethroid insecticide; NPYT, natural pyrethroid insecticide; PHC, phenoxy-carboxylate herbicide; CHL, chlorophenoxy herbicide; BEN, benzoate herbicide; GLY, glycine herbicide; NEO, neonicotinoid insecticide; PHT, phthalimide fungicide; OB, organobromine insecticide; QUI, quinone algicide; IMI, imidazole fungicide; DIC, dicarboximide fungicide; MoV, metabolite of vinclozolin; CHP, chlorinated pyridine herbicide; ORG, organoarsenic herbicide; TRI, triazine herbicides; Mo2, metabolite of 2,4-D; Mod, metabolite of dinoseb dinitrophenol herbicide; αCH, α-chloroacetamides herbicide; IMZ, imidazolinones herbicides; CAR, carbamate insecticide/nematicide; BEZ, benzamide, pyramide fungicide; BED, benzimidazole fungicide; TRZ, triazole fungicide; HYD, hydroxyanilides fungicides.

Table 3.
List of Pesticides Suggesting Carcinogenicity in different types of studies.

7.2.2.1 Non-Hodgkin lymphoma and Hodgkin lymphoma

Non-Hodgkin lymphoma (NHL) is a diverse group malignancies and its incidence has increased worldwide. Patients with immune dysfunction are at a high risk to develop NHL. Studies have reported an elevated risk of NHL with exposure to several classes of pesticides. Terbufos (OP nematicide), dimethoate, malathion and chlorpyrifos (OP insecticide), and 2,4-D and dichlorprop (chlorophenoxy herbicides) have been associated with significant risk of developing HL.

7.2.2.2 Leukemia

Leukemia has been associated with occupational exposure with a higher risk in livestock farmers and golf course superintendents. The risk of chronic myelocytic leukemia (CML) and acute myeloblastic leukemia (AML) was found to be higher in women. Children whose parents used garden and indoor insecticides, or whose mothers had been exposed while pregnant had increased rates of all types of leukemia. Children living on farms and those exposed to household pesticides have increased risk of leukemia. Association between occupational exposure to pesticides and chronic lymphocytic leukemia (CLL) has been reported from Spain. A nationwide study in France showed a moderate increase in incidence of childhood AL in municipalities where viticulture is common.

7.2.2.3 Brain cancer

Brain tumors are the most common solid tumors in children and the leading cause of cancer-related mortality during childhood. A positive association has been reported between parental occupational, prenatal or residential exposure, living on a farm, mothers living on farms, rural activity and childhood brain tumors. Increased risk for primitive neuroectodermal tumors (PNETs) was associated with maternal exposure living on pig or poultry farms. Exposure to pyrethroid formulations used to control mosquitoes and cockroaches at home also increase the risk of brain tumors.

7.2.2.4 Breast cancer

Breast cancer is the leading cause of cancer-related deaths among women. About 650 pesticides out of the 800 used worldwide can affect the functioning of the endocrine system and are called endocrine disrupting pesticides (EDPs). EDPs have the potential ability to act as tumor promoters and increasing risk of breast cancer. All women diagnosed with breast cancer between 1995 and 2005 in the city of Arica (geographic area that received massive aerial applications of malathion in 1980) were 5.7 times more likely to suffer from breast cancer compared to women diagnosed during the same period in the city of Iquique, Chile [14]. Several chemical classes of insecticides, fungicides and herbicides have been associated with breast cancer in women (**Table 3**).

7.2.2.5 Prostate cancer

Prostate cancer is the second most common cancer in men globally, and accounts for 7% of all cancers. More than 95% of cases of prostate cancer are androgen-dependent. The higher incidence of prostate cancer, at least in part, has been associated with the hormone disrupting pesticides and consistent positive associations between prostate cancer and pesticide exposure have been reported.

7.2.2.6 Hepatocellular carcinoma

Hepatocellular carcinoma (HCC) is the 6th most common cancer, and the 4th most common cause of cancer-related mortality. The major risk factors include hepatitis B virus (HBV), hepatitis C virus (HCV), alcohol, aflatoxin contaminated foods, obesity, smoking and type 2 diabetes besides pesticides. Pesticides exposure has been associated with increased risk of developing HCC.

7.2.3 Reproductive effects

EDCs are emerging as one of the leading risks and are recognized as serious and urgent threats to public health. In laboratory studies, EDCs are reported to shorten gestation, alter intrauterine growth, and disrupt metabolic programming.

Prenatal exposure to EDCs can affect fetal neurodevelopment through disruption of peroxisome proliferator activated receptors, mainly estrogen receptors, and thyroid hormone receptors.

Failure of testosterone production in Leydig cells leads to failure of testosterone-bound androgen receptor-mediated gene transcription necessary for spermatogenesis. Many studies have shown that various pesticides decrease testosterone levels. Testosterone is required for the final stages of sperm maturation, so a decrease in intra-testicular testosterone is likely to impair fertility. Vinclozolin (fungicide) and chlorpyrifos (OP) can reduce testosterone production. Exposure to higher concentrations of OP and dialkyl phosphates (metabolites of OPs), p,p'-DDE, fenvalerate and atrazine (chlorotriazine herbicide) have been consistently associated with lower semen quality (sperm concentration, motility, and morphology).

A study of male children from a village of cashew plantations, where endosulfan (OC, EDC) had been aerially sprayed for more than 20 years, showed a delay in sexual maturity and an alteration in sex hormone synthesis. Endosulfan, in exposed mothers, can move via trans-placental route and breast feeding to children. Exposure during critical periods of development might contribute to decline conception rates and increased incidence of female reproductive disorders, such as altered cyclicity, endometriosis, fetal growth retardation, and pregnancy loss [60].

A high incidence of spontaneous abortions 81.02 / 1000 live newborns was reported in Valparaíso Region (agricultural area) compared to 9.5 /1000 live newborns in the rest of Chile. A 28% incidence of congenital malformations in live newborns was reported in the O'Higgins Region (agricultural area) compared to only 15% of cases in non-agricultural in Chile [14].

8. Conclusion

Pesticides are used in managing pests of agricultural and public health importance, and their use will continue in future because of food security and vector control. Additionally, pesticides are used at home in fumigation for structural pests and to mitigate household pest using aerosols or sprays. It is difficult to eliminate pesticides in the near future, but they should be used with care and caution. Most pesticides are potentially toxic to human beings resulting in severe health consequences including cancers.

Epidemiological evidence suggests that there is an increased incidence of different diseases including leukemia, lymphoma, and several other types of cancers in farmers, and those who are associated with application of pesticides. There is also evidence that parental exposure, as well as, exposure in early life or adolescence could increase the longer-term risks.

Since animal studies are problematic, expensive and often generate ethical problems, cell cultures are increasingly used as a model of research. Correctly conducted and properly selected, the cell culture is an excellent experimental model reflecting human exposure to different xenobiotics through all relevant routes. The cell cultures are also becoming more widely used to study the effect of pesticides on the human body at a molecular level, which is necessary to understand the hazards and determine the level of exposure.

Some pesticides (OCs) are no longer used worldwide due to their persistence and toxicity. However, their residues or metabolites are still found in food and water samples. The use of OPs and carbamate insecticides has been reduced since the arrival of newer chemistries in different parts of the world but most of them are still use around the world.

The workplace safety standards and proper pesticide management and storage must be implemented to reduce the risks posed to human health. Pesticide users should be aware of their risks and proper handling, as well as must use personal protective equipment which are effective in reducing damage to human health. To ensure healthy childhood growth, efforts should be made to develop comprehensive pesticides risk mitigation strategies and interventions to reduce children's exposure.

It is critical to achieve sustainable development in agricultural systems. Newer approaches in pest management have been developed which should be encouraged. For example, RNA interference- (RNAi-) based pesticides are emerging as a promising new biorational control strategy [61] and steam treatment at temperature of 150.56°C can kill 93.99% of nematode 97.49% of bacteria [62].

Future research need in the context of minimizing the impact on human health due to exposure to pesticides include an urgent need to eliminate the use of carcinogenic pesticides and to develop environmentally sound integrated pest management (IPM) strategies that use the minimum amount of pesticides. Such IPM strategies should aim at reducing the pesticides residues on food products and pesticides-free water and air.

Acknowledgements

The author acknowledges the financial support by the Sultan Qaboos University, Muscat, Oman. This work was funded through an Internal Grant # IG/AGR/CROP/18/02.

Author details

Riaz Shah
Sultan Qaboos University, Muscat, AlKhod, Oman

*Address all correspondence to: riazshah@squ.edu.om

References

[1] University of Kentucky. PESTICIDE FORMULATIONS; Kentucky Pesticide Safety Education Program [Internet]. 2020. Available from: https://www.uky.edu/Ag/Entomology/PSEP/3formulations.html

[2] Pimentel D. Integrated Pest Management. In: Peshin, Rajinder, Dhawan AK, editor. 1st ed. Springer; 2009.

[3] USAID. Integrated Vector Management Programs for Malaria Vector Control. 2007. p. 524.

[4] HRAC. HRAC MODE OF ACTION CLASSIFICATION 2020 MAP [Internet]. 2020. Available from: https://www.hracglobal.com

[5] Muñoz-Quezada MT, Iglesias V, Lucero B, Steenland K, Barr DB, Levy K, et al. Predictors of exposure to organophosphate pesticides in schoolchildren in the Province of Talca, Chile. Environ Int [Internet]. 2012 Oct;47:28-36. Available from: https://linkinghub.elsevier.com/retrieve/pii/S0160412012001225

[6] Petraitis J, Jarmalaite I, Vaičiunas V, Uščinas R, Jankovskiene G. A review of research studies into pesticide residues in food in Lithuania. Zemdirbyste. 2013;100(2):205-12.

[7] Commission of the European Communities. Monitoring of Pesticide Residues in Products of Plant Origin [Internet]. 2008. p. 40. Available from: http://ec.europa.eu/food/fvo/specialreports/pesticide_residues/report_2006_en.pdf

[8] Evoy R, Kincl L. Evaluation of Pesticides Found in Oregon Cannabis from 2016 to 2017. Ann Work Expo Heal. 2020;64(7):770-4.

[9] Agarwal A, Prajapati R, Singh OP, Raza SK, Thakur LK. Pesticide residue in water—a challenging task in India. Environ Monit Assess. 2015;187(2).

[10] Koçyiğit H, Sinanoğlu F. Method validation for the analysis of pesticide residue in aqueous environment. Env Monit Assess. 2020;192(9):567.

[11] Bennett B, Workman T, Smith MN, Griffith WC, Thompson B, Faustman EM. Characterizing the Neurodevelopmental Pesticide Exposome in a Children's Agricultural Cohort. Int J Environ Res Public Health [Internet]. 2020 Feb 25;17(5):1479. Available from: https://www.mdpi.com/1660-4601/17/5/1479

[12] Benka-Coker W, Hoskovec L, Severson R, Balmes J, Wilson A, Magzamen S. The joint effect of ambient air pollution and agricultural pesticide exposures on lung function among children with asthma. Environ Res [Internet]. 2020 Nov;190:109903. Available from: https://linkinghub.elsevier.com/retrieve/pii/S0013935120307982

[13] Benka-Coker W, Loftus C, Karr C, Magzamen S. Association of Organophosphate Pesticide Exposure and a Marker of Asthma Morbidity in an Agricultural Community. J Agromedicine [Internet]. 2020 Jan 2;25(1):106-14. Available from: https://www.tandfonline.com/doi/full/10.1080/1059924X.2019.1619644

[14] Zúñiga-Venegas L, Saracini C, Pancetti F, Muñoz-Quezada MT, Lucero B, Foerster C, et al. Exposición a plaguicidas en Chile y salud poblacional: urgencia para la toma de decisiones. Gac Sanit [Internet]. 2020 Jul; Available from: https://linkinghub.elsevier.com/retrieve/pii/S0213911120301291

[15] Latifovic L, Freeman LEB, Spinelli JJ, Pahwa M, Kachuri L,

Blair A, et al. Pesticide use and risk of Hodgkin lymphoma: results from the North American Pooled Project (NAPP). Cancer Causes Control [Internet]. 2020;31(6):583-99. Available from: https://doi.org/10.1007/s10552-020-01301-4

[16] Alavanja MCR, Hofmann JN, Lynch CF, Hines CJ, Barry KH, Barker J, et al. Non-Hodgkin Lymphoma Risk and Insecticide, Fungicide and Fumigant Use in the Agricultural Health Study. Akiba S, editor. PLoS One [Internet]. 2014 Oct 22;9(10):e109332. Available from: https://dx.plos.org/10.1371/journal.pone.0109332

[17] Kachuri L, Beane Freeman LE, Spinelli JJ, Blair A, Pahwa M, Koutros S, et al. Insecticide use and risk of <scp>non-Hodgkin</scp> lymphoma subtypes: a subset meta-analysis of the North American Pooled Project. Int J Cancer [Internet]. 2020 Jun 23;ijc.33164. Available from: https://onlinelibrary.wiley.com/doi/abs/10.1002/ijc.33164

[18] Moura LTR de, Bedor CNG, Lopez RVM, Santana VS, Rocha TMB da S da, Wünsch Filho V, et al. Exposição ocupacional a agrotóxicos organofosforados e neoplasias hematológicas: uma revisão sistemática. Rev Bras Epidemiol [Internet]. 2020;23. Available from: http://www.scielo.br/scielo.php?script=sci_arttext&pid=S1415-790X2020000100428&tlng=pt

[19] Smith AM, Smith MT, La Merrill MA, Liaw J, Steinmaus C. 2,4-dichlorophenoxyacetic acid (2,4-D) and risk of non-Hodgkin lymphoma: a meta-analysis accounting for exposure levels. Ann Epidemiol [Internet]. 2017 Apr;27(4):281-289.e4. Available from: https://linkinghub.elsevier.com/retrieve/pii/S1047279717302466

[20] McDuffie HH, Pahwa P, McLaughlin JR, Spinelli JJ, Fincham S, Dosman JA, et al. Non-Hodgkin's lymphoma and specific pesticide exposures inmen: Cross-Canada study of pesticides and health. Cancer Epidemiol Biomarkers Prev. 2001;10(11):1155-63.

[21] Pahwa P, Karunanayake CP, Spinelli JJ, Dosman JA, McDuffie HH. Ethnicity and incidence of Hodgkin lymphoma in Canadian population. BMC Cancer [Internet]. 2009 Dec 11;9(1):141. Available from: https://bmccancer.biomedcentral.com/articles/10.1186/1471-2407-9-141

[22] Ellsworth RE, Kostyniak PJ, Chi L-H, Shriver CD, Costantino NS, Ellsworth DL. Organochlorine pesticide residues in human breast tissue and their relationships with clinical and pathological characteristics of breast cancer. Environ Toxicol [Internet]. 2018 Aug;33(8):876-84. Available from: http://doi.wiley.com/10.1002/tox.22573

[23] He T-T, Zuo A-J, Wang J-G, Zhao P. Organochlorine pesticides accumulation and breast cancer: A hospital-based case–control study. Tumor Biol [Internet]. 2017 May;39(5):101042831769911. Available from: http://journals.sagepub.com/doi/10.1177/1010428317699114

[24] Arrebola JP, Belhassen H, Artacho-Cordón F, Ghali R, Ghorbel H, Boussen H, et al. Risk of female breast cancer and serum concentrations of organochlorine pesticides and polychlorinated biphenyls: A case-control study in Tunisia. Sci Total Environ. 2015;520:106-13.

[25] Zárate L V., Pontillo CA, Español A, Miret N V., Chiappini F, Cocca C, et al. Angiogenesis signaling in breast cancer models is induced by hexachlorobenzene and chlorpyrifos, pesticide ligands of the aryl hydrocarbon receptor. Toxicol Appl Pharmacol [Internet]. 2020 Aug;401:115093. Available from: https://

linkinghub.elsevier.com/retrieve/pii/S0041008X20302179

[26] Yang KJ, Lee J, Park HL. Organophosphate Pesticide Exposure and Breast Cancer Risk: A Rapid Review of Human, Animal, and Cell-Based Studies. Int J Environ Res Public Health [Internet]. 2020 Jul 13;17(14):5030. Available from: https://www.mdpi.com/1660-4601/17/14/5030

[27] Montes-Grajales D, Olivero-Verbel J. Structure-based Identification of Endocrine Disrupting Pesticides Targeting Breast Cancer Proteins. Toxicology [Internet]. 2020 Jun;439:152459. Available from: https://linkinghub.elsevier.com/retrieve/pii/S0300483X20300986

[28] Silva AMC, Campos PHN, Mattos IE, Hajat S, Lacerda EM, Ferreira MJM. Environmental exposure to pesticides and breast cancer in a region of intensive agribusiness activity in brazil: A case-control study. Int J Environ Res Public Health. 2019;16(20):1-10.

[29] Mills PK, Dodge JL, Bush J, Thompson Y, Shah P. Agricultural Exposures and Breast Cancer Among Latina in the San Joaquin Valley of California. J Occup Environ Med [Internet]. 2019 Jul;61(7):552-8. Available from: http://journals.lww.com/00043764-201907000-00003

[30] Thongprakaisang S, Thiantanawat A, Rangkadilok N, Suriyo T, Satayavivad J. Glyphosate induces human breast cancer cells growth via estrogen receptors. Food Chem Toxicol. 2013;59:129-36.

[31] Band PR, Abanto Z, Bert J, Lang B, Fang R, Gallagher RP, et al. Prostate cancer risk and exposure to pesticides in British Columbia Farmers. Prostate. 2011;71(2):168-83.

[32] Wang Y, Guo Y, Hu Y, Sun Y, Xu D. Endosulfan triggers epithelial-mesenchymal transition via PTP4A3-mediated TGF-β signaling pathway in prostate cancer cells. Sci Total Environ [Internet]. 2020 Aug;731:139234. Available from: https://linkinghub.elsevier.com/retrieve/pii/S0048969720327510

[33] Alavanja MC. Use of Agricultural Pesticides and Prostate Cancer Risk in the Agricultural Health Study Cohort. Am J Epidemiol [Internet]. 2003 May 1;157(9):800-14. Available from: https://academic.oup.com/aje/article-lookup/doi/10.1093/aje/kwg040

[34] Potti A, Sehgal I. Exposure to pesticides increases levels of uPA and uPAR in pre-malignant human prostate cells. Environ Toxicol Pharmacol [Internet]. 2005 Feb;19(2):215-9. Available from: https://linkinghub.elsevier.com/retrieve/pii/S1382668904001383

[35] Pardo LA, Beane Freeman LE, Lerro CC, Andreotti G, Hofmann JN, Parks CG, et al. Pesticide exposure and risk of aggressive prostate cancer among private pesticide applicators. Environ Heal [Internet]. 2020 Dec 5;19(1):30. Available from: https://ehjournal.biomedcentral.com/articles/10.1186/s12940-020-00583-0

[36] Chien J-M, Liang W-Z, Liao W-C, Kuo C-C, Chou C-T, Hao L-J, et al. Ca 2+ movement and cytotoxicity induced by the pyrethroid pesticide bifenthrin in human prostate cancer cells. Hum Exp Toxicol [Internet]. 2019 Oct 17;38(10):1145-54. Available from: http://journals.sagepub.com/doi/10.1177/0960327119855129

[37] Lee H-H. Ca^(2+) Movement Induced by Deltamethrin in PC3 Human Prostate Cancer Cells. Chin J Physiol [Internet]. 2016 Jun 30;59(3):148-55. Available from: http://www.airitilibrary.com/Publication/alDetailedMesh?DocID=03044920-201606-201606290012-201606290012-148-155

[38] Thomas P, Dong J. Novel mechanism of endocrine disruption by fungicides through binding to the membrane androgen receptor, ZIP9 (SLC39A9), and antagonizing rapid testosterone induction of the intrinsic apoptotic pathway. Steroids [Internet]. 2019 Sep;149:108415. Available from: https://linkinghub.elsevier.com/retrieve/pii/S0039128X19300996

[39] Ansbaugh N, Shannon J, Mori M, Farris PE, Garzotto M. Agent Orange as a risk factor for high-grade prostate cancer. Cancer [Internet]. 2013 Jul 1;119(13):2399-404. Available from: http://doi.wiley.com/10.1002/cncr.27941

[40] Hu K, TIAN Y, DU Y, HUANG L, CHEN J, LI N, et al. Atrazine promotes RM1 prostate cancer cell proliferation by activating STAT3 signaling. Int J Oncol [Internet]. 2016 May;48(5):2166-74. Available from: https://www.spandidos-publications.com/10.3892/ijo.2016.3433

[41] Jones RR, Barone-Adesi F, Koutros S, Lerro CC, Blair A, Lubin J, et al. Incidence of solid tumours among pesticide applicators exposed to the organophosphate insecticide diazinon in the Agricultural Health Study: an updated analysis. Occup Environ Med [Internet]. 2015 Jul;72(7):496-503. Available from: http://oem.bmj.com/lookup/doi/10.1136/oemed-2014-102728

[42] Huang F, Chen Z, Chen H, Lu W, Xie S, Meng QH, et al. Cypermethrin Promotes Lung Cancer Metastasis via Modulation of Macrophage Polarization by Targeting MicroRNA-155/Bcl6. Toxicol Sci [Internet]. 2018 Jun 1;163(2):454-65. Available from: https://academic.oup.com/toxsci/article/163/2/454/4870159

[43] Lerro CC, Koutros S, Andreotti G, Hines CJ, Blair A, Lubin J, et al. Use of acetochlor and cancer incidence in the Agricultural Health Study. Int J Cancer [Internet]. 2015 Sep 1;137(5):1167-75. Available from: http://doi.wiley.com/10.1002/ijc.29416

[44] Koutros S, Silverman DT, Alavanja MCR, Andreotti G, Lerro CC, Heltshe S, et al. Occupational exposure to pesticides and bladder cancer risk. Int J Epidemiol. 2016;45(3):792-805.

[45] Zhao B, Shen H, Liu F, Liu S, Niu J, Guo F, et al. Exposure to organochlorine pesticides is an independent risk factor of hepatocellular carcinoma: A case–control study. J Expo Sci Environ Epidemiol [Internet]. 2012 Nov 14;22(6):541-8. Available from: http://www.nature.com/articles/jes201129

[46] Peyre L, Zucchini-Pascal N, de Sousa G, Luzy A-P, Rahmani R. Potential involvement of chemicals in liver cancer progression: An alternative toxicological approach combining biomarkers and innovative technologies. Toxicol Vitr [Internet]. 2014 Dec;28(8):1507-20. Available from: https://linkinghub.elsevier.com/retrieve/pii/S0887233314001246

[47] Huang T, Huang Y, Huang Y, Yang Y, Zhao Y, Martyniuk CJ. Toxicity assessment of the herbicide acetochlor in the human liver carcinoma (HepG2) cell line. Chemosphere [Internet]. 2020 Mar;243:125345. Available from: https://linkinghub.elsevier.com/retrieve/pii/S0045653519325858

[48] Tinwell H, Rouquié D, Schorsch F, Geter D, Wason S, Bars R. Liver tumor formation in female rat induced by fluopyram is mediated by CAR/PXR nuclear receptor activation. Regul Toxicol Pharmacol [Internet]. 2014 Dec;70(3):648-58. Available from: https://linkinghub.elsevier.com/retrieve/pii/S0273230014002165

[49] Lerro CC, Hofmann JN, Andreotti G, Koutros S, Parks CG, Blair A, et al. Dicamba use and cancer incidence in the agricultural health study: an updated analysis. Int J

Epidemiol [Internet]. 2020 May 1; Available from: https://academic.oup.com/ije/advance-article/doi/10.1093/ije/dyaa066/5827818

[50] Van Leeuwen JA, Waltner-Toews D, Abernathy T, Smit B, Shoukri M. Associations between stomach cancer incidence and drinking water contamination with atrazine and nitrate in Ontario (Canada) agroecosystems, 1987-1991. Int J Epidemiol [Internet]. 1999 Oct 1;28(5):836-40. Available from: https://academic.oup.com/ije/article-lookup/doi/10.1093/ije/28.5.836

[51] Lerro CC, Koutros S, Andreotti G, Friesen MC, Alavanja MC, Blair A, et al. Organophosphate insecticide use and cancer incidence among spouses of pesticide applicators in the Agricultural Health Study. Occup Environ Med [Internet]. 2015 Oct;72(10):736-44. Available from: http://oem.bmj.com/lookup/doi/10.1136/oemed-2014-102798

[52] Perdichizzi S, Mascolo MG, Silingardi P, Morandi E, Rotondo F, Guerrini A, et al. Cancer-related genes transcriptionally induced by the fungicide penconazole. Toxicol Vitr [Internet]. 2014 Feb;28(1):125-30. Available from: https://linkinghub.elsevier.com/retrieve/pii/S0887233313001574

[53] Brasil VLM, Ramos Pinto MB, Bonan RF, Kowalski LP, da Cruz Perez DE. Pesticides as risk factors for head and neck cancer: A review. J Oral Pathol Med [Internet]. 2018 Aug;47(7):641-51. Available from: http://doi.wiley.com/10.1111/jop.12701

[54] Pan H, Xuan Q, Han D, Yin J, Wu S, Wang Q. [The change of cancer-related genes expression profile in Nthy-ori-3-1 cell induced by the pesticide amitrole]. Wei Sheng Yan Jiu [Internet]. 2016 Jul;45(4):558-62. Available from: http://www.ncbi.nlm.nih.gov/pubmed/29903322

[55] Tusha Sharma, Basu D Banerjee, Gaurav K Thakur, Kiran Guleria DM. Polymorphism of xenobiotic metabolizing gene and susceptibility of epithelial ovarian cancer with reference to organochlorine pesticides exposure. Exp Biol Med. 2019;244(16):1446-53.

[56] Go R-E, Kim C-W, Choi K-C. Effect of fenhexamid and cyprodinil on the expression of cell cycle- and metastasis-related genes via an estrogen receptor-dependent pathway in cellular and xenografted ovarian cancer models. Toxicol Appl Pharmacol [Internet]. 2015 Nov;289(1):48-57. Available from: https://linkinghub.elsevier.com/retrieve/pii/S0041008X15300752

[57] Song L, Liu J, Jin X, Li Z, Zhao M, Liu W. p, p′-Dichlorodiphenyldichloroethylene Induces Colorectal Adenocarcinoma Cell Proliferation through Oxidative Stress. Liu C, editor. PLoS One [Internet]. 2014 Nov 11;9(11):e112700. Available from: https://dx.plos.org/10.1371/journal.pone.0112700

[58] Lee WJ, Sandler DP, Blair A, Samanic C, Cross AJ, Alavanja MCR. Pesticide use and colorectal cancer risk in the agricultural health study. Int J Cancer [Internet]. 2007 Jul 15;121(2):339-46. Available from: http://doi.wiley.com/10.1002/ijc.22635

[59] KAUR P, RADOTRA B, MINZ R, GILL K. Impaired mitochondrial energy metabolism and neuronal apoptotic cell death after chronic dichlorvos (OP) exposure in rat brain. Neurotoxicology [Internet]. 2007 Nov;28(6):1208-19. Available from: https://linkinghub.elsevier.com/retrieve/pii/S0161813X07001659

[60] Milesi MM, Durando M, Lorenz V, Gastiazoro MP, Varayoud J. Postnatal exposure to endosulfan affects uterine development and fertility. Mol Cell Endocrinol [Internet].

2020 Jul;511:110855. Available from: https://linkinghub.elsevier.com/retrieve/pii/S0303720720301556

[61] Mysore K, Hapairai LK, Sun L, Li P, Wang C-W, Scheel ND, et al. Characterization of a dual-action adulticidal and larvicidal interfering RNA pesticide targeting the Shaker gene of multiple disease vector mosquitoes. Abd-Alla AMM, editor. PLoS Negl Trop Dis [Internet]. 2020 Jul 20;14(7):e0008479. Available from: https://dx.plos.org/10.1371/journal.pntd.0008479

[62] Huh D-A, Chae WR, Lim HL, Kim JH, Kim YS, Kim Y-W, et al. Optimizing Operating Parameters of High-Temperature Steam for Disinfecting Total Nematodes and Bacteria in Soil: Application of the Box–Behnken Design. Int J Environ Res Public Health [Internet]. 2020 Jul 13;17(14):5029. Available from: https://www.mdpi.com/1660-4601/17/14/5029

Chapter 5

Impacts of Agricultural Toxicity on Non-Target Organisms in Aquatic Ecosystem

Marlyete Chagas de Araújo, Caio Rodrigo Dias de Assis, Rafael Artur de Queiroz Cavalcanti de Sá, Jordany Gomes da Silva, Bruno Oliveira de Veras, Kaline Catiely Campos Silva, Glauber Pereira Carvalho dos Santos, Vagne de Melo Oliveira, Juliana Ferreira dos Santos, Ana Lúcia Figueiredo Porto and Maria Betânia Melo de Oliveira

Abstract

The present review addresses the impacts of pesticides used in crops on non-target organisms in aquatic ecosystems. In recent decades, these ecosystems have received large amounts of these compounds, which are released by urban communities, rural and industrial properties. Pesticides reach the aquatic environment through different routes (leaching, irrigation, drainage, and surface runoff) and can easily reach non-target organisms, such as fish, mollusks, as well as other benthic organisms. Usually, these animals tend to undergo bioaccumulation. Exposure to these pesticides can cause numerous physiological changes by direct influence on certain cellular structures, such as on the lysosomal membrane, which can be degraded. Also, they can even react with nucleic acids resulting in several genetic injuries, thus causing adverse reactions to the body. There is a need for more incentives for the adoption of sustainable agroecological practices, as well as a ban on active ingredients harmful to the environment, in addition to strict inspection by competent environmental agencies.

Keywords: environmental impact, pesticides, aquatic organisms, non-target organisms, aquatic ecosystems

1. Introduction

In the last decades, with the machinery modernization and the consolidation of the sector of modern inputs, agriculture has been growing at a fast pace, with pesticides being one of the main instruments that drive the agricultural sector in productivity gains [1]. However, the indiscriminate use of these substances has easily reached non-target organisms and their effects on the environment are varied, ranging from the reduction in the availability and quality of water to the compromise of air and food quality, harming human health. Also, it can directly affect cellular structures of aquatic or terrestrial organisms resulting in damage to biodiversity [2].

In the early 1960s, society began to worry about the adverse effects and potential risk that these pesticides posed to human health and the environment. In several countries, production, marketing, and use of many of these compounds, in particular those considered persistent organic pollutants (POPs), such as organochlorines, were banned [3]. With the ban on most organochlorine compounds (less toxic, but with greater bioaccumulation in the environment), after the Second World War, carbamate and organophosphate pesticides had their use intensified. Also, it became the most used pesticides worldwide, being widely used in developing countries with a predominantly agricultural economy [4].

These toxic substances have the potential to cause various biochemical and genetic injuries to non-target organisms. Carbamates and organophosphates, for example, are potent inhibitors of the acetylcholinesterase enzyme, which damages the nervous system of an exposed organism [5, 6]. This enzyme acts in the hydrolysis of the acetylcholine neurotransmitter in cholinergic synapses. Its inhibition can lead the individual to death due to cholinergic hyperstimulation. Pesticides are also known for their mutagenic and carcinogenic effects. They react with nucleic acids causing adverse reactions in the body. Thus, monitoring and controlling the presence of these substances in the environment are necessary, since these compounds have become a human health and environmental problems [7].

2. Pesticides in aquatic environments

Pesticides or agrochemicals are defined as:

> *"Products and agents of physical, chemical or biological processes, intended for use in the sectors of production, in the storage and processing of agricultural products, in pastures, in the protection of forests, native or implanted, and of other ecosystems and also of urban environments, whose purpose is to change the composition of flora or fauna, in order to preserve them from the adverse action of living beings considered harmful" [8].*

According to the harmful species that intend to eliminate, these compounds are classified as insecticides, fungicides, herbicides, acaricides, rodenticides, molluscicides, among others. Herbicides represent 48% of the total pesticides, which is followed by insecticides (25%) and fungicides (22%) [9]. Depending on the chemical class, they can be grouped into pyrethroids, organochlorines, organophosphates, carbamates, benzoylureas, neonicotinoids, among others [10].

Pesticides arrive in the environment carried by runoff and leaching of rainwater, irrigation, and drainage or by spraying, as shown in **Figure 1**. Among these processes, runoff and leaching can contaminate reservoirs, lakes, and rivers. Also, they expose aquatic organisms at levels of pesticides that can be toxic to many species. Once present in the aquatic environment, these compounds can penetrate the organisms orally - through the ingestion of contaminated food, respiratory - through the gills, and dermal - through the surface of the body. In most cases, these organisms tend to suffer bioaccumulation [1, 13]. Pesticide exposure can cause numerous physiological changes by direct influence on

DOI: http://dx.doi.org/10.5772/intechopen.93941

certain cellular structures, for example, on the lysosomal membrane, which can be degraded or can react with nucleic acids, resulting in several genetic injuries that cause adverse reactions in the body [2].

Currently, there is a growing concern about the exacerbated use of pesticides since, in recent decades, aquatic ecosystems have received alarmingly large amounts of these compounds, which are released by urban communities, rural properties, and industries. Thus, society started to worry about the adverse effects of these substances and their potential risk [14]. According to Silva et al. [4], carbamates and organophosphates are the most used pesticides worldwide. They together account for more than 50% of what is marketed.

Organophosphate pesticides (OPs) comprise a large number of substances classified chemically as esters, amides, or derivatives of pentavalent phosphoric acids. Carbamates (CBs) are esters, or N-substituted derivatives of carbamic acid (carbamic acid monoamide) (**Figure 2**). Both have low water solubility and are, in general, easily hydrolyzable in alkaline environments [10, 15, 16]. In general, OPs need biotransformation to become toxicologically active, unlike CBs that are already bioactive.

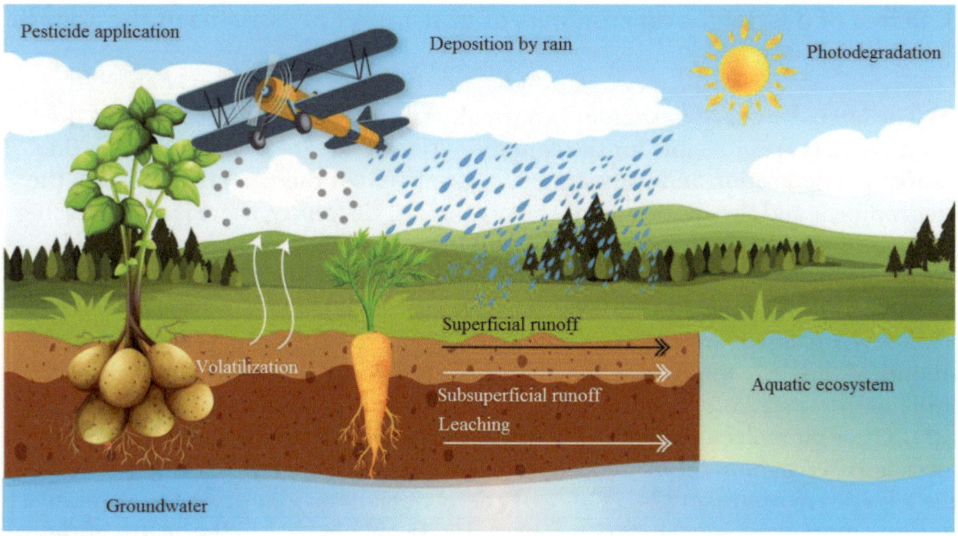

Figure 1.
Pesticide paths to the aquatic environments. Source: [11, 12].

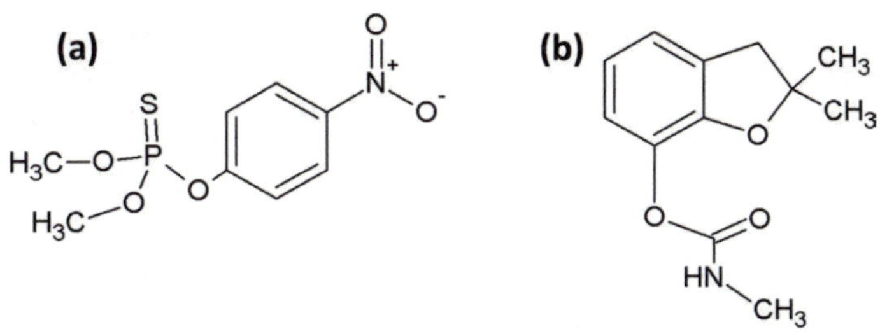

Figure 2.
Organophosphate pesticides: (a) methyl-paration (triesters of phosphoric acid), (b) carbamate carbofuran (esters derived from carbamic acid).

Carbamates and organophosphates affect the nervous system of organisms. They inhibit the activity of the enzyme acetylcholinesterase (AChE), as demonstrated by Wang et al. [17]. In their study, AChE inhibition in carp (*Cyprinus carpio*) exposed to various concentrations of organophosphates, malathion, and triazophos, as well as carbamates fenobucarb and carbosulfan, was evaluated. In equitoxic mixtures, the authors noted that AChE activity was inhibited by the combination of triazophos and malathion, as well as triazophos and carbosulfan, with synergism occurring. The effects of organophosphates on the behavior and activity of the AChE of zebrafish larvae have also been studied, through exposure to chlorpyrifos and malathion, and changes in swimming speed (hypoactivity and hyperactivity), rest and tigmotatism have been found [18].

Recently, benzoylurea, a class of pesticide that in the past was not considered an acetylcholinesterase inhibitor, since its main mode of action is the inhibition of chitin biosynthesis in insects (which interrupts the incorporation of N-acetylglycosamine monomers), demonstrated anticholinesterase potential [19]. In 2011, this class of pesticides represented 3.6% of the world's pesticide market. Since then, its commercial importance has grown over the years [20].

3. Effect of pesticides on enzymes of non-target organisms in the aquatic environment

The intensive use of pesticides in agricultural cultivation has been one of the main problems responsible for the contamination of aquatic ecosystems. It is due to both the deposition and consequent accumulation of these contaminants in the environment and the sensitivity of the organisms. Currently, there is an increasing number of studies in which fish, for example, are used as indicators of pesticides in the aquatic ecosystem, since these substances, even in low concentrations, can affect their physiology and survival capacity [17, 21, 22].

These organisms are sources of biologically active molecules. When their functioning is altered, compromise the organism's physiological functions, which culminates in genetic, biochemical, morphological, ecological, or behavioral changes [23]. These biomolecules are considered as biomarkers, and their measurement has been used in biomonitoring programs to detect exposure to toxic substances in the aquatic environment [24]. This early detection allows identifying the presence of the contaminant, even before it causes significant changes in the health of the exposed individuals.

Among exposure biomarkers, recent studies showed great interest in enzyme biomarkers as an alternative for monitoring impacted aquatic environments due to their high specificity and speed in responding to changes from target substances [4, 6, 17, 21, 25, 26]. The use of enzymes as biomarkers is based on inhibitory or inductive interference caused by contaminants in their catalytic activity. Most of these toxic compounds have a high affinity for electron pairs found in the amino acids that form the enzymes, such as the sulfhydryl - SH groups and other functional groups from the catalytic site [5, 27]. Among the main enzymes used extensively for this purpose, cholinesterase enzymes stand out (ChEs; EC 3.1.1.x).

3.1 Influence of pesticides on cholinesterase enzyme activity

Two distinct cholinesterases are found in vertebrate and invertebrate aquatic organisms, acetylcholinesterase (AChE, EC 3.1.1.7) and butyrylcholinesterase (BChE, EC 3.1.1.8). AChE is a hydrolase that predominates mainly in erythrocytes, neurons, ganglia of the autonomic nervous system, and terminal motor plates. Its main function is to promote the hydrolysis of the neurotransmitter acetylcholine. It releases acetate and choline in the cholinergic synapses. Due to its key function in the control of synaptic transmission, this enzyme becomes one of the most vulnerable molecular targets to the action of neurotoxic agents. For this reason,

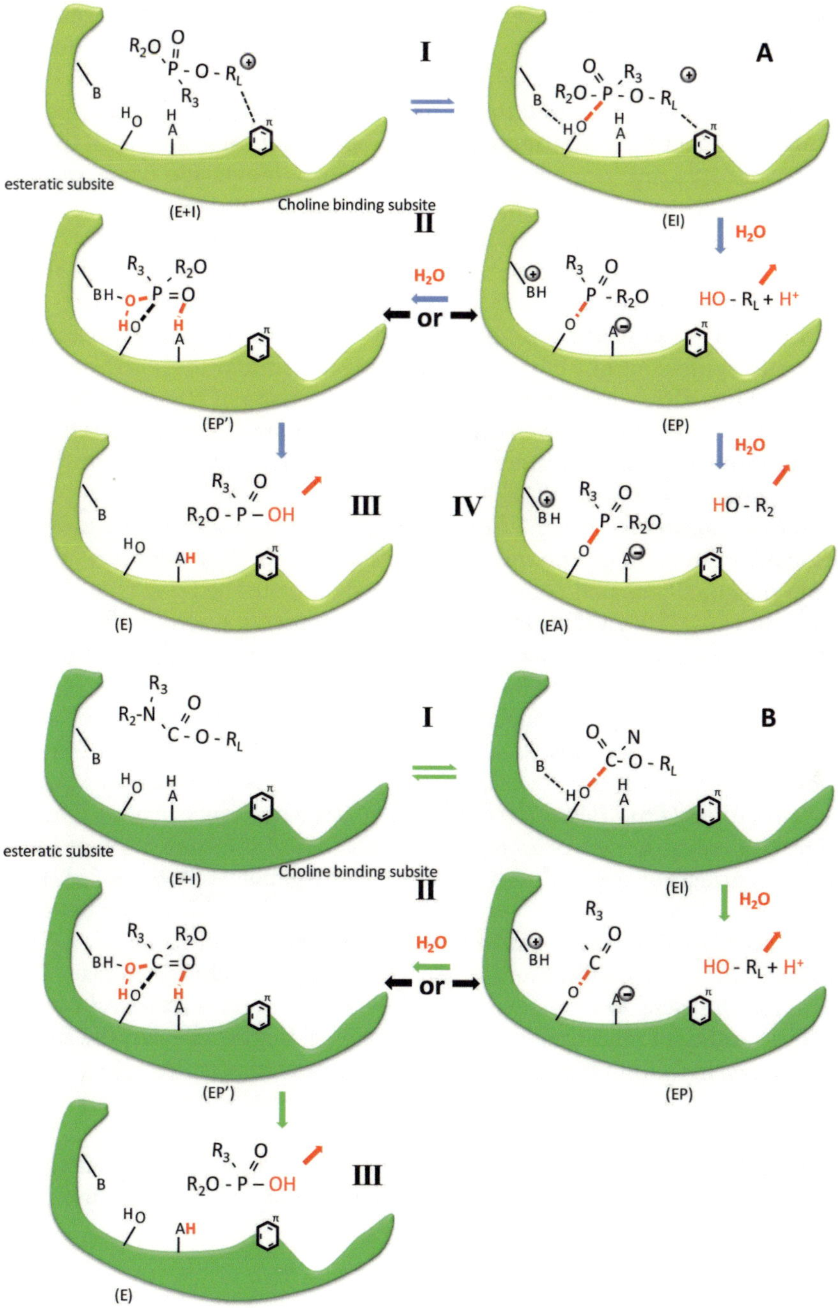

Figure 3.
Steps of inhibition by organophosphorus (A) and carbamates (B): I – Approaching of the organophosphorus (OP) or carbamate (CB) pesticide into the bottom of the catalytic cavity attracted by the choline binding sub-site (for OP only) and transition state in the interaction between enzyme and the pesticide. In particular, the bonds involved; II - Scheme representing the two occurrence possibilities during the existence of the enzyme-OP complex: spontaneous reactivation (left) or aging (right – only OP); III – Free enzyme; IV - Before undergoing aging (only OP), R2 was attracting electrons from the phosphorus atom. After the removal of R2, these electrons are shared with "O"-Serine, strengthening the binding, which cannot be hydrolyzed.

it has been widely studied in aquatic organisms and proposed for use in monitoring programs, given its sensitivity [5, 6]. On the other hand, BChE predominates in plasma, liver, neuroglia, pancreas, and digestive tract walls. It has not fully clarified its function, and the absence of its activity has been reported in the brains of several fish species [4, 28].

These enzymes are widely used in biomonitoring of aquatic ecosystems. They are used as biomarkers of the presence of two specific classes of pesticides: carbamates and organophosphates, which generally have low environmental persistence, especially when compared to organochlorines, but with greater toxicity. These substances act by inhibiting enzymatic activity. It interacts with the steratic site by phosphorylation (organophosphates) or carbamoylation (carbamates) (**Figure 3**) [12, 29].

Inhibition, once initiated, tends to generate acute or chronic intoxication. Depending on the degree of exposure to the toxic substance, the individual may die, due to over-stimulation of his nervous system, since with AChE inhibition, acetylcholine accumulates in neuromuscular junctions and cholinergic synapses [22, 30]. The signs and symptoms of carbamate poisoning are similar to those of organophosphates. They differ only in the duration and intensity of toxicity. The moderate effects of carbamates compared to organophosphates are due to the fact that they reversibly inhibit acetylcholinesterase (hydrolysis with enzyme regeneration) and are rapidly metabolized *in vivo* [31].

The anticholinesterase action of these pesticides, simultaneously, causes AChE inhibition of central and peripheral nervous tissue. Also, they inhibit erythrocyte AChE and plasma BChE [29]. According to the data from the Food and Agriculture Organization (FAO, 2007) [32], inhibition of cholinesterase activity from 20% characterizes the action of anticholinesterase agents. After 50% inhibition, clinical signs are visualized, and after 90% inhibition, the organism dies.

The *in vitro* study of acetylcholinesterase activity in various fish species, such as arapaima (*Arapaima gigas*), peacock bass (*Cichla ocellaris*), tambaqui (*Colossoma macropomum*), zebrafish (*Danio rerio*), jaguar cichlid (*Parachromis managuensis*), streaked prochilod (*Prochilodus lineatus*), cobia (*Rachycentron canadum*), and tilapia (*Oreochromis niloticus*), have been proposed to be used in the detection of harmful physiological effects of pesticides to these aquatic organisms [4, 6, 33–35]. In addition to these, we can mention the works of GHAZALA et al. [26], who tested the effect of three sublethal concentrations of the profenofos and carbofuran pesticides on the activity of acetylcholinesterase (AChE) and butyrylcholinesterase (BChE) in the brain, gills, muscle, kidney, liver, and blood of the species *Labeo rohita* (Indian carp). These authors found that exposure to both pesticides affected the functions of these organs, including metabolism and neurotransmission. Araújo et al. [6] also reported *in vitro* inhibition of acetylcholinesterase by carbamate and organophosphate pesticides in the brain of the Jaguar cichlid, showing high degree of toxicity.

These studies have confirmed fish as a practical and economically viable source of acetylcholinesterase, which is capable of making water resource biomonitoring procedures routine.

4. Genotoxic effects of pesticides

Pesticides, in general, are known to have genotoxic, mutagenic, and carcinogenic action, since they interact chemically with the genetic material, promoting changes in the DNA molecule. These alterations in the organisms' DNA can cause serious consequences, since, at the individual level, they damage cells and organs and can even affect their reproductive function [36]. Among the most used methodologies for assessing DNA damage in aquatic organisms, the micronucleus (MN) test stands out, which allows the observation of macrolesions in the genome quickly, simply, and minimally invasive [37–39]. This test consists of a blood smear on a slide (**Figure 4A**) and is commonly applied to fish erythrocytes, oysters hemocytes, and crabs as an alternative in the detection of genotoxic agents, such as pesticides, in environmental biomonitoring programs [36, 40–42].

Currently, it is one of the most used cytogenetic tests in the field of toxicological genetics since it is a sensitive test for detecting structural either-or numerical chromosomal changes [43, 44]. In addition to the micronucleus test, nuclear morphological changes (NMC) can also be analyzed. Several studies describe the presence of these changes in fish cells as a result of exposure to genotoxic substances [7, 44, 45].

In addition to these tests, the Comet Assay (single cell gel electrophoresis assay) is also one of the most used in the evaluation of genomic damage caused by pesticides. It presents high sensitivity in detecting pre-mutagenic lesions in individual cells. It is a technique capable of detecting microlesions in DNA, which are genomic lesions that can be repaired [46]. In this technique, cells that have damages in their DNA, form different fragments which tend to migrate at different speeds during the electrophoretic run, forming a comet under fluorescence microscopy (**Figure 4B**).

Among the studies that demonstrate the action of pesticides in aquatic organisms, we can mention the study by Silva et al. [7] that evaluated the genotoxic potential of the herbicide trifluralin (one of the herbicides most used in weed control) on *Colossoma macropomum* (tambaqui). The mutagenic and genotoxic effects of different concentrations of trifluralin (0.25, 0.5, 0.75, 1.0 mg L^{-1}) in peripheral erythrocytes of *C. macropomum*, were investigated using the micronucleus test (MN), assay comet, and apoptosis. After an exposure period of 96 h, the results showed a significant rate of micronuclei and nuclear abnormalities in erythrocytes from *C. macropomum* exposed to 0.5, 0.75, 1.0 mg L^{-1} of trifluralin compared to the group control, thus confirming the genotoxicity of the herbicide trifluralin in the investigated species.

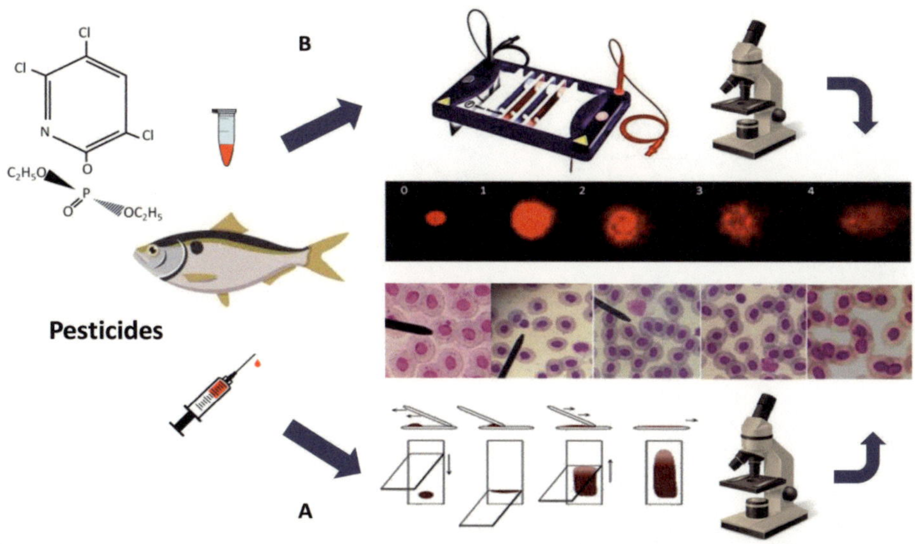

Figure 4.
Micronucleus test (A) and Comet assay (B) to evaluate DNA da mages in aquatic organisms.

5. Conclusion

In the search for high productivity, the use of pesticides has been intensified in agricultural crops. Also, the indiscriminate use of these substances has reached non-target organisms, causing deleterious effects on biodiversity, especially in the aquatic ecosystem.

To mitigate these impacts, several methodologies have been used to detect exposure to these toxic substances in aquatic environments. Among them, the methodologies that aim to evaluate the exposed organism at the biochemical and genetic level, as described in this review, show efficiency. It allows the early identification of the presence of the contaminant even before it causes significant changes in the health of the exposed individual, as well as before higher levels of biological organization are reached. It is worth mentioning that the pesticides present in aquatic ecosystems can accumulate in high concentrations in the organisms throughout the trophic level reaching the human being.

Monitoring and controlling the presence of these substances in the environment is necessary since these compounds have become a human and environmental health problem. Allied to this, there is a need for more incentives for the adoption of sustainable agroecological practices, as well as the prohibition of harmful active ingredients to the environment, added to the strict inspection by competent environmental agencies.

Acknowledgements

The authors would like to thank the Laboratory of Comparative Physiology and Animal Behavior (LabFCCA), the Laboratory of Evolutionary and Environmental Genomics (LAGEA) and the Directorate of Environmental Management (DGA) of the Infrastructure Superintendence of the Federal University of Pernambuco/UFPE for their support and collaboration in the execution of this work.

Author details

Marlyete Chagas de Araújo[1,2], Caio Rodrigo Dias de Assis[2,7], Rafael Artur de Queiroz Cavalcanti de Sá[1], Jordany Gomes da Silva[1], Bruno Oliveira de Veras[1], Kaline Catiely Campos Silva[3], Glauber Pereira Carvalho dos Santos[4], Vagne de Melo Oliveira[5], Juliana Ferreira dos Santos[6], Ana Lúcia Figueiredo Porto[5] and Maria Betânia Melo de Oliveira[1*]

1 Laboratory of Molecular Biology (BIOMOL), Biochemistry Department, Federal University of Pernambuco - UFPE, Recife, Pernambuco, Brazil

2 Laboratory of Enzymology (LABENZ), Biochemistry Department, Federal University of Pernambuco - UFPE, Recife, PE, Brazil

3 Education Department, Biology Collegiate, University of the State of Bahia – UNEB, Paulo Afonso, BA, Brazil

4 Institute of Technology of Pernambuco - ITEP, Recife, PE, Brazil

5 Laboratory of Bioactive Products Technology (LABTECBIO), Department of Morphology and Animal Physiology, Federal Rural University of Pernambuco - UFRPE, Recife, PE, Brazil

6 Laboratory of Physioecology in Aquaculture (LAFAq) and Laboratory of Aquaculture Production Systems (LAPAq), Department of Fisheries and Aquaculture, Federal Rural University of Pernambuco - UFRPE, Recife, PE, Brazil

7 Laboratory of Organic Compounds in Coastal and Marine Ecosystems, Department of Oceanography, Federal University of Pernambuco, Brazil

*Address all correspondence to: maria.bmoliveira@ufpe.br

References

[1] Belchior DCV, Saraiva AS, López AMC, Scheidt GN. Impactos de agrotóxicos sobre o meio Ambiente a a saúde Humana. Cadernos de Ciência & Tecnologia. 2014; 34: 135-151.

[2] Cogo AJD, Siqueira AF, Ramos AC, Cruz ZMA, Silva AG. Utilização de enzimas do estresse oxidativo como biomarcadoras de impactos ambientais. Natureza on line. 2009; 7: 37-42. [on line] http://www.naturezaonline.com.br

[3] França TCC, Silva GR, Castro AT. Defesa química: uma nova disciplina no ensino de química. Revista Virtual de Química. 2010; 2: 84-104.

[4] Silva KCC, Assis CRD, Oliveira VM, Carvalho Jr LB, Bezerra RS. Kinetic and physicochemical properties of brain acetylcholinesterase from the peacock bass (*Cichla ocellaris*) and in vitro effect of pesticides and metal ions. Aquatic Toxicology. 2013; 126: 191-197.

[5] Assis CRD, Linhares AG, Oliveira VM, França RCP, Santos JF, Marcuschi, M, Maciel Carvalho EVM, Bezerra RS, Carvalho Jr LB. Characterization of catalytic efficiency parameters of brain cholinesterases in tropical fish, Fish Physiology and Biochemistry. 2014; 40: 1659-1668.

[6] Araújo MC, Assis CRD, Silva LC, Machado DC, Silva KCC, Lima AVA, Carvalho Jr LB, Bezerra RS, Oliveira MBM. Brain acetylcholinesterase of jaguar cichlid (*Parachromis managuensis*): From physicochemical and kinetic properties to its potential as biomarker of pesticides and metal ions. Aquatic Toxicology. 2016; 177: 182-189.

[7] Silva JM, Santos FLB, Santos RV, Barreto EO, Santos EL, Goulart Santana AE, Rodarte RS, Machado SS, Abreu FC. Determination of genotoxic effect of trifluralin on *Colossoma macropomum* (Teleostei: Characidae: Serrasalminae, Cuvier, 1816) using a multibiomarker approach. Ecotoxicology and Environmental Contamination. 2017; 12: 85-93 doi: 10.5132/eec.2017

[8] BRASIL. Lei Federal n° 7.802 de 11/07/89. Dispõe sobre o tratamento de embalagens e resíduos de agrotóxicos e dá outros tratamentos. Disponível em: http://www.planalto.gov.br/ccivil_03/leis/L9974.htm. Acessso em: 22 de Julho de 2015.

[9] Pelaez V, Terra FHB, Silva LR. A regulamentação dos agrotóxicos no Brasil: entre o poder de mercado e a defesa da saúde e do meio ambiente. Revista de Economia. 2010; 36: 27-48. DOI: 10.5380/re.v36i1.20523.

[10] WHO. WORLD HEALTH ORGANIZATION. Organophosphorus insecticides: a general introduction. Environmental Health Criteria 63. Genebra, 1986.

[11] Tomita RY, Beyruth Z. Toxicology of pesticides in the aquatic environment. O Biológico, 64, 135-142, 2002.

[12] Matos KS. Aspectos moleculares da reativação da acetilcolinesterase inibida por ciclosarin e carbofurano / Karina Silvia Matos. – Lavras: UFLA, 2012. 148 p. : il. Dissertação (mestrado) – Universidade Federal de Lavras, 2012.

[13] Rodrigues SR, Caldeira C, Castro BB, Gonçalves F, Nunes B, Antunes SC. Cholinesterase (ChE) inhibition in pumpkinseed (*Lepomis gibbosus*) as environmental biomarker: ChE characterization and potential neurotoxic effects of xenobiotics. Pestic Biochem Physiol. 2011; 99:181-188.

[14] Carneiro FF, Pignati W, Rigotto RM, Augusto LGS, Rizollo A, Muller NM,

Alexandre VP, Friedrich K, Mello MSC. Dossiê ABRASCO – Um alerta sobre os impactos dos agrotóxicos na saúde. ABRASCO, Rio de Janeiro, abril de 2012. 1ª Parte. 98p.

[15] Dauterman WC. Biological and nonbiological modifications of organophosphorus compounds. Bulletin of the World Health Organization, 44, 133-150, 1971.

[16] Vale JA. Toxicokinetic and toxicodynamic aspects of organophosphorus (OP) insecticide poisoning. Toxicology Letters. 1998; 102-103: 649-652.

[17] Wang Y, Chen C, Zhao X, Wang Q, Qian Y. Assessing joint toxicity of four organophosphate and carbamate insecticides in common carp (*Cyprinus carpio*) using acetylcholinesterase activity as an endpoint. Pesticide Biochemistry and Physiology. 2015; 122: 81-85.

[18] Richendrfer H, Pelkowski SD, Colwill RM, Créton R. Developmental sub-chronic exposure to chlorpyrifos reduces anxiety-related behavior in zebrafish larvae. Neurotoxicology and Teratology. 2012; 34: 458-465. ISSN 0892-0362

[19] Araújo MC, Assis CRD, Silva KCC, Souza KS, Azevedo RS, Alves MHE, Silva LC, Silva VC, Adam ML, Carvalho Jr LB, Bezerra RS, Oliveira MBM. Characterization of brain acetylcholinesterase of bentonic fish *Hoplosternum littorale*: Perspectives of application in pesticides and metal ions biomonitoring. Aquatic Toxicology. 2018; 205: 213-226.

[20] Sun R, Liu C, Zhang H, Wang Q. Benzoylurea chitin synthesis inhibitors. Journal of Agricultural and Food Chemistry. 2015; 63: 6847-6865.

[21] Nigam AK, Srivastava N, Rai AK, Kumari U, Mittal AK, Mittal S. The first evidence of cholinesterase in skin mucus of carps and its applicability as biomarker of organophosphate exposure. Environmental Toxicology. Doi 10.1002/tox.2012.

[22] Nunes B, Barbosa AR, Antunes SC, Gonçalvez F. Combination effects of anticholinesterasics in acetylcholinesterase of a fish species: effects of a metallic compound, an organophosphate pesticide, and a pharmaceutical drug. Environmental Science and Pollution Research. 2014; 21: 6258-6262.

[23] Mansouri B, Ebrahimpour M, Babaei H. Bioaccumulation and elimination of nickel in the organs of black fish (*Capoeta fusca*). Toxicology and Industrial Health. 2012; 28: 361-368.

[24] Coimbra RSC, Santos CRS, Saraiva VB, Oliveira MM. Biomarcadores como ferramentas na avaliação da qualidade do pescado contaminado com metais traço. Boletim do Observatório Ambiental Alberto Ribeiro Lamego. 2013; 7: 153-172.

[25] Botte ES, Jerry DR, Codi King S, Smith– Keune C, Negri AP. Effects of chlorpyrifos on cholinesterase activity and stress markers in the tropical reef fish *Acanthochromis polyacanthus*. Marine Pollution Bulletin. 2012; 65: 384-393.

[26] Ghazala, Mahboob S, Ahmad L, Sultana S, Alghanim K, Al-Misned F, Ahmad Z. Fish Cholinesterases as biomarkers of sublethal effects of organophosphorus and carbamates in tissues of *Labeo rohita*. Journal of Biochemistry and Molecular Toxicology. 2014; DOI 10.1002/jbt.2014.

[27] Ivanina AV, Habinck E, Sokolova IM. Differential sensitivity to cadmium of key mitochondrial enzymes in the eastern oyster, *Crassostrea*

virginica Gmelin (Bivalvia: Ostreidae). Comparative Biochemistry and Physiology Part C. 2008; 148: 72-79.

[28] Pezzementi L, Chatonnet A. Evolution of cholinesterases in the animal kingdom. Chemico-Biological Interactions. 2010; 187: 27-33.

[29] Assis CRD, Bezerra RS, Carvalho Jr LB. Fish cholinesterases as biomarkers of organophosphorus and carbamate pesticides. In: Pesticides in modern world, Book 5 - Pests control and pesticides exposure and toxicity assessment. Stoytcheva M. (Ed). Intech, Rijeka, Croatia, 614 p. 2011.

[30] Valbonesi P, Brunelli F, Marttioli M, Rossi T, Fabbri E. Cholinesterase activities and sensitivity to pesticides in different tissues os silver European eel, *Anguilla anguilla*. Comparative Biochemistry and Physiology Part C. 2011; 154: 353-359.

[31] Jebali J, Khedher SB, Sabbagh M, Kamel N, Banni M, Boussetta H. Cholinesterase activity as biomarkers of neurotoxicity: utility in the assessment of aquatic environment contamination. Journal of Integrated Costal Zone Management. 2013; 13: 525-537.

[32] FAO, Food Agriculture Organization, 2007. Pesticides in Food Report 2007 FAO Plant Production and Protection Paper 191. Food and Agriculture Organization, Rome, Italy.

[33] Maduenho LP, Martinez CBR. Acute effects of diflubenzuron on the freshwater fish *Prochilodus lineatus*. Comparative Biochemistry and Physiology Part C. 2008; 148: 265-272.

[34] Assis CRD, Linhares AG, Oliveira VM, França RCP, Maciel Carvalho EVM, Bezerra RS, Carvalho Jr LB. Comparative effect of pesticides on brain acetylcholinesterase in tropical fish. Science of the Total Environment. 2012; 441: 141-150.

[35] Dai Y-J, Jia Y-F, Chen N, Bian W-P, Li Q-K, Ma Y-B, Chen Y-L, Pei D-S. Zebrafish as a model system to study toxicology. Environmental Toxicology and Chemistry. 2014; 33: 11-17. ISSN 1552-8618. Available at: < http://dx.doi.org/10.1002/etc.2406

[36] Bolognesi C, Hayashi M. Micronucleus assay in aquatic animals. Mutagenesis. 2011; 26: 205 213.

[37] Adam ML, Torres RA, Sponchiado G, Motta TS, Oliveira CMR, Carvalho-Filho MA, Correia MTS. Environmental degradation at a public park in southern Brazil as revealed through a genotoxicity test (MN) on peripheral blood cells from *Poecilia vivipara* (Teleostei). Water, Air and Soil Pollution. 2010; 211: 61-68.

[38] Ansari RA, Rahman S, Kaur M, Anjum S, Raisuddin S. *In vivo* cytogenetic and oxidative stress-inducing effects of cypermethrin in freshwater fish, *Channa punctata* Bloch. Ecotoxicology and Environmental Safety. 2011; 74: 150-156.

[39] Pinheiro DA, Dias MT, Dias MKR, Santos EF, Marinho RGB. Primeiro registro da ocorrência de protozoários em Tamoatá *Hoplosternum littorale* no Brasil. Boletim do Instituto de Pesca. 2013; 39: 169-177.

[40] Obiakor M, Okonkwo J. Eco-genotoxicology: Micronucleus assay in fish erythrocytes as in situ aquatic pollution biomarker: a review. Journal of Animal Science Advances. 2012; 2: 123-133.

[41] Braham RP. Micronuclei and other erythrocyte nuclear abnormalities in fishes from the Great Lakes Basin, USA. Environmental and Molecular Mutagenesis. DOI: 10.1002/em. 2017.

[42] Lima ARB, Torres RA, Jacobina UP, Pinheiro MAA, Adam ML. Genomic damage in *Mugil curema*

(Actinopterygii: Mugilidae) reveals the effects of intense urbanization on estuaries in northeastern Brazil. Marine Pollution Bulletin. 2019; 138: 63-69. DOI: 10.1016/j.marpolbul.2018.07.037.

[43] Gutierrez JM, Villar S, Plavan AA. Micronucleus test in fishes as indicators of environmental quality in subestuaries of the Rio de la Plata (Uruguay). Marine Pollution Bulletin. 2015; 91: 518-523. http://dx.doi.org/10.1016/j.marpolbul.2014.10.027. 2014.

[44] Hussain B, Sultana T, Sultana S, Masoud MS, Ahmed Z, Mahboob S. Fish eco-genotoxicology: Comet and micronucleus assay in fish erythrocytes as in situ biomarker of freshwater pollution. Saudi Journal of Biological Sciences. 2017; 25: 393-398. DOI: 10.1016/j.sjbs.2017.11.048

[45] Thomé RG, Silva PM, Santos HB. Avaliação de genotoxidade da água de um rio urbano utilizando estudo de células sanguíneas de *Danio rerio*. Revista Conexão Ciência. 2016; 11: 9-16.

[46] ScalonMCS,RechenmacherC,Kayser M C, Rodrigues MTC, Maluf SWD, Rodrigues MAS, SILVA LBF. Evaluation of Sinos River water genotoxicity using the comet assay in fish. Brazilian Journal Biology. 2010; 70: 1217-1222.

Chapter 6

Limited Knowledge and Unsafe Practices in Usage of Pesticides and The Associated Toxicity Symptoms among Farmers in Tullo and Finchawa Rural Kebeles, Hawassa City, Sidama Regional State, Southern Ethiopia

Lana MHD Jamal Alshalati

Abstract

The insufficient knowledge regarding safe and proper pesticide handling by farmers in developing countries has led to extensive agricultural expansions at the expense of the health of farmers. The objective of this study was to assess the knowledge and field practices of farmers regarding pesticide handling, and to determine the prevalence of acute and chronic health-related problems in Finchawa and Tullo rural Kebeles of Hawassa City Administration. A cross-sectional mixed methods research design was employed to capture the fuller image of the issue. Farmers' knowledge regarding pesticide handling and toxicity found to be on average. The odds of the knowledge concerning proper pesticide handling was positively influenced by the factor of age, access to training; and years of experience; Field practices adopted by farmers were disappointing and intentional suicide incidents among teenagers were the result of farmers' unsafe storage. Nearly all the farmers did not use any means of PPE, and the lack of awareness about the dermal route presented a high risk of exposure. The common self-reported toxicity symptoms experienced by the participants included a headache (84.93%) and slow heartbeats (72.60%). The odds of prevalence of long-term toxicity symptoms found to be negatively correlated with the training factor while the same was positively influenced by the working hours in the farm. The study revealed that there is a high risk of exposure among farmers and their families in the study area.

Keywords: farmers, knowledge, unsafe practices, awareness, Toxicity symptoms, exposure

1. Introduction

1.1 Background of the study

In parallel with the line of feeding 9.8 billion people by 2050, agricultural productions are largely expanding and intensifying [1]. The agricultural sector that made a living for 65% of poor working adults living in the developing world [2], includes the largest segment of society afflicted with the greatest amount of pesticide exposure [3]. Farmers and farmworkers put themselves at risk of pesticide poisoning during the production process [4]. They are routinely exposed to toxic pesticides via spray, drift or direct contact with pesticide residues on treated crops or soil [5].

Excessive and indiscriminate use of the most infamous synthetic organic chemicals, specifically chlorinated hydrocarbons like DDT, left a long-lasting imprint on every atom on earth since the beginning of the green revolution to date [6]. Despite the early alarming signs on the gruesome effects of these chemicals, such as the sharp decline of the bald eagle and other birds in the highest trophic levels of the food chain [7], the ascendency of using these chemicals pressed onward. This entailed unintentional pesticide exposure to the general population via food consumption [3, 8].

In Ethiopia, the agricultural sector is a source of livelihood for more than 80% of the population and represents more than three quarters of national exports [9], which impacts that almost 80% of the population in Ethiopia contributes to posing negative health issues to both humans and the environment, leading to the prevalence of acute and chronic health issues among people in the whole population [10, 11]. Hence, addressing this problem will have practical benefits for Ethiopia and will contribute to understanding of this widespread phenomenon among farmers almost in the whole country.

Unsafe practices exhibited by farmers in Ethiopia included the frequent mixing of highly toxic pesticides, unsafe storage and unsafe transport methods [12], unsafe disposal of empty pesticide containers [10], spraying while barefooted, using obsolete pesticides [13], selling illegal, expired, hazardous, unknown, repacked pesticides in small containers without any labeling, manufacturing or expiry date by unauthorized and untrained people in shops and local village in open markets [13], reusing empty insecticide containers for food and potable water, washing pesticide-contaminated work clothing with the family clothes and using highly toxic insecticides to treat head lice, fleas, bedbugs, and even to try to cure open wounds using malathion and (DDT), sometimes with fatal results [14].

Few researches in Ethiopia were conducted concerning the same area of study [10–12, 15], and further studies about pesticide use and pesticide-related illnesses are needed to develop more effective approaches for protecting farmers from pesticide exposure and moving the targets towards organic farming, resulting in having a healthy community free of disabilities and other dysfunctional diseases.

2. Material and methods

2.1 Study area

The study was conducted at Finchawa and Tullo rural kebeles in Hawella Tula sub city, which is located in Hawassa City Administration, Sidama, Ethiopia. Tullo and Finchawa are considered the catchment area of Lake Hawassa. The area lies on 1728 meters above sea level. The land form is plain [16], and has tropical savanna climate with two seasons [17]. The mean annual temperature is 20.9 °C with mean rainfall 997.6 mm [17]. The type of soils around Lake Hawassa in general is Andisols. A source of information about the type of soil in the study area is Hawassa University, Agricultural Campus. The two main crops cultivated in both areas are corn and ensete.

2.2 Study design and data collection tools

A cross-sectional survey with both qualitative and quantitative data collection research design were employed. The study was conducted between June and September 2020. The study also applied concurrent triangulation in data collection, where both qualitative and quantitative data were collected in one phase of the research study and equally weighted. A concurrent triangulation procedure is the model mostly employed when researchers first consider mixed methods as it results in a shorter data collection time period [18]. The three different primary data sources used in this study were a standardized questionnaire, key informant interview and a document review.

2.3 Methods of data collection

A supportive letter was given from the College of Natural and Computational Science, Biology Department, Hawassa University, to get the permission of the respective directions to select the study participants and conduct the interviews in extension offices, health institutions and hospitals. Two days of training were allocated to train four experienced data collectors. The training focused on explaining the purpose of the study, the meaning and interpretation of some scientific terminologies in each question, and obtaining consent from every single participant. The data collectors were experienced and capable of speaking the local languages (Amharic and Sidamingia). The four data collectors conducted door-to-door visits based on a list of members of the households to get responses and fill in the questionnaire. The list of householders was coded and their names were not mentioned for anonymity and confidentiality. In-depth interviews with officials and physicians were conducted by the researchers with the help of a professional translator who spoke both Amharic and English languages.

2.4 Data analysis

All data was coded and analyzed using SPSS version 25. Descriptive statistics were used to summarize frequencies and proportions, and results were presented in tables and charts. A multiple logistics regression model was employed to determine the effect of the independent variables on farmers' knowledge and the prevalence of self-reported toxicity symptoms. Multiple logistic regression was employed due to its powerful statistical way of modeling a binomial outcome for categorical data [19]. Chi-square, as well as Hosmer and Lemeshow tests, were firstly used before running the logistic regression test to measure the association between the independent and dependent variables and to check whether the model fits the data or not respectively. The data was summarized using the odds ratio, 95% confidence interval at .05 alpha levels.

2.5 Sampling technique and sample size determination

The study employed a multi-stage sampling technique due to the advantage it gives of using more than one stage and combine several sampling techniques. The multi-stage sampling in this study entitled four stages. In the first stage, the

Tula sub-city was purposively selected as it is relatively accessible by scientists. In the second stage, the Finchawa and Tullo rural kebeles were also purposively selected because of the considerable number of farmlands available in both kebeles, the extensive usage of pesticides in their farmlands, and their strategic location around Lake Hawassa. Both rural kebeles are considered the catchment area of Lake Hawassa. In the third stage, the study applied a simple random sampling to select farmers from both rural kebeles. All participants agreed to participate in the research study by signing informed consent forms. In the fourth stage, a convenience sampling was employed to select one official from the extension office in Finchawa, one official from the extension office in Tullo, one physician from the Bushullo Health Institution and one physician from the Referral Hospital.

The farmers' representative of both Finchawa and Tullo rural Kebeles estimated the number of farmers that use pesticides in their farmland as 100 farmers distributed as follows: Finchawa 49% and Tullo 51%. The sample size was determined by using the formula of Kothari [20]; at 95% level of confidence. Accordingly, the total sample including 10% of the contingency is 73.

2.6 Pilot testing

The questionnaire was piloted with 20 farmers (10 participants from Finchawa and 10 participants from Tullo) who did not participate in the study. Hence, all the forwarded comments regarding the wording of sentences, vague sentences and unclear scientific ideas were amended to ensure the validity of the items. The research was also expected to be reliable on its findings. Reliability of binary items were tested using Kuder–Richardson 20. The KR-20 can be applied to any test item responses that are dichotomously scored [21]. The value of internal consistency tests suggested a good level of reliability. Further, the internal consistency of the Likert scale items, was also tested using Chronbach's alpha. Cronbach's (1951) alpha was developed based on the necessity to evaluate items scored in multiple answer categories [21]. The value of internal consistency tests indicated a good level of reliability.

3. Results and discussion

3.1 Demographic characteristics of the respondents

As depicted in item 1 of **Table 1** below regarding male to female ratio, about 91.8% of the participants were males, while only 8.2% were females. The dominance of males over females in this study might be due to the nature of work as men usually are more involved in pesticide handling than women [22]. The sex ratio presented in this study is in line with the finding of [23].

Regarding the age of the respondents, about 50.7% had their ages between 36–45 years, while 20.5% and 16.4% were within the range of 46–55 and 25–35, respectively. The average age of the farmers was 44.42 years. The obvious decrease in the proportion of farmers on the both sides of the age spectrum might be due to youths' lack of interest in farming, in addition to their tendency to shift to urban areas for better employment and higher income [24]. The decrease may also be due to the health deterioration of aged farmers, as this limits their abilities to put up with more physical-consuming tasks that might add more health burdens to the bunch they already have.

Concerning the education status of the participants, the majority 65.75% had a formal education, mainly primary education 10.96%, secondary education 49.32%,

Variables	Category	Freq.	Percentage
Gender	Female	6	8.2%
	Male	67	91.8%
Age of the Respondent	25.00–35.00	12	16.4%
	36.00–45.00	37	50.7%
	46.00–55.00	15	20.5%
	56.00–70.00	9	12.3%
Educational Level	No formal education	25	34.25%
	First cycle primary (Grade 1–4)	8	10.96%
	Second cycle primary (Grade 5–8)	22	30.14%
	Secondary (Grade 9–10)	14	19.18%
	Preparatory and above level	4	5.48%
Years of Experience	Less than 5 years	12	16.4%
	5–10 years	25	34.2%
	10–20 years	20	27.4%
	20 years and above	16	21.9%
Average Monthly Income	<1000 birr	46	63.0%
	1001–1500 birr	14	19.2%
	1500 birr and above	13	17.8%
Residential Area	Living on the farm	48	65.8%
	Within 5 km distance from the farm	18	24.7%
	5–12 km away from the farm	7	9.6%
Is Farming	Your only source of income	32	43.8%
	You have another source (other job)	41	56.2%
Working Hours Per Day on the Farm	Part-timer <8 hours	41	56.2%
	Full-timer >8 hours	32	43.8%

Table 1.
Demographic characteristics of sample respondents (N = 73).

preparatory and above level 5.48%, while 34.25% of the participants were illiterates. Therefore, there is a considerable proportion of educated farmers in both kebeles and, the highest observations were categorized in the secondary education level. In comparison with similar studies that were previously conducted, there is a leap of improvement in the achievements of the educational sector in Ethiopia. A study done by [25] reported that only 24.3% of the sample participants completed their secondary level of education. Accordingly, the investment of the Ethiopian government in education through a sustainable increase in national expenditure and aids to the educational sector [26] is well translated in this study. Yet, more efforts are still needed to eradicate illiteracy completely, especially in the country sides.

With respect to farmers' work experience, most of the respondents 34.2% had 5–10 years of farming experience, followed by 27.4% and 21.9% of the same with 10–20 years and over 20 years of experience, in that order. Conversely, only 16.4% of the study participants had less than 5 years of experience. This clearly shows that most of the farmers in the study area had quite adequate experience which, in other studies, proved to have a significant contribution to good pesticide knowledge and safe practices [23, 27].

Majority of the respondents 63% earned less than 1000 Ethiopian birr, followed by 19.2% who earned 1001–1500 birr, while 17.8% of the respondents were found to earn 1500 birr and above. The low monthly income was also reflected by the study of [16] in Ethiopia.

About 65.8% of the respondents reported to be living on the farm, while 24.7% of the same replied that they are living within 5 km distance from the farm. On the contrary, 9.6% of the participants were living 5–12 km away from the farm. However, residing in or close to agricultural lands might increase the potential risk of pesticide exposure on farmers and their families through non-occupational pathways via spray drift and volatilization of pesticides beyond the treated area [28].

Majority 56.2% of the sample farmer respondents indicated that they have additional sources of income other than farming, while 43.8% of the same mentioned that farming is their only source of income.

Concerning the working hours per day, the majority of the respondents 56.2% indicated that they work as part-timers and they spend less than 8 hours per day on the farm, while 43.8% of the participants made known that they work as full-timers and they spend more than 8 hours on the farm. The low income of the majority of the respondents justifies the steep decrease in the proportion of youth in farming lands found in this study, and shows the modern-day slavery lifestyle which is portrayed in cheap wages beside the long working hours.

Accordingly, the socio-economic profile of the participants in both kebeles in this study indicated a kind of harsh lifestyle that swings in a range of difficult circumstances, including poverty, modern life slavery and illiteracy.

3.2 Farmers' knowledge regarding pesticides handling and toxicity

Participants were presented with eight questions that they could answer either "Yes or No". For the purpose of analysis, data was coded as (Yes = 1 and No = 0). The result showed that the total sum of the knowledge score was 288 with the mean and SD of knowledge score of 3.95 ± 1.07. The range of the knowledge score was 0 to 8 where: <4 = poor knowledge while ≥ 4 = good knowledge [23].

The sample farmer respondents were asked whether pesticides make people feel sick or not. Accordingly, the majority 56.2% of the farmers had a lack of awareness about pesticide use posing some potential risk to human health, while 43.8% of the respondents perceived that pesticides make people feel ill or sick. Similarly, the lack of awareness regarding the adverse health effects of highly toxic pesticides among farmers in the study area was also assured by the emergency physician in the Referral Hospital, who indicated during the interview that farmers would not store highly toxic pesticide in their homes if they were well aware about pesticide's fate in the environment and the negative health effects that pesticides might pose on humans. This finding is in line with the study of [29] which reported that 71% of the farmers had limited knowledge about pesticides posing a health problem in their community.

When inquired as to whether water gets polluted from pesticide runoff or not, about 35.61% of the respondents believed that the pesticides could pollute the aquatic environment, while the majority 64.39% of the respondents did not consider that the pesticide may affect the water bodies. The limited knowledge of farmers about the end fate of pesticides in the environment justifies the finding of [30] regarding the high concentration of DDT found in the *Barbus intermedius* fish, which represents the highest trophic level of the food chain in Lake Hawassa. However, the majority 67.12% of the farmers reported that contaminated water makes people sick or ill. This finding is in line with a previous study reported by [29] which stated that 91% of the farmers knew that water might get polluted from pesticide runoff, and people can get sick from water contaminated by pesticide runoff.

In item 4 of **Table 2**, participants were asked about the routes of which pesticides can enter their body. According to the data in **Table 2**, Inhalation 56.16%, followed by Oral 36.99%, Ocular 32.88%, and Dermal 12.76%. Conversely, considerable fraction 42.47% of the respondents did not know about the exposure route by which the pesticide could enter the body. However, despite the fact that dermal absorption is the main exposure route for pesticide appliers [31], it was the least route reported by respondents 24.66%. This rationalizes the high proportion of participants who reported to clean the contaminated area of skin after finishing their shift, rather than cleaning it immediately. The depiction of this action is illustrated in **Figure 1**. Moreover, this study noted that the majority of farmers were aware of the important entry routes of pesticide exposure, including inhalation and ingestion. These findings are in line with similar studies carried out by [32, 33].

Regarding farmers' knowledge of the effect of pesticides on food quality and quantity, about 71.2% of the respondents replied that pesticide application affects

Items	Category	Freq.	%
Do pesticides make people feel ill or sick?	No	41	56.2%
	Yes	32	43.8%
Does the water get polluted from pesticide runoff?	No	47	64.39%
	Yes	26	35.61%
Does contaminated water make people sick?	No	24	32.9%
	Yes	49	67.1%
Routes of which pesticides can enter your body? [Multiple responses possible]	Skin	18	24.66%
	inhalation	41	56.16%
	oral	27	36.99%
	Eye	24	32.88%
	Do not know	31	42.47%
Will the food be of the same quality without using pesticides?	No	52	71.2%
	Yes	21	28.8%
Can a farmer obtain the same yields without pesticides? (Quantity)	No	31	42.5%
	Yes	42	57.5%
Would people get sick if they entered the farm after a few hours of spraying (2–3 hours)	No	52	71.2%
	Yes	21	28.8%
Do you know about biological and natural control?	No	18	24.66%
	Yes	55	75.34%
What kind of pesticides do you use? [Multiple responses possible]	Insecticide	43	58.90%
	Herbicide	57	78.08%
	Fungicide	14	19.18%
	Rodenticides	3	4.11%
Knowledge status	Poor	24	32.12%
	Good	49	67.12%

Table 2.
Farmers' knowledge of pesticide handling and toxicity (n = 73).

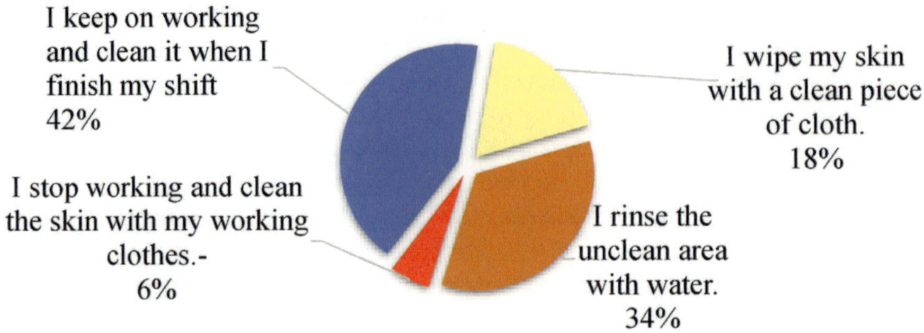

Figure 1.
Actions farmers usually take when their skin gets in contact with pesticides.

food quality. In contrast, 28.8% of the study participants disagreed with the stated statement. However, studies proved that pesticide application brings a primary benefit of better quality on crops, and this quality brings a benefit that outweighs the potential risk of human's exposure to very low residues of pesticides, especially in a diet containing fresh fruits and vegetables [3].

Regarding the effect of pesticides on crop quantity, about 42.5% of the farmers indicated that pesticides application is essential for high crop yield and productivity, while the majority 57.5% contradicted the correlation between pesticide application and crop quantity.

Farmers' tendency to rely on pesticides application to speed up the eradication of pest infestation is evident in this study. This reflects the high influence of the government on their officials perception, as the Ethiopian government extension program encourages the use of pesticides in agriculture and supports the widely-accepted perception that there is no other alternatives to pesticides [13].

Concerning whether people get sick if they entered the farm after a few hours of spraying (2–3 hours) or not, about 28.8% of the respondents perceived that pesticides make people feel ill or sick if they entered the farm after a few hours of spraying (2–3 hours), while 71.2% of the respondents did not relate any significant health effect to the exposure of pesticides. Farmers who re-enter treated fields soon after pesticide application might be in serious risk of exposure, especially when safety measures are poorly adopted [4].

Participants were also asked whether they know about biological and natural control or not. Majority 75.34% of the farmers indicated that they know about biological and natural control of pesticides, while only 24.66% of the participants did not know about natural pest control. In similar vein, officials in the agricultural extension office mentioned during the interview that farmers had used tobacco leaves as pest repellant and added healthy soil to the infested one when they are confronted with a shortage of pesticide availability. The aforementioned findings concerning farmers' knowledge about biological and natural control correlate with similar statements made by their authorities.

Insecticides 58.90% and herbicides 78.08% are the most common types of pesticides utilized in this study, while fungicides 19.18% and rodenticides 4.11% were the least pesticides used. Authorities from the agricultural office also indicated that the three main pesticides used in both kebeles are endosulfan to eradicate aphids, diazinone to combat American bollworm, and flazasulfuron to control the growth of the unwanted weeds. These responses strongly suggest that farmers and local authorities are in agreement on the use of the above-mentioned agents.

Regarding the training that is regularly given for the participants in both Kebeles, the experts in the extension offices reported during the interview that:

'We visit farmers in their farmlands about three times per week, and in addition, we keep on giving them professional training and instructions regularly.'

Despite the efforts excreted by the experts in the extension offices in both Kebeles regarding the proper communication and the regular training they give to farmers in their farmlands, there is a limited knowledge among the participants about the effects of pesticides on humans' health and the environment, in addition to the lack of awareness regarding the dermal route of exposure. Accordingly, the training given to the sample participants in the study area is in insufficient, and different strategies should be taken to improve the process of learning and further mitigate their risk of exposure.

3.3 Factors that influence farmers' knowledge regarding pesticide handling and toxicity

The study also Investigated the factors that influenced farmers' good knowledge regarding pesticide handling and toxicity. Accordingly, item 2 of **Table 3** revealed that the odds of the knowledge are positively influenced by the age factor; therefore, older farmers are 1.377 times more likely to have a good knowledge regarding pesticide handling and toxicity, than those who are younger. The estimated odds ratio is statistically significant since ($p = 0.002$) which indicates that ($p < 0.05$) is within a 95% confidence interval (OR = 1.377, CI = 1.125–1.685). Therefore, acquiring domain knowledge through aging might be due to daily observations and interaction with people who are knowledgeable and very well experienced in farming work.

Dependent variable: Good knowledge **Independent Variables**	B	S.E.	Sig.	OR	95% C.I. Lower	95% C.I. Upper
Gender (1 = Male), Ref. = Female	0.584	1.342	0.664	1.792	0.129	24.864
Age of the respondent	0.320	0.103	0.002	1.377	1.125	1.685
Reference = no formal education			0.176			
First cycle primary (Grade 1–4))	1.207	1.330	0.364	3.344	0.246	45.377
Second cycle primary (Grade 5–8)	2.503	1.203	0.037	12.222	1.157	129.162
Secondary level (Grade 9–10)	2.799	1.230	0.023	16.436	1.476	183.046
Preparatory and above level (11+)	3.677	3.117	0.238	39.522	0.088	17769.305
Training in pesticides/ safety (1 = yes)	2.549	1.241	0.040	12.799	1.124	145.789
Experience (Ref. = > 5 years)			0.071			
5–10 years	3.455	1.434	0.016	31.649	1.903	526.455
10–20 years	2.726	1.333	0.041	15.278	1.121	208.284
Above 20 years	5.015	2.059	0.015	150.681	2.666	516.792
Constant	–18.148	5.348	0.001	0.000		

Table 3.
The multiple logistic regression analysis of factors that influence farmers' knowledge of pesticide handling and toxicity.

According to item 3, **Table 3**, the odds of knowledge is also positively influenced by farmers' educational level, especially with the second cycle primary (grade 5–8) and secondary level (grade 9–10) of education. Hence, farmers who achieved second cycle primary (grade 5–8) education level are (12.222) times more likely to have a good knowledge in pesticide handling and toxicity, than those who did not achieve this level of education since ($p = 0.037$) which indicates ($p < 0.05$). The estimated odds ratio is statistically significant within a 95% confidence interval (OR = 12.222, CI = 1.157–129.162). A similar analysis showed that farmers who attended secondary level (grade 9–10) of formal education are 16.436 times more likely to have good knowledge in pesticide handling and toxicity than those who did not achieve this level of education since ($p = 0.023$) which indicates ($p < 0.05$). The estimated odds ratio is statistically significant within a 95% confidence interval (OR = 16.436, CI = 1.476–183.046).

Regarding farmers' access to training which is presented in item 4 of **Table 3**, the odds of knowledge status is positively influenced by the training factor since ($p = 0.040$) which indicates that ($p < 0.05$). Thus, farmers who attended field training are 12.799 times more likely to have a good knowledge in pesticide handling and toxicity than those who did not take any training. The estimated odds ratio is statistically significant within 95% confidence interval (OR = 12.799, CI = 1.124–145.789).

Item 5 of **Table 3** also showed that the odds of the knowledge were positively influenced by the experience factor, Accordingly, farmers who had 5–10 years of work experience are 31.64 times more likely to have a good knowledge in pesticide handling and toxicity than those who had less years of work experience since ($p = 0.016$) which indicates ($p < 0.05$). The estimated odds ratio is statistically significant with 95% confidence interval (OR = 31.64, CI = 1.903–526.455). In addition, farmers who had 10–20 years of work experience are 15.278 times more likely to have a good knowledge in pesticide handling and toxicity than those who had less years of work experience since ($p = 0.041$) which indicates ($p < 0.05$). The estimated odds ratio is statistically significant with 95% confidence interval (OR = 15.278, CI = 1.121–208.284). The result also indicated that farmers who have 20 years of work experience and above are 150.681 times more likely to have a good knowledge in pesticide handling and toxicity, compared with the reference category (Ref. < 5 years) since ($P = 0.015$) which indicates ($p < 0.05$). The estimated odds ratio is statistically significant within 95% confidence interval (OR = 150.681, CI = 2.666–516.792), while gender and first cycle primary (grade 1–4) failed to be significant predictors under the given conditions.

3.4 Farmers' field practices including storage, disposal, transportation and pesticide preparation

The study evaluated farmers' field practices by assessing the way they disposed of empty pesticide containers, the way they transported pesticides from the vendors' shop to the field, the way they stored pesticides and their spraying equipment, and their methods of pesticide preparation.

3.4.1 Disposing empty chemical containers by farmers

About 12.33% of the participants reported burning empty pesticide containers as a method of disposal. The guidelines on the management options for emptying pesticide containers by [34] warned against and manifested on the prohibition of such practice. Open burning of pesticide containers generates environmentally-persistent toxic fumes resulting from chemical traces lining the container, chemicals which are used to make the body of the container, or emissions of incomplete combustion. Such toxic fumes might be inhaled by animals or humans that exist in the burning area, causing irreversible damage to health.

The majority 64.38% of the farmers reported reusing empty pesticide containers for other purposes, like water or food storage. Reusing empty pesticide containers might increase the risk of non-occupational exposure via pesticide residuals ingestion as it is impossible to remove all traces of pesticide chemicals' residue from empty plastic or metal pesticide containers [35]. According to [36], empty pesticide containers should be shaken clean and triple-rinsed before disposing them in an environmental-friendly manner.

The fact that the majority of the farmers reuse empty pesticide contains for food and water storage is worrisome, and might be a result of a wrong perception that once the container is washed, it becomes pesticide-free and poses a zero negative health effect on them. This perception might be acquired from their daily observations as they did not suffer or saw anyone complain of any negative health effects on the short-term of ingestion. Hence, it is very obvious that farmers are unaware of the long-term risks of pesticide exposure and the adverse health effects these residuals might pose when they accumulate in body tissue. Reusing empty pesticide containers by majority of the participants was also reported by the study of [12].

About 10.96% of the participants disposed their empty pesticide containers anywhere on the farm. Such improper disposal method contaminates the soil with chemicals that might leach to both ground and surface water, or reach the latter via surface runoffs, posing a threat to aquatic organisms, as well as humans that use ground water for drinking.

Dropping pesticide containers in a public dump was reported by 5.4% of the sample participants. Such practice might expose solid waste collectors in the formal and informal sector to unintentional exposure via skin contact, inhalation or ingestion, especially those workers do not use any means of protection measures to avoid several safety issues they experience on a daily basis. Moreover, a considerable proportion of the workers in the informal sector are children who might take the containers, wash them improperly, and reuse them for water drinking or sell them to people in the local market.

In Finchawa, reusing pesticide containers was not limited to illiterate people. The official from the agriculture office who had 7 years of work experience in agriculture, a university degree and a robust background in farming, recommended and instructed farmers to reuse empty pesticide containers. Further, officials in Tullo kebele instructed farmers to dispose empty pesticide containers in pit latrines. When the latrines are full, they are backfilled and new pits are dug again.

Pit latrines usually lack a physical barrier that separate human's excreta or chemicals thrown in the pit and soil or ground water. Therefore, contaminants from latrines potentially leach into ground water and seeps into other water surfaces, like lakes or rivers, posing a threat to humans using ground water or aquatic organisms living in water bodies [37].

Encouraging farmers to reuse empty pesticide containers, or throw them in latrines by officials, might be a part of the kebeles' waste management strategy to reduce environmental contamination and save water sources due to the lack of availability of proper disposal facilities in Hawassa. Yet, it is strong evidence of poor understanding and a lack of awareness about the effects of pesticide residues that come in contact with food and drinking water, the long-term health effects on humans, and the negative impacts of pesticides on the environment.

Throwing empty pesticide containers in the lake by 6.85% of the participants is presented in item 5 of **Table 4**. According to [38], once pesticides reach water

No	Items	Freq.	Percent
1	Burn	9	12.33%
2	Reuse them for food and water storage	47	64.38%
3	Throw them anywhere in the farm	8	10.96%
4	Drop in the public dump	4	5.4%
5	Throw in the lake	5	6.85%
Total		73	100%

Table 4.
Disposal methods of empty pesticide containers by farmers.

bodies, they can impact the whole ecological food chain, since other animals, including humans, feed on aquatic animals that may be contaminated.

Back to the literature review, recalling the Lake Michigan incident about children who had an intellectual impairment due to their in-utero exposure to organochlorines [39], in addition to the finding of [40], several cases of learning disabilities, autism, ADHD, child cancer and juvenile diabetes among the generation living around the lake, and consuming food and water contaminated with highly toxic pesticides in the 30–40 coming years, is expected by this study. The increment in the rate of suicide incidents among teenagers that was reported by [40] was also reported and manifested by the physician of the emergency room in the Referral Hospital.

3.4.2 The way farmers transport pesticides from the shop to the farm

Respondents were asked about the way they transported pesticides from the store to the farm. Accordingly, the majority 42% of respondents indicated that they carry it and walk, while the rest used the bed of a truck alone with no other items 22%, the backs of donkeys 18% and bajaj with other passengers 18% (**Figure 2**).

Using the backs of donkeys, public transport or self-carrying methods to deliver pesticides from shop to field are inappropriate pesticide transport practices that might expose human beings, animals and the environment to danger, in the case of unexpected accidents during trips. Such accidents might cause a container puncture, break or torn which increase the risk of spillage. However, spillage or leakage of highly toxic pesticides might be absorbed through inhalation or directly through unbroken or broken skin. Unfortunately, delivering pesticide containers using the bed of the truck alone with no other items was reported only by 22% of the participants.

3.4.3 Storage of pesticides and spraying equipment by farmers

Farmers' practices on storing pesticides and spraying equipment were assessed through 5 items. Almost all of the farmers 98.63% reported that they frequently take the spraying equipment after the field work to their home. A considerable proportion of farmers 79.45% reported storing pesticides in the bedrooms, 60.27% in the living room, and 63.01% in the kitchen. However, storing pesticides beside food, potable water or seeds may increase the risk of their contamination with vapors, dust or spills, and increase the likelihood of accidental human exposure (**Table 5**).

Frequent intentional suicide incidents among teenagers and youths in their early twenties in both kebeles were reported by the physician in the emergency room at the Referral Hospital during the interview due to the free availability of pesticides

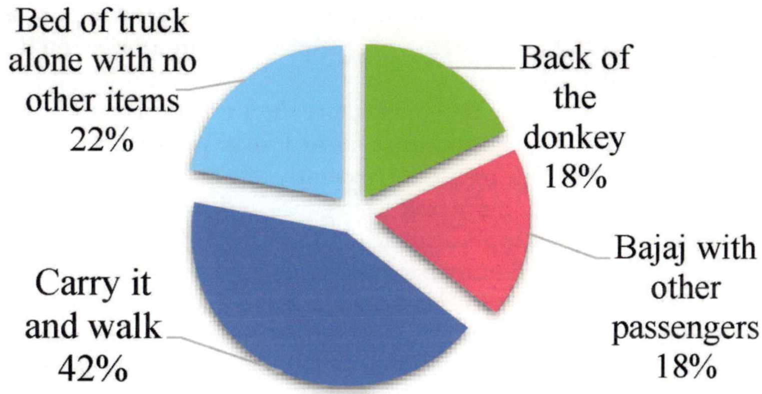

Figure 2.
Means of transporting pesticides from the store by farmers.

Items	No	Yes
Do you usually take the spraying equipment after the field work to your home?	1 1.37%	72 98.63%
Have you ever stored pesticides in the bedroom?	15 20.55%	58 79.45%
Have you ever stored pesticides in the living room?	29 39.73%	44 60.27%
Have you ever stored pesticides in the kitchen?	27 36.99%	46 63.01%
Do you store pesticides in a locked and separate place that is specified for pesticide storage?	36 49.32%	37 50.68%

Table 5.
Storage of pesticides and spraying equipment by farmers.

in farmers' houses. The rate of suicide cases among the mentioned segments is 3–4 cases per month (around one case per week). The physician at the Referral Hospital indicated that organophosphate is the most commonly used insecticide in both kebeles.

The official in Finchawa reported that farmers use diazinone and endosulfan to eradicate bed bugs in their homes, which exposes them and their families to a high risk of exposure via direct inhalation, especially that Endosulfan evaporates rapidly after spraying and poses high toxicity if inhaled on the long-term of exposure [41].

Unfortunately, only 50.68% reported that they store pesticides in a locked and separate place that is specified for pesticide storage.

Methods of pesticides' storage among farmers in both kebeles found to be inappropriate in this study. Moreover, it increased the risk of exposure on farmers and their families through direct and indirect ingestion, inhalation or dermal absorption.

3.4.4 Preparation of pesticides by farmers

The sample participants were asked about the likelihood of using a measuring cup or measuring tool to add the exact amount of pesticide mentioned on the label.

Accordingly, the majority 97.25% of the farmers reported that they never or rarely use a measuring tool to add the exact amount of pesticide mentioned on the label, while 2.73% responded sometimes.

The failure of using a gauging tool to measure the amount of pesticide needed for the application, results in using a less or more dose of pesticide than the one recommended on the label. The result of frequently low or heavy doses of the application is more resistance, more pest resurgence and more secondary outbreaks. Moreover, heavy doses of the application lead to an unacceptable environmental contamination and high risk of human exposure. One study reported that only 0.0000001% of DDT reach the target pest, while more than 99.99% are dispersed into the environment through volatilization, surface runoff, infiltration and drift [42]. This finding is in line with the result of [11] which indicated that none of the farmers used scaled measuring equipment.

Practices	Never	Rarely	Some times	Often	Always	Mean	SD
Do you usually use a measuring cup or measuring tool to add the exact amount of pesticide mentioned on the label?	66 90.41%	5 6.84%	2 2.73%	0 0%	0 0%	1.12	0.41
How often do you check the defect (inadequacy) of the spraying equipment you are using before you start applying?	23 31.5%	9 12.3%	38 52.1%	3 4.1%	0 0.0%	2.25	1.32
Do you usually check the defect (inadequacy) of the PPE before dealing with pesticides?	56 76.72%	15 20.55%	2 2.73%	0 0%	0 0%	2.29	0.96
Do you use special tools (only for pesticide usage) to mix and apply pesticides?	32 43.8%	8 11.0%	23 31.5%	3 4.1%	7 9.6%	1.26	0.50
Average	177 60.62%	37 12.67%	65 22.26%	6 2.05%	7 2.40%	1.73	0.50

Table 6.
Farmers' pesticides handling and preparation.

Participants also were asked about the likelihood of checking the inadequacy of the spraying equipment before they start applying pesticides. About 43.8% of the farmers reported that they never or rarely checked their spraying equipment before they start applying, while 52.1% responded sometimes, with 4.1% reporting that they often check the inadequacy of the spraying equipment before they start applying. However, punctures, breaks and cracks in the tank of the spraying equipment for liquid formulation results in a high risk of exposure to workers via dermal absorption.

Majority 97.72% of the participants reported that they rarely or never checked the defect of their PPE before dealing with pesticides because PPE are mostly unavailable, while 2.73% only reported doing such practice sometimes.

Participants were asked if they usually use special tools to mix and apply pesticides. The majority 54.8% reported that they never or rarely used special tools to mix and apply pesticides, rather, they use their hands or any available stick in the farm for the purpose of mixing. About 31.5% indicated that they sometimes used special tools to mix and apply. The rest of the participants reported often 4.1% or always 9.6%. Mixing pesticides with hands increases the risk of exposure via dermal absorption or ingestion as farmers can easily carry traces of pesticides from their hands to their mouth, especially in the case of poor usage of PPE that was reported by this study. This finding is in line with the result of [13].

Statistically, it is also observed from the results in **Table 6** above that the mean average score of the participants' responses 2.27, 2.25, 2.29, and 1.26 regarding the methods of pesticide preparation is below the average Likert scale 3. This indicates that the farmers are not in a position to reduce the risk of pesticide exposure.

As depicted in **Figure 3** below, respondents were asked about the source of information before and during mixing, applying or loading pesticides. Accordingly, the majority of the farmers 72% got information from kebeles' agricultural experts, while the minority of the farmers 17% reported that they get such information from their neighbors or 11% vendor. Despite the considerable proportion of the participants receiving information from kebeles' agricultural experts, it is disquieting to have 27% of the respondents seeking information from improper sources. Studies proved that there is a big difference between experts and novices in the way they perceive, remember and express their observations through the language they use [43]. Hence, seeking information from neighbors or vendors about the kind of

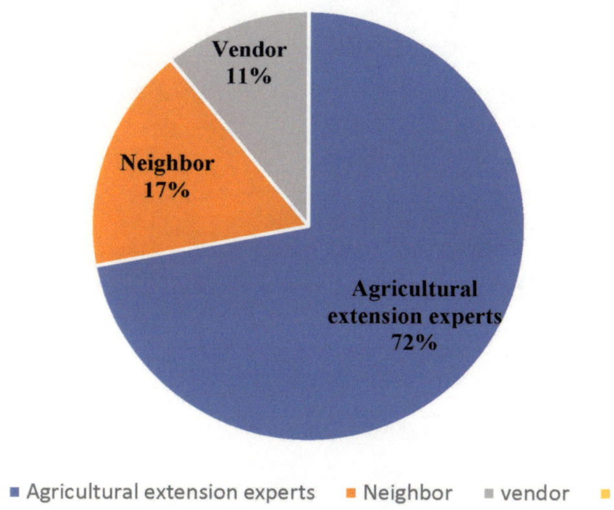

Figure 3.
Farmers' sources of information while mixing, applying or loading pesticides.

pesticide should be used, or any technical issue regarding pesticide handling results in farmers' information inadequacy.

Officials in the agricultural office in both kebeles assured, during the interview, that farmers resort to them when they observe any pest infestation in their farms to get advice about the right type of pesticide that should be applied in such cases. This implies that participants are somehow on the same page with their authorities. This finding is in line with the result of [44] which reported that 57.2% of the farmers seek information from woreda agricultural extension experts.

As depicted in **Figure 4** below, respondents were also asked about the area they usually use to mix and load pesticides. The majority 56% of the respondents reported that they mix and load pesticides within their residential area in the garden, while 27% reported that they mix and load in the field and 17% reported that they never prepare pesticides and they use the one that is already prepared by someone else. Mixing and loading pesticides within the house garden increases the risk of exposure of farmers' family members via dermal absorption, inhalation or ingestion, especially among small children who spend most of their time outdoors playing with the mud.

The inadequate knowledge of farmers regarding pesticide handling and toxicity was reflected in their field practices and found to be unsatisfactory. Moreover, it implied a high potential of pesticide exposure for them and their families.

3.5 Safety precaution and protective measures adopted by farmers

3.5.1 The usage of personal protective equipment by farmers

The usage of PPE by farmers while dealing with pesticides was assessed in this study. Nearly all the participants 93.15% did not use any means of PPE while dealing with pesticides and 6.85% of the respondents only reported using gloves.

This indicates that farmers are in a high potential risk of exposure while preparing, handling and spraying pesticides through all the exposed parts of their bodies,

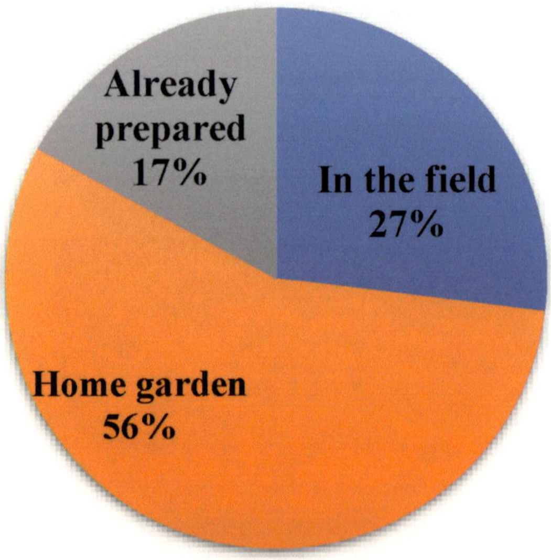

Figure 4.
Pesticide preparation area by farmers.

especially their hands, which are considered the main carriers of pesticide, traces to every single part of their bodies. Wearing proper PPE during pesticide application proved to have a significant effect on mitigating farmers' risk of exposure. According to [45], an increase in the use of protective measures decreases the probability of poisoning by 44% to 80% (**Table 7**).

Farmers were also asked to list the factors that stops them from using PPE while handling pesticides. As depicted in **Figure 5** below, the answers were as follows: too expensive 42%, not comfortable in the tropical climate 5%, not available when needed 22% and no health challenges from using pesticides 31% are some of the cited reasons.

The officer in Finchawa Kebele reported during the interview that farmers do not use any kind of personal protective equipment while dealing with pesticides because the government does not provide them with any. However, the officer in Tullo indicated that farmers tie their clothes on their nose and mouth while spraying as a kind of precautionary measure to protect themselves from pesticide exposure, which increases the risk of their dermal exposure after wearing their clothes again.

3.5.2 Instant actions taken by farmers when their skin gets in contact with pesticides

The study assessed the actions that farmers take instantly when their skins get contaminated with pesticides. As depicted in **Figure 1**, sample respondents were asked about the actions they take when their skin gets in contact with pesticides. Accordingly, about 18% of the respondents reported that they wipe the unclean area of skin with a clean piece of clothing, while 6% indicated that they stop working and clean their skin with working cloths, and 34% replied that they rinse the pesticide-contaminated area of skin with water. However, the majority of the respondents 42% reported taking no action until they finish their shift **Figure 1**.

Ignoring the contaminated area of skin with pesticides until the shift is finished, rather than cleaning it immediately, is worrisome. According to [46], the extent of skin absorption increases positively with the duration of exposure. This implies that the sooner the cleaning of the contaminated area of skin is performed, the greater the decrease in the dermal absorption is achieved, especially in the case of liquid formulations. Besides, the longer the operators ignore the stains of pesticide on their skin, the greater the risk of their exposure will become via the inhalation of pesticide volatiles.

No	PPE Items	Freq.	percent
1	Gloves	5	6.85%
2	Mask/Respirator	0	0%
3	Protective eyewear	0	0%
4	Special shoes	0	0%
5	Overall	0	0%
6	Hat/Cap	0	0%
7	Wear all of them at the time of handling pesticide	0	0%
8	Do not use any of them	68	93.15

Table 7.
The usage of personal protective equipment by farmers.

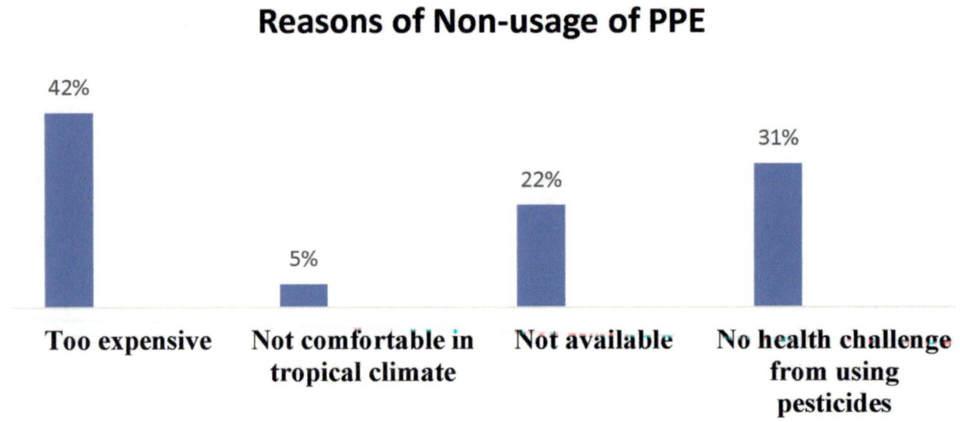

Figure 5.
Factors that hindered the usage of PPE by farmers.

Rinsing the contaminated area of skin with water is considered a proper action in the case of using water soluble pesticide formulation, like flazasulfuron, based on the general rule of like dissolves like. Yet, both organochlorine and organophosphate insecticides reported to be used in this study are lipid-soluble. They are very well absorbed through the skin as they dissolve easily in the sebum that is released by the sebaceous glands; therefore, rinsing the contaminated area with water is an inappropriate instant action performed by farmers and implies a high risk of exposure.

Wiping the contaminated area of skin with a work cloth is also worrying. Cloths soaked with pesticides increase the risk of dermal exposure and volatiles inhalation. Also, the longer the time the operator wears the contaminated clothes, the greater the extent of absorption and inhalation will be.

The level of self-protection among farmers regarding the proper usage of adequate PPE while dealing with pesticides, and the instant action taken when their skin gets contaminated with pesticides in this study, is found to be disappointing and presents a potential risk to pesticide exposure, especially via the dermal route.

3.6 Self-reported toxicity symptoms associated with pesticide exposure among farmers

Sample respondents were asked about the acute toxicity symptoms they experienced in 24 hours after mixing, loading or applying pesticides. Significant number 94.52% of the farmers reported at least one symptom of acute pesticide poisoning in the previous year immediately after applying or handling pesticides, while 5.48% of the respondents did not ascribe any health problems encountered to pesticide exposure. The most frequently symptoms reported by the participates were headaches 84.93%, skin rash 60.27%, slow heartbeats 72.60%, chest wheezing 67.12%, change in their mood 71.23%, dizziness 42.46%, burning in the skin or eyes 61.64%, lacrimation 17.81% and day/night coughing 23.29%. Other symptoms reported by respondents were pain in the hands or on the feet, excessive sweating and chest tightness (**Table 8**).

Participants were also asked about the actions they take following an incident of poisoning. The majority 53% reported that they resort to traditional methods like drinking milk, applying creams and washing the affected area, 32% reported that they do not take any action as long as the incident is minor or required only self-medication. Only 15% of respondents reported a serious poisoning incident that required medical attention in a clinic (**Figure 6**).

Items	Freq.	Percentage
Skin rash	44	60.27%
Headache	64	84.93%
Slow heartbeats	53	72.60%
Chest wheezing	49	67.12%
Burning in the skin or eyes	45	61.64%
Change in the mood	52	71.23%
Day/night cough	17	23.29%
Dizziness	31	42.46%
Excessive sweating	24	32.88%
Pain in the hands or in the feet	14	19.18%
Chest tightness	14	19.18%
Pain in the hands or in the feet	14	19.18%
Eye tears	13	17.81%
No health impairment	4	5.48%

Table 8.
Toxicity symptoms reported by the participants on the short-term of exposure.

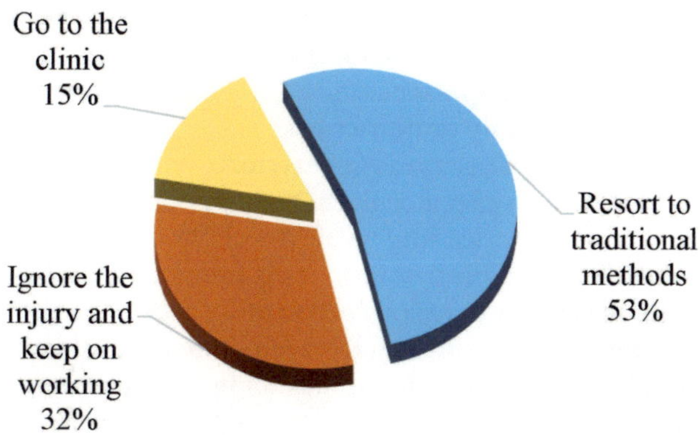

Figure 6.
Action farmers usually take following an incident of poisoning.

Self-medication methods that farmers resort to in the case of injury were manifested by the physician in the Referral Hospital during the interview.

> *Farmers usually drink milk after swallowing bleach (Barakina) to reduce the risk of injury.*

In general, actions taken by farmers following an incident of poisoning are unsatisfactory. Only a few of participants have visited a health institution after incidents of pesticide poisoning, and others resorted to traditional-based care methods or did not take any action. This finding is also in line with what the physician has indicated in the health institution. Similar findings were reported by the study conducted in Tanzania [47].

On the long-term of exposure, symptoms reported by farmers are illustrated in **Figure 7** below. The majority 46% of the sample respondents reported libido, whereas, the remaining reported poor memory 24%, diabetes 10% and others 20%.

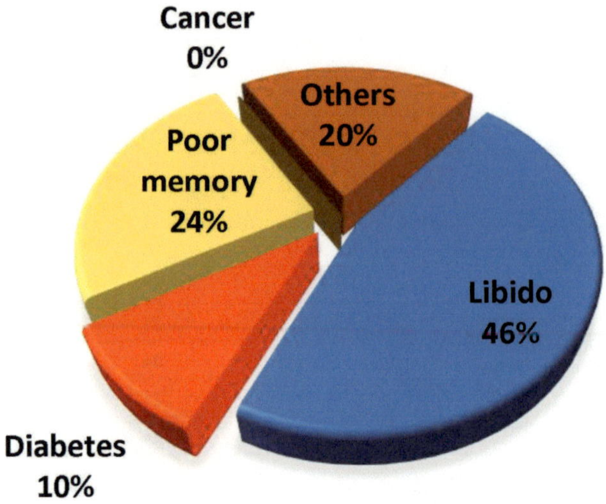

Figure 7.
Long-term health effects reported by farmers.

Through increasing the reactivity of toxic xenobiotics by converting them into electrophiles, free radicals or nucleophiles, chemicals cause damage to major biological systems leading to the development of various diseases, such as diabetes, neurodegeneration, schizophrenia, respiratory disorders, aging, cancer, immunodeficiency syndromes, and hypertension [48]. The ultimate toxicant may bind to the target molecules covalently or non-covalently, or may alter it by hydrogen abstraction, electron transfer or enzymatic reaction.

Endocrine disruptors, such as endosulfan, were proved to change the levels of insulin secretion in the body, leading to the development of diabetes [49]. Regarding men infertility, including libido, it was found that exposure to organochlorines is the main reason for hyperprolactinemia (a severe decrease in serum prolactin) which is the main reason for decreased libido and infertility [50]. Furthermore, a developmental exposure to pesticides may generate oxidative stressors that result in irreversible damage in the brain cells, followed by reducing the ability of the cells to communicate with each other. In time, chemical connections between brain cells are lost, and cells begin to die, resulting in poor memory [51].

The finding in the study revealed that the prevalence of toxicity symptoms among farmers is quite high.

3.7 Factors influence the prevalence of long-term toxicity symptoms among farmers

A multiple logistics regression model was employed to determine the effect of the independent variables (working hours per day, residential area, source of income, experience and training) on the prevalence of long-term toxicity symptoms among farmers (**Table 9**).

The result illustrated that the odds of the prevalence of long-term toxicity symptoms is negatively correlated with the training factor; therefore, farmers who have access to training are 9% less likely to develop long-term toxicity symptoms than farmers who did not attend training in pesticide handling and toxicity. Hence, the estimated odds ratio is statistically significant since ($p = 0.042$) which indicates that ($p < 0.05$) within 95% confidence interval (OR = .090, CI = 0.009–.0.920).

The result also revealed that the odds of the prevalence of long-term toxicity symptoms is positively influenced by working hours per day; therefore, farmers

Dependent variable: self-reported toxicity	B	S.E.	Sig.	Exp(B)	95% CI for EXP(B)	
					Lower	Upper
Training(1 = yes)	−2.406	1.185	.042	.090	.009	.920
Experience			.990			
Experience (1 = 5–10 years)	−.341	1.062	.748	.711	.089	5.697
Experience (2 = 10–20 years)	−.311	1.128	.783	.733	.080	6.682
Experience (3 = above 20 years)	−.362	1.257	.773	.696	.059	8.169
Working hours per day on the farm (1 = Full time)	2.681	.841	.001	14.599	2.809	75.881
Knowledge	−1.566	.828	.059	.209	.041	1.059
Constant	−.884	.838	.291	.413		

a. Variable(s) entered on step 1: training, Years of using pesticides (experience), working hours per day, and knowledge.

Table 9.
The factors that influenced the prevalence of long-term toxicity symptoms among farmers.

who work more than 8 hours per day in the farm (full-timers) are 14.599 times more likely to develop long-term toxicity symptoms than farmers who spend less than 8 hours in the farm (part-timers). Hence, the estimated odds ratio is statistically significant since ($p = .001$) which indicates that ($p < 0.05$) within 95% confidence interval (OR = 14.599, CI = 2.809–75.881), while knowledge and years of experience failed to be significant under the given conditions.

3.8 Data extracted from the interview with the physicians in Bushullo health center and referral hospital

The physician in the Bushullo health institution reported during the interview that:

> *"Only one acute pesticide intoxication case was reported for the last five years. The case was a female farmer and wasn't officially registered because she refused to pay the card fee which was 20 birr. The farmer patient was excessively salivating and dizzy when she arrived to the health center. Symptoms were quickly reversible and the patient returned back healthy in 15-30 minutes. Severe cases of acute pesticide intoxication are not treated in the health center and usually are transferred to the Referral Hospital because antidotes for poisons are not available in the health institution. The health institution treats patients with minor symptoms only by providing them with oxygen and fluids. Cases of pesticide intoxication were never under-estimated, and all the crew in the health institution resort to the Ethiopian hospital guidelines to diagnose all cases. In addition, doctors ask about the history of using pesticides as a part of the diagnosis."*

Both hard and soft documents where reviewed in the health center by the interviewed physician to obtain the accurate number of toxicity cases registered before.

The physician also indicated that under-reporting the cases by farmers is due to a lack of knowledge as most of the farmers are illiterates and the government's help would be appreciated by providing more educational programmes and training to farmers, and their families.

The interviewed physician in the emergency room at the Referral Hospital declared that:

"All pesticide intoxication incidents reported before were intentional suicide cases among teenagers and youths in their early twenties. Most of suicidal cases resort to the hospital when they reach the brink of death. Symptoms experienced by patients are vomiting, diarrhea, sweating, breathing difficulties, uncontrollable defecation and too much fluid around the lungs. The antidote usually given in these cases is atropine as it rapidly dries up the body and reduces secretions. In the case of total respiratory failure, patients are treated in the intensive care units with the help of a machine that helps them breathe properly. There is a limited number of these machines in the hospital and the patient might pass away in the case of all the machines being occupied. Besides, when the farmers experience slight symptoms of pesticide intoxication, they resort to nurses living in the same residential area, which results in misdiagnosing the cases properly due to nurses' lacking the adequate experience. In addition, there is a poor registration system in general, and registration only matters for patients. All highly toxic pesticides should be officially banned and the free availability is a serious issue."

Documents associated with pesticide intoxication were all reviewed in the Referral Hospital. The employee in the registration room reported that the ICD 10 system is the one that has been implemented for a long time now (more than 10 years) and to this present date. This system was implemented by the World Health Organization (WHO) in 1993 to replace ICD-9, which was developed by the WHO in the 1970s. ICD-10 is used in almost every country in the world, except the United States [52].

The employee also clarified that the registration system is not really efficient and the Ethiopian government will develop and start using its own system soon; however, only 56 cases of poisoning were officially registered in the past two years and labeled as poisoned due to unspecified drugs and biological substances. Therefore, data inserted in the system found to be not properly categorized, and the exact number of pesticide intoxication cases and intentional suicide trials among farmers and their families is unknown. This finding is in line with what the physician in the Referral Hospital declared about the poor registration system. However, the poor categorization of disease causals might be due to the registrars' lack of awareness about the importance of the accuracy of these numbers which are definitely a solid clue for the authorities to check the improvement of their performance.

The finding of this study regarding the registration of pesticide intoxication cases among farmers in Finchawa and Tullo rural kebeles found to be poor and in line with the study of [14] that was previously conducted.

4. Conclusion and recommendations

In conclusion, there is no gap of communication between farmers, and their authorities, as answers from both parties were perfectly matched. Statistically, the level of knowledge among the sample participants was found to be on average and was reflected in their field practices. However, while going into deeper details to address the presented and absent areas of knowledge among the participants, it was revealed that they were knowledgeable about the daily tasks that should be performed in the field, while the information about the effect of pesticides on the environment, on humans especially in the long-term, as well as the dermal route of exposure, were absent. Accordingly, the knowledge that farmers acquired from their experience, practices, field training and daily observations were insufficient to fill the hiatus of knowledge that is known to be obtained from the accumulation of information through education, and this was the gap that hindered farmers from mitigating their risk of exposure and had a joint significance on influencing the prevalence of pesticide intoxication among farmers and their families.

Therefore, knowledge-based training programs with practical classes and related courses are essential to improve farmers' level of knowledge about the adverse health effects of pesticides on human and environmental health, and help them address the simple protective methods to protect themselves and the environment around them. In addition, a specific budget should be dedicated by the government to provide farmers with adequate personal protective equipment to reduce their risk of exposure. Since the existence of highly toxic pesticides in farmers' residential area increased the risk of exposure among their family members, an official banning of highly toxic pesticides and replacing them with less toxic ones should be seriously considered. In the same respect, Pesticide application should be restricted to certified people who are trained, experienced and adequately equipped. Besides, improving the registration system in governmental hospitals is pivotal, and physicians should not prescribe any type of medication to their patients until they are registered and the disease is well categorized. Finally, construct hazardous waste collection units in Hawassa City for the proper disposal of empty chemical containers, rather than disposing of them in an inappropriate way.

Acknowledgements

Main advisor Dr. Daniel Fitamo. Co-advisor Prof. Solomon Sora. Statistical Analysis: Mr. Bereket Mekonenn.

Author details

Lana MHD Jamal Alshalati
Hawassa University, Hawassa, Southern Ethiopia

*Address all correspondence to: lshalati82@gmail.com

References

[1] FAO and IWMI. More people, more food, worse water? a global review of water pollution from agriculture [Internet]. 2018. 224 p. Available from: http://www.fao.org/3/CA0146EN/ca0146en.pdf

[2] Castañeda A, Doan D, Newhouse D, Cong M, Hiroki N, João U, et al. Who Are the Poor in the Developing World? Poverty and Shared Prosperity Report 2016: Taking on Inequality Background Paper. World Bank Policy Res Work Pap. 2016;7844(October).

[3] Aktar W, Sengupta D, Chowdhury A. Impact of pesticides use in agriculture: Their benefits and hazards. Interdiscip Toxicol. 2009;2(1):1-12.

[4] Damalas CA, Koutroubas SD. Farmers' exposure to pesticides: Toxicity types and ways of prevention. Toxics. 2016;4(1):1-10.

[5] Farmworker Justice. Exposed and Ignored: How pesticides are endagering our nation's farmworkers. 2013; Available from: http://www.farmworkerjustice.org/sites/default/files/aExposed and Ignored by Farmworker Justice singles compressed.pdf

[6] Hopper S, McSherry B. The insanity defence and international human rights obligations. Psychiatry, Psychol Law. 2001;8(2):161-73.

[7] EarthTalk. How Important Was Rachel Carson's Silent Spring in the Recovery of Bald Eagles and Other Bird Species? Sci Am [Internet]. 2012; Available from: https://www.scientificamerican.com/article/rachel-carson-silent-spring-1972-ddt-ban-birds-thrive/?print=true

[8] Saeedi Saravi SS, Shokrzadeh M. Role of Pesticides in Human Life in the Modern Age: A Review. Pestic Mod World - Risks Benefits. 2011;

[9] Ochilo KW, Musebe R, Bateman M. Study on crop protection where the "Green Innovation Centres for the Agriculture and Food Sector" (GIAE) initiative is being implemented. 2018;(April).

[10] Gesesew HA, Woldemichael K, Massa D, Mwanri L. Farmers knowledge, attitudes, practices and health problems associated with pesticide use in rural irrigation villages, Southwest Ethiopia. PLoS One. 2016;11(9):1-13.

[11] Negatu B, Kromhout H, Mekonnen Y, Vermeulen R. Use of chemical pesticides in Ethiopia: A cross-sectional comparative study Onknowledge, attitude and practice of farmers and farm workers in three farming systems. Ann Occup Hyg. 2016;60(5):551-66.

[12] Aliyi N, Sota SS, Deribe E. Pesticide Usage and Safety Measures Awareness of Small Scale Farmers in Gera District, Jimma Zone, Western Ethiopia. 2018;18-30.

[13] Mengistie BT, Mol APJ, Oosterveer P. Pesticide use practices among smallholder vegetable farmers in Ethiopian Central Rift Valley. Environ Dev Sustain. 2017;19(1):301-24.

[14] Williamson. Understanding the Full Costs of Pesticides: Experience from the Field, with a Focus on Africa. Pestic - Impacts Pestic Expo. 2011;(March):24-48.

[15] Kalayou Hiluf. and Amare Ayalew. ASSESSMENT OF PESTICIDE USE , PRACTICE AND ENVIRONMENTAL EFFECTS ON THE SMALL HOLDER FARMERS IN THE NORTH SHOA ZONE OF AMHARA NATIONAL REGIONAL STATE OF ETHIOPIA Kalayou Hiluf . G and Amare Ayalew . A , Sample Size and Sampling Procedures. 2015;2(2):16-24.

[16] Bekele F. The Status and Quality of Water Supply and Its Related Problems. 2017;7(9).

[17] Wikipedia. Awasa. 2020;55(March):2020.

[18] Ten C. Mixed Methods Procedures. 2003;

[19] Fang J. Why Logistic Regression Analyses Are More Reliable Than Multiple Regression Analyses. 2013;4(7):620-33.

[20] C.R.KOTHARI. Research Methodology. 1390.

[21] Ritter N. Understanding a Widely Misunderstood Statistic: Cronbach's. Online Submiss. 2010;

[22] Mrema EJ, Ngowi AV, Kishinhi SS, Mamuya SH. Pesticide Exposure and Health Problems Among Female Horticulture Workers in Tanzania. 2017;

[23] Zyoud SH, Sawalha AF, Sweileh WM, Awang R, Al-Khalil SI, Al-Jabi SW, et al. Knowledge and practices of pesticide use among farm workers in the West Bank, Palestine: Safety implications. Environ Health Prev Med. 2010;15(4):252-61.

[24] May D, Arancibia S, Behrendt K, Adams J. Preventing young farmers from leaving the farm: Investigating the effectiveness of the young farmer payment using a behavioural approach. Land use policy [Internet]. 2019;82(September 2018):317-27. Available from: https://doi.org/10.1016/j.landusepol.2018.12.019

[25] Mekonnen Y, Agonafir T. Pesticide sprayers ' knowledge , attitude and practice of pesticide use on agricultural farms of Ethiopia. 2002;52(6):311-5.

[26] ESC. Ethiopian Education Development Roadmap An integrated Executive Summary. 2018;(July):1-101.

[27] Jallow MFA, Awadh DG, Albaho MS, Devi VY, Thomas BM. Pesticide knowledge and safety practices among farm workers in Kuwait: Results of a survey. Int J Environ Res Public Health. 2017;14(4).

[28] Nigatu AW, Bråtveit M, Moen BE. Self-reported acute pesticide intoxications in Ethiopia. BMC Public Health [Internet]. 2016;16(1):1-8. Available from: http://dx.doi.org/10.1186/s12889-016-3196-5

[29] Ncube NM, Fogo C, Bessler P, Jolly CM, Jolly PE. Factors associated with self-reported symptoms of acute pesticide poisoning among farmers in Northwestern Jamaica. Arch Environ Occup Heal. 2011;66(2):65-74.

[30] Deribe E, Rosseland BO, Borgstrøm R, Salbu B, Gebremariam Z, Dadebo E, et al. Organochlorine pesticides and polychlorinated biphenyls in fish from Lake Awassa in the Ethiopian rift valley: Human health risks. Bull Environ Contam Toxicol. 2014;93(2):238-44.

[31] Commission E. Guidance Document on Dermal Absorption. 2004;1-15.

[32] Damalas CA, Eleftherohorinos IG. Pesticide exposure, safety issues, and risk assessment indicators. Int J Environ Res Public Health. 2011;8(5): 1402-19.

[33] Lekei EE, Ngowi A V, London L. Farmers ' knowledge , practices and injuries associated with pesticide exposure in rural farming villages in Tanzania. BMC Public Health [Internet]. 2014;14(1):1-13. Available from: BMC Public Health

[34] WHO. International Code of Conduct on the Distribution and Use of Pesticides Guidelines on Management Options for Empty Pesticide Containers. 2008.

[35] FAO. Guidelines for the management of small quantities of unwanted and obsolete pesticides. 1999;

[36] WHO. Guidelines for personal protection when handling and applying pesticides: International Code of Conduct on Pesticide Management [Internet]. 2020. 75 p. Available from: http://www.fao.org/3/ca7430en/CA7430EN.pdf

[37] Graham JP, Polizzotto ML. Review Pit Latrines and Their Impacts on Groundwater Quality : A Systematic Review. 2013;121(5):521-30.

[38] Souza RM, Beatrizquesada H. Occurrence , impacts and general aspects of pesticides in surface water : A review. 2020;

[39] Jacobson JL, Jacobson SW. Intellectual impairment in children exposed to polychlorinated biphenyls in utero. N Engl J Med. 1996;335(11):783-9.

[40] Colborn T. Neurodevelopment and endocrine disruption. Environ Health Perspect. 2004;112(9):944-9.

[41] Watts M. Pesticide Action Network Asia and the Pacific Endosulfan A PAN AP Factsheet Series. Down to Earth Mag [Internet]. 2001;9(19). Available from: http://www.panna.org/sites/default/files/PAN AP pesticides-factsheet-hhps-endosulfan.pdf

[42] Duke SO. Pesticide Dose – A Parameter with Many Implications Pesticide Dose – A Parameter with Many Implications. 2020;(September 2017).

[43] Kim K, Bae J, Nho MW, Lee CH. How do experts and novices differ? Relation versus attribute and thinking versus feeling in language use. Psychol Aesthetics, Creat Arts. 2011;5(4):379-88.

[44] Brhane G, Mammo Y, Negusse G. Sources of information and information seeking behavior of smallholder farmers of Tanqa Abergelle Wereda, central zone of Tigray , Ethiopia. 2017;9(April):47-52.

[45] Matthews GA. Attitudes and behaviours regarding use of crop protection products - A survey of more than 8500 smallholders in 26 countries Attitudes and behaviours regarding use of crop protection products — A survey of more than 8500 smallholders in 26 countries. 2008;(February).

[46] Sandt JJMVANDE, Dellarco M, Hemmen JJVAN. From dermal exposure to internal dose. 2007;(December 2006).

[47] Lekei EE, Ngowi A V., London L. Farmers' knowledge, practices and injuries associated with pesticide exposure in rural farming villages in Tanzania. BMC Public Health. 2014;14(1):1-13.

[48] Pandey AK, Nagpure NS, Trivedi SP, Kumar R, Kushwaha B. Profenofos induced DNA damage in freshwater fish, *Channa punctatus* (Bloch) using alkaline single cell gel electrophoresis. Mutat Res - Genet Toxicol Environ Mutagen [Internet]. 2011;726(2):209-14. Available from: http://dx.doi.org/10.1016/j.mrgentox.2011.09.011

[49] Howard SG. Developmental Exposure to Endocrine Disrupting Chemicals and Type 1 Diabetes Mellitus. 2018;9(September):1-6.

[50] Kaur RP, Gupta V, Christopher AF, Bansal P. Asian Paci fi c Journal of Reproduction. Asian Pacific J Reprod [Internet]. 2015;4(4):322-30. Available from: http://dx.doi.org/10.1016/j.apjr.2015.07.012

[51] Alzheimer's Society. What is dementia? Factsheet 400LP. 2017;(January). Available from: https://www.alzheimers.org.uk/sites/default/files/migrate/downloads/what_is_dementia.pdf

[52] WHO. ICD-10. 2010;2.

Chapter 7

Analysis of Exogenous Poisoning by Pesticide in the State of Bahia-Brazil during the Period from 2007 to 2017

Karolayne Silva Souza, Flávia Steffany Leite Miranda, Milena Roberta Freire da Silva, Caio Rodrigo Dias de Assis, Rafael Artur de Queiroz Cavalcanti de Sá, Marcilio de Oliveira Santos, Jordany Gomes da Silva, Kelma Sirleide de Souza, Bruno Oliveira de Veras and Maria Betânia Melo de Oliveira

Abstract

Brazil is one of the main exporters of agricultural products and is one of the largest consumers of pesticides in the world. Bahia stands out in irrigated agriculture, with growth at the national and international level exports. The aim of this study is to describe and analyze the spatial distribution, and the epidemiological profile of the population with confirmed cases of pesticides intoxication in the state of Bahia-Brazil, from 2007 to 2017. It is an ecological and retrospective study of a series, with secondary data collected at the Department of Informatics of the System Of Health (DATASUS), on notifications and confirmed cases of pesticide poisoning in the state in 2007 to 2017. In Bahia, about 1,632 notifications of exogenous pesticide poisoning in the investigated period, confirming about 1137 cases. The main age group affected by these intoxications were related to 20-39 years of age. The results of this study show the increase in the number of notified and confirmed cases in Bahia during years from 2007 to 2017, thus suggesting the carrying out of research on the subject in this and other Brazilian states, in order to monitor and evaluate the causes of these pesticide poisoning, and improve health promotion.

Keywords: agriculture, notifications, healthy

1. Introduction

Brazil is one of the main countries in the world in the production and export of products from the agricultural sector. It uses fertilizers and standing out as the main world consumer of pesticides [1]. This fact is due to its extensive planting area, which results in large areas dedicated to monocultures. In turn, it results in increasing pests in plantations, cultivation of transgenics, the imposition of the policy of the green revolution, and the absence of reduction policies and monitoring the use of pesticides and encouraging ecological agriculture [2].

The use of pesticides started to become popular during the Second World War when it emerged as a revolution in pest control. This product is classified as efficient and with an advantageous cost/benefit ratio, which caused new organosynthetic compounds to be produced, thus strengthening the agrochemical industry until today. The use of these chemical products added to the process of developing modern varieties with a high capacity to use these inputs became known as "the Green Revolution". Recently, the increase in the consumption of pesticides is closely related to the use of transgenic cultivars [1, 3, 4].

This process, linked to the modernization of agriculture, inserted pesticides and fertilizers into the country's technological package, where the expansion of capital in the countryside forced small producers and farmers to enter the new model of agricultural production. In this perspective, the use of chemical pesticides and fertilizers has become a constant, since the land was forced to produce more in less time, paying attention only to market needs, exponentially increasing productivity per hectare and the frequency of harvests [5, 6].

Pesticides interfere with the physiological mechanisms of human beings and, therefore, are harmful to health. According to the World Health Organization (WHO), there are three to five million acute poisonings in the world each year, with approximately 200 thousand deaths annually, mainly in developing countries. The high toxicity of some products, the ease of access, and the increase in the number of products containing these substances are factors related to pesticide poisoning, causing serious public health [7–11]. Exogenous autointoxications are directly involved (about 70%) with pesticides and medications, since hospital admissions, according to the Sistema Único de Saúde (SUS), is related to self-inflicted violence, by pesticides, chemicals, and medicines [12].

In agricultural terms, the state of Bahia stands out nationally and internationally, mainly in irrigated agriculture, showing an increase in exports and favoring population growth. However, with this highlight, it was also possible to observe an increase in the number of notifications for poisoning with pesticides [13]. Therefore, this study aimed to describe and analyze the spatial distribution and epidemiological profile of confirmed cases of pesticide poisoning in the state of Bahia - Brazil, from 2007 to 2017.

2. Methodology

This is an ecological and retrospective study of a time series, based on secondary data collected through SINAN (Sistema de Informação de Agravos de Notificação) linked to DATASUS. For this purpose, an analysis of confirmed cases of pesticide poisoning was carried out and data collection of estimates and projections of the population of the state of Bahia in the period from 2007 to 2017. To search for the data, we used the option of Disease and Notifiable grievances from SINAN: Exogenous pesticide infection, with the following variables: sex, age group, education, and area of residence. Data processing was performed using Excel software, in which the calculations of absolute and relative frequencies were performed, in addition to the arithmetic mean.

This research follows the rules of the National Ethics Council of Resolution no. 466/2012, using data available to the public through the DATASUS digital platform.

Therefore, authorization from the Research Ethics Committee was not required, as no identification data was used for individuals.

3. Results

In Bahia, there recorded about 1671 notifications of exogenous poisoning by agricultural pesticides between 2007 and 2017. Where, there confirmed about 1155, of which 96 died. Of the investigated periods, the year 2013 had the highest number of reported cases (240), unlike in the year 2008, where the lowest number was registered (59). The average number of cases during the analyzed period was also verified, which was 151.9 notifications. Regarding the number of confirmed cases (1155) of pesticide poisoning, 2013 had the highest number (195) and 2008 the lowest (42). Also, the investigated period showed an average of 105 confirmed cases (**Table 1**).

In **Table 2**, it is possible to observe the overview of cases of exogenous poisoning by pesticides in cities (health regions) in the state of Bahia, during the years 2007 to 2017. The highest percentage of notifications and confirmed cases were in the city of Salvador, in which there were 268 notifications and about 238 confirmed cases, followed by Juazeiro with 234 notifications and 165 confirmed cases, Feira de Santana with 155 notifications and 140 confirmed cases, and Paulo Afonso with 114 notifications and 93 confirmed cases. The other health regions can also be seen in **Table 2**.

The epidemiological profile of intoxication confirmed by pesticides in the state of Bahia (2007 to 2017) was divided according to the following criteria: sex (Male and female), age group (<1 to 9 years, 10 to 19 years, 20 to 39 years, 40 to 59 years, > 60 years and white/ignored), area of residence (Urban, Rural, peri-urban, and ignored), and education (Illiterate, complete/incomplete elementary school, complete/incomplete high school, complete/incomplete higher education,

Years	Population	Notified cases	Confirmed case
2007	14.080.670	103	71
2008	14.502.575	59	42
2009	14.637.364	89	66
2010	14.021.432	107	62
2011	14.097.534	149	106
2012	14.175.341	181	138
2013	15.044.137	240	195
2014	15.126.371	234	158
2015	15.203.934	203	136
2016	15.276.566	168	109
2017	15.344.447	138	72
TOTAL	—	**1.671**	**1.155**
MEAN[1]	—	**151,9**	**105**

[1]*Calculated arithmetic mean. Source: Authors based on data from DATASUS (2019).*

Table 1.
Distribution in the period from 2007 to 2017 of notified and confirmed cases in the state of Bahia.

Notification health regions	Total of notifications	Total of confirmed cases
Alagoinhas	18	7
Barreiras	101	53
Brumado	81	46
Camaçari	2	1
Cruz das Almas	11	7
Feira de Santana	155	140
Guanambi	92	68
Ibotirama	1	—
Ilhéus	13	9
Irecê	57	33
Itaberaba	27	17
Itabuna	29	20
Itapetinga	7	2
Jacobina	29	11
Jequié	80	37
Juazeiro	234	165
Paulo Afonso	114	93
Porto Seguro	51	29
Ribeira do Pombal	23	17
Salvador	268	238
Santa Maria da Vitória	78	55
Santo Antônio de Jesus	35	16
Seabra	8	6
Senhor do Bonfim	24	20
Serrinha	21	11
Teixeira de Freitas	56	21
Valença	38	26
Vitória da Conquista	18	7
TOTAL	**1.671**	**1.155**

Table 2.
Distribution of cases of exogenous poisoning by pesticides in the state of Bahia, during the period from 2007 to 2017.

white/ignored and does not applied). Of the confirmed cases, the male sex had the highest number of cases, about 756 (65%), when compared to the female sex who presented 399 (35%) (**Figure 1**).

The age range with the highest number of confirmed cases was between aged individuals 20 to 39 years old, which represented 588 cases, following by the age range 40 to 59 years old (279) and the age range 10 to 19 years (148). For those aged >60 years, there identified 78 cases, and the age range < 1 to 9 years (61), lastly, the ignored cases presented only one confirmed case. **Figure 2** shows the percentages by age range of cases confirmed by pesticide poisoning.

According to the area of residence of individuals with confirmed cases of pesticide poisoning in the state of Bahia from 2007 to 2017, the urban area had the highest number, approximately 647 confirmed cases, followed by the rural area (473) and ignored (27), lastly, the peri-urban area (8). **Figure 3** shows the percentages by area of residence.

The last epidemiological profile observed was education. For this parameter, the ignored criterion was the one that had the highest number of confirmed cases (622),

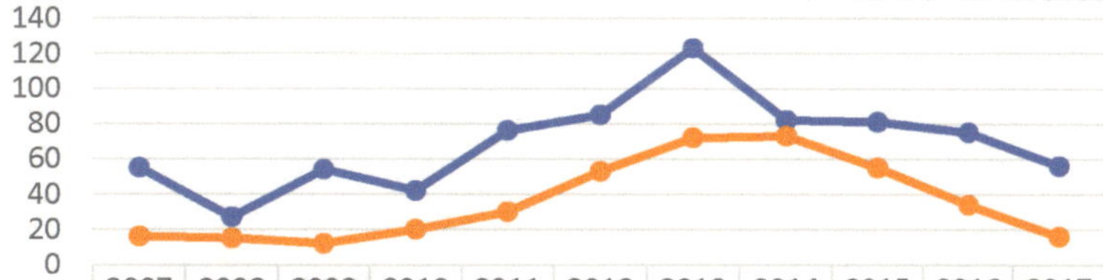

	2007	2008	2009	2010	2011	2012	2013	2014	2015	2016	2017
Male	55	27	54	42	76	85	123	82	81	75	56
Female	16	15	12	20	30	53	72	73	55	34	16

Figure 1.
Epidemiological profile of pesticide poisoning distributed by sex in the state of Bahia from 2007 to 2017.

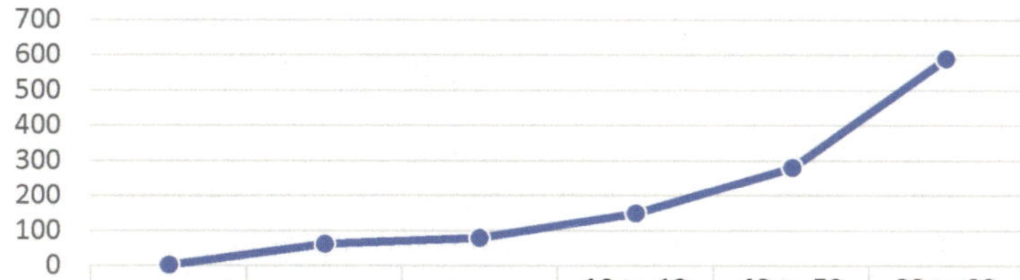

	White (0%)	< 1 to 9 years (5%)	> 60 years (7%)	10 to 19 years (13%)	40 to 59 years (24%)	20 to 39 years (51%)
Age range	1	61	78	148	279	588

Figure 2.
Epidemiological profile of pesticide poisoning distributed by age group in the state of Bahia from 2007 to 2017.

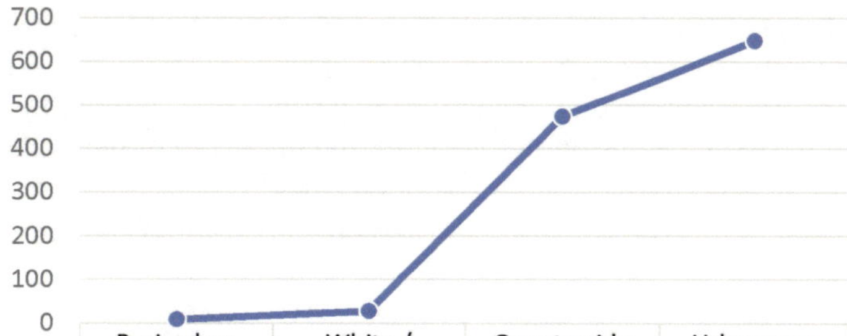

	Peri-urban area (1%)	White / ignored (2%)	Countryside (41%)	Urban area (51%)
residence area	8	27	473	647

Figure 3.
Epidemiological profile of pesticide poisoning distributed by area of residence in the state of Bahia from 2007 to 2017.

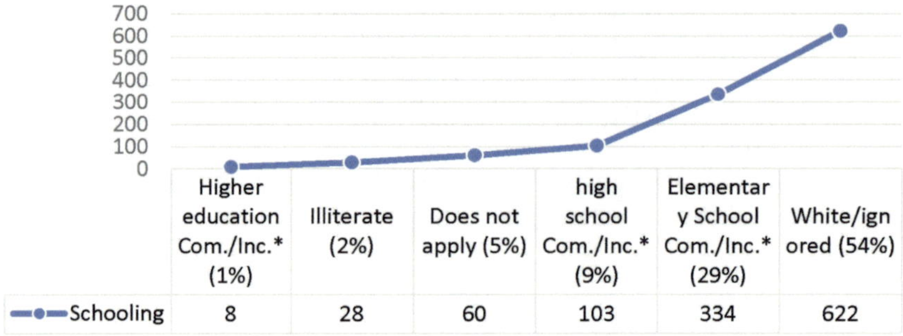

Figure 4.
Percentage by schooling of confirmed cases by pesticides poisoning in the state of Bahia from 2007 to 2017.
**Complete/incomplete.*

followed by complete/incomplete elementary school (334), complete/incomplete high school (103), illiteracy (28), and higher education (8). **Figure 4** addresses the percentage by the education of the aforementioned data.

4. Discussion

The use of pesticides is a serious problem for human health and the environment. Brazil, in turn, increased the consumption of pesticides in agricultural sectors and insect vector control programs. With these determining factors, the number of records of human poisoning by pesticides is increasing [3, 14]. The release of pesticides in 2019 by the government of Brazil totaled, until July, 290 active ingredients. Despite these alarming decisions, the present study refers to a retrospective study in the state of Bahia (2007 to 2017). At the time, there was still such a large number of licenses to use these compounds. However, this information does not make it any less relevant, as the number of pesticides released and cases of reported poisoning were already alarming.

According to data from the main cities in the state of Bahia, the largest number of cases of pesticide poisoning occurred in Salvador, Juazeiro, Feira de Santana, and Paulo Afonso. The findings of the present study suggest that males are the main victims of pesticide poisoning in the state of Bahia, with about 65% of cases confirmed in the period studied, which corroborates those of Rebelo et al., (2011) and Matos, (2013), who found that males represented a higher percentage of pesticide poisoning.

This data is directly linked to the figure of the man in the handling and use of pesticides in agricultural work, thus being more exposed to these chemicals [8, 10, 11] Regarding the age range, the most affected are concentrated mainly between 20 and 39 years (51%), followed by aged 40 to 59 years (24%). According to these data, the number of pesticide poisonings is increasing mainly in individuals of economically active age groups, as previously described [14, 15].

The results referring to the area of residence (urban, rural, peri-urban, and ignored) in the state of Bahia, demonstrate an important aspect mainly between the rural and urban areas, in which the urban obtained a percentage of 56% of confirmed cases and the rural (41%). It is worth mentioning that the number of confirmed cases in agricultural (rural) regions can be even higher. However, structural problems in health services hamper the efficiency of services, actions, and even the notifications of these individuals. According to LONDON, (2012), these problems have been previously reported, which reappear in the present work demonstrating that they persist, and that is necessary more effective administrative action by the health sectors to change and prevent this situation.

The data resulting from schooling address mainly that white/ignored had a higher percentage (54%) of confirmed cases, and complete/incomplete elementary education (29%). Complete/incomplete higher education a have a smaller number of individuals affected by pesticide poisoning, since this data is possibly related to the age of education, since, individuals with low education do not have the habit of using Personal Protective Equipments (PPEs) [16].

It is important to address the limitations of the present study of a large number of ignored, mainly in the epidemiological profile by education, which produces a possible bias regarding the notifications of confirmed cases of pesticide poisoning. Also, notifications must be made insightfully and correctly when diagnosing these individuals, since, the WHO makes the following estimate that for each case record of exogenous poisoning by pesticides, there are about 50 other unregistered cases [14].

5. Conclusion

The results of the present study of exogenous poisoning by pesticides show that there was an increase in reported and confirmed cases in the state of Bahia from 2007 to 2017, mainly from the year 2011. The health regions most affected were Salvador, Feira de Santana, Juazeiro, and Paulo Afonso that cover other respective municipalities in which the affected individuals reside. According to the epidemiological profile of the risk group with the highest percentages of confirmed cases, the male sex, the age groups of 20 to 39 and 40 to 59 years old, individuals with low schooling, and individuals living in the urban area are the most expressive. However, it is worth mentioning that these data are directly related to deficiencies in the structuring of the health system and the registration of notifications in the rural area when compared to the urban area.

Finally, it is suggested to expand research on the topic in the investigated state, as well as in other states in the region, in order to assess the causes of these pesticide poisonings, besides to encouraging the training of health professionals to carry out the diagnosis, clinical notification, and treatment of the individual affected by pesticide poisoning. Also, it is extremely important to develop and strengthen public policies with surveillance actions to reduce the indiscriminate use of pesticides and, thus, improve the health promotion of the population and the environment.

Author details

Karolayne Silva Souza[1], Flávia Steffany Leite Miranda[1], Milena Roberta Freire da Silva[1], Caio Rodrigo Dias de Assis[1,2], Rafael Artur de Queiroz Cavalcanti de Sá[1], Marcilio de Oliveira Santos[3], Jordany Gomes da Silva[1], Kelma Sirleide de Souza[4], Bruno Oliveira de Veras[1] and Maria Betânia Melo de Oliveira[1]*

1 Laboratory of Molecular Biology (BioMol), Biochemistry Department, Federal University of Pernambuco, Recife, Pernambuco, Brazil

2 Laboratory of Organic Compounds in Coastal and Marine Ecosystems, Department of Oceanography, Federal University of Pernambuco, Recife, PE, Brazil

3 Post-Graduation in Human Rights, Federal University of Pernambuco, Recife, Pernambuco, Brazil

4 Laboratory of Enzymology (LABENZ) Biochemistry Department, Federal University of Pernambuco - UFPE, Recife, PE, Brazil

*Address all correspondence to: maria.bmoliveira@ufpe.br

References

[1] Carneiro FF, Pignati W, Rigotto RM, Augusto LGS, Rizollo A, Muller NM, Alexandre VP, Friedrich K, Mello MSC Dossiê ABRASCO – Um alerta sobre os impactos dos agrotóxicos na saúde. ABRASCO, Rio de Janeiro, abril de 2012. 1ª Parte. 98p.

[2] Pignati WA et al. Distribuição espacial do uso de agrotóxicos no Brasil: uma ferramenta para a Vigilância em Saúde. Ciência & Saúde Coletiva, 2017; 22: 3281-3293.

[3] Dias CITR. Incorporação e implicações do uso de agrotóxico no ambiente e na saúde e a possibilidade de transição agroecológica. 2017.

[4] Soares WL. Uso dos agrotóxicos e seus impactos à saúde e ao ambiente: uma avaliação integrada entre a economia, a saúde pública, a ecologia e a agricultura. 2010. 163 f. 2010. Tese de Doutorado. Tese (Doutorado em Saúde Pública e Meio Ambiente)–Escola Nacional de Saúde Pública, Fundação Oswaldo Cruz, Rio de Janeiro.

[5] Silva RB, Garcia MF. Riscos à saúde dos trabalhadores rurais: o cultivo de maracujá em pequenas comunidades rurais na Paraíba. Procedings of XIII Jornada do Trabalho Procedings of XIII Jornada do Trabalho, 2012.

[6] Severo FE et al. Agrotóxicos: Riscos e Implicações na Saúde da População do Vale do Ribeira/SP. Anais do Encontro Nacional de Pós Graduação, 2019; 3[1]: 318-322.

[7] Eddleston M, Buckley NA, Eyer P, Dawson AH. Management of acute organophosphorus pesticide poisoning. Lancet, 2008; 371[9612]: 597-607.

[8] Costa MB. Análise de expressão dos genes de reparo da lesão de fita simples do DNA de trabalhadores rurais expostos à agrotóxicos. 2017.

[9] Xavier FG, Righi DA, Spinosa HS. Toxicologia do praguicida aldicarb (" chumbinho"): aspectos gerais, clínicos e terapêuticos em cães e gatos. Ciência Rural, 2007; 37[4]: 1206-1211.

[10] Pereira JN et al. Assistência de enfermagem realizada a agricultores em uso de agrotóxico. 2019.

[11] Oliveira LGB et al. Levantamento sobre uso de defensivos agrícolas na região de Andradina-SP. 2018.

[12] Veloso C et al. Violência autoinfligida por intoxicação exógena em um serviço de urgência e emergência. Revista Gaúcha de enfermagem, 2017; 38[2].

[13] Souza CDF, Costa KF, Ramos LS. Distribuição espacial das intoxicações exógenas por agrotóxicos em trabalhadores rurais no estado da Bahia-brasil, de 2007 A 2011. Hygeia, 2016; 12[23]: 133-141.

[14] Matos AS. Análise das intoxicações exógenas por agrotóxicos no Brasil, entre 2007 a 2012 [trabalho]. Brasília: Universidade de Brasília, 2013.

[15] Rebelo FM et al. Intoxicação por agrotóxicos no Distrito Federal, Brasil, de 2004 a 2007-análise da notificação ao Centro de Informação e Assistência Toxicológica. Ciência & Saúde Coletiva, 2011; 16: 3493-3502.

[16] Rangel CF, Rosa ACS, Sarcinelli PN. Uso de agrotóxicos e suas implicações na exposição ocupacional e contaminação ambiental. Cad. saúde colet.,(Rio J.), 2011.

Chapter 8

Potential Use of Agro/Food Wastes as Biosorbents in the Removal of Heavy Metals

Faizan Ahmad and Sadaf Zaidi

Abstract

The production of large quantities of agro/food wastes from food processing industries and the release of pollutants in the form of heavy metals from various metallurgical industries are the grave problems of the society as well as serious threats to the environment. It is estimated that approximately one–third of all food that is produced goes to waste, meaning thereby that nearly 1.3 billion tonnes of agro/food wastes are generated per year. This readily available and large amount waste can be utilized for the removal of toxic metals obtained from metallurgical industries by converting it into the adsorbents. For example, mango peel showed adsorption capacity of 68.92 mg/g in removing cadmium II ions. Similarly, coconut waste showed a higher adsorption capacity of 285 and 263 mg/g in removing cadmium and lead ion, respectively. Biosorption and bioaccumulation are recommended as novel, efficient, eco-friendly, and less costly alternative technologies over the conventional methods such as ion exchange, chemical precipitation, and membrane filtration, etc. for the removal of toxic metal ions. Because of the presence of metal-binding functional groups, the industrial by-products, agro-wastes and microbial biomass are considered as the potential biosorbents. Thus they can be used for the removal of toxic metal ions. This chapter highlights the available information and methods on utilizing the agro/food waste for the eradication of toxic and heavy metal ions. Furthermore, this chapter also focuses on the sorption mechanisms of different adsorbents as well as their adsorbing capacities.

Keywords: agro/food wastes, heavy metals, adsorption, biosorption, wastewater

1. Introduction

The generation of large quantities of agro/food wastes by various means of food processing industries and the release of high amount of heavy metals into the environment through various industrial activities such as refining ores, metal plating, fertilizer industries, batteries, mining, and tanneries, etc. are very serious increasing problems of the society as well as they are grave threats to the environment [1–3]. It is estimated that worldwide, around 39% of food is wasted in the food manufacturing industries and this is expected to rise to about 126 million tonnes by 2020 if proper prevention polices are not put in place [2, 4]. Similarly in India a large amount of agro/food wastes are generated every year. Nearby about 20% of the produced fruits and vegetables are wasted because, India is the second largest producer of fruits, vegetables, groundnut, sugarcane, rice, wheat and cotton.

As the production increased, it also increased the percentage of wastes generated from them [5, 6]. This readily available, cheap, and large amount of waste could be utilized for the removal of heavy metals from the effluents as bioadsorbents either in their natural form or in modified form. So the idea of utilization of agro/food wastes for the removal of heavy metals can suppress both the aforementioned problems to a great extent. Some of the industries such as paper, electroplating, metallurgy, textiles, batteries, metal plating, dyes, pesticides, and fertilizers, etc. are the major contributors of heavy metals into the wastewater [7]. These industries directly or indirectly discharge the heavy metals into the environment, especially in developing countries. The discharge of heavy metals like mercury (Hg), iron (Fe), chromium (Cr), lead (Pb), nickel (Ni), cadmium (Cd), cobalt (Co), zinc (Zn), copper (Cu), arsenic (As) etc. from industries is hazardous to humans [8]. These heavy and toxic metals are a serious concern and are threats to environment due to their toxicity, persistence in nature, and bio-accumulation tendency [9]. Several reputed and standard organizations have set the limits for the release of toxic metals into the water streams. But the addition of heavy metals into the aquatic stream at a higher concentration is increasing day by day by the various industrial activities, thus it increases the human health hazards and environmental pollution. Two tragedies namely, Minamata and Itai-Itai (Jintsu River) occurred in Japan because of the contamination of methyl-mercury and cadmium in aquatic streams [3, 10].

In the past several years various conventional techniques have been used for the eradication of toxic metal ions such as ultrafiltration, reverse osmosis, oxidation, ion exchange, chemical precipitation, reduction, and electrodialysis [11]. But these conventional methods are mainly associated with certain limitations, such as they produce large amount of sludge, are less efficient, are sensitive in operating conditions and are of costly disposal [12]. Thus, the use of agro/food wastes material and by-products of fruits and vegetables such peel, pomace and seeds as a bioadsorbents for the eradication of toxic metals is an emerging and potentially alternative method and in the recent years this method has gained much attention. This technique has several advantages over the conventional method such as low cost, high efficiency, produces less sludge, and regenerates biosorbents [13]. Various types of agro/food waste materials such as sugarcane bagasse, wheat bran, rice bran, rice husk, orange peels, coffee beans, hazelnut shells, groundnut shells, wheat husk, waste tea leaves, maize corn cob, apple peels, banana peels, coconut shells, sugar beet pulp, soybean hulls, cotton stalks etc. have been tried by several researchers for the eradication of heavy metals [14–16]. The agro/food waste materials, especially those containing cellulose, show a high biosorption capacity. Their components such as lignin, hemicellulose, starch, lipids, proteins, hydrocarbons and other functional groups expedite metal complexation that helps in the removal of heavy metals [17]. Because of the several advantages that are associated with the agricultural waste as mentioned above it seems to be a viable option for the removal of heavy metals. These readily available, low cost, promising agro/food waste materials can be effectively utilized for the eradication of various toxic metal ions either in their natural form or after some modifications [18].

2. Conventional methods for the eradication of toxic metal ions

Several conventional methods are readily available for the removal of heavy metal ions present in effluents that come from different sources. These conventional methods can be classified as physical, chemical, and biological [19]. In the past several years, various industries were extensively using physical and chemical methods, but nowadays, they do not prefer these traditional methods due to several reasons associated with them such as high cost and disposal problems. Membrane filtration, gravity concentration, adsorption, flotation, mechanical screening, and magnetic separation are some of the physical methods. Membrane filtration can be further classified as electrodialysis, reverse osmosis, ultrafiltration, and nanofiltration. In comparison to membrane filtration and adsorption, adsorption has been considered the appropriate and effective method for the removal of heavy metals because most of the times membrane separation is associated with several problems such as fouling. Chemical methods include electrochemical processes, coagulation, electro kinetic coagulation, irradiation, and electroflotation [20]. These chemical methods are efficient for removing the heavy metals from the effluent but they are very expensive and commercially unattractive. Like physical methods, these chemical methods are also associated with several limitations such as; they require high energy and high consumption of chemical reagents. Several studies have reported that both the physical and chemical methods are not much effective and economical towards the removal of heavy metals especially when the metals are present in low concentration [21, 22]. Some of the researchers have also used metal and metal oxide nanoparticles for water decontamination and purification [23]. Biological methods or bio removal is one of the alternative methods which can effectively reduce the concentration of heavy metals to environmentally tolerable levels at a reasonable cost [24]. In the past several years a large number of studies were conducted for the eradication of toxic metal ions from the waste water by utilizing the various products and by-products of plants [18, 25, 26]. They have also confirmed the efficacy of agricultural products or by-products as sorbents for the removal of metals. Further, they have recommended that this process is a good and efficient alternative for the removal of heavy metal but only if the adsorbents are inexpensive and do not require any pre-treatment before the application. Besides the several advantages of biological method, various other authors have also highlighted its drawbacks such as, it requires large area and has less design flexibility and lesser modes of operation. Further, it has also been recommended that this process is not favourable for the handling of large amount of effluent [27]. In order to overcome the drawbacks associated with the above mentioned techniques, several researchers have recommended the biosorption method as an economical and environmental friendly method. Biosorption is not only a method for the eradication of heavy metals but it is also an example of the potential use of bio waste (agro/food waste). Formation of adsorbent by using agro/food waste is a cheap, economical, environmental friendly, and simple to design method and moreover it does not produce any toxic and dangerous material [18].

3. Biosorption and its classification

It is a process that utilizes the biological material or components as adsorbents for the eradication of toxic metal and non-metal ions, and small particulates from the wastewater, which come from numerous industries such as ore refining, metallurgic, fertilizer, paper, batteries, etc. The biological materials that are used as adsorbents in the process of biosorption are called as biosorbents. They can be classified as natural, biological, and waste oriented as shown in **Figure 1**. Some of the natural adsorbents that are having the good properties of an adsorbent are clay, zeolite, and siliceous material and they are easily available in the ecosystem [28]. Most of the biological adsorbents emanate from microorganisms which include bacteria, fungi, algae, and yeast [20]. Other than these, some of the biological adsorbents obtained from biological sources such as chitin and chitosan, peat, and

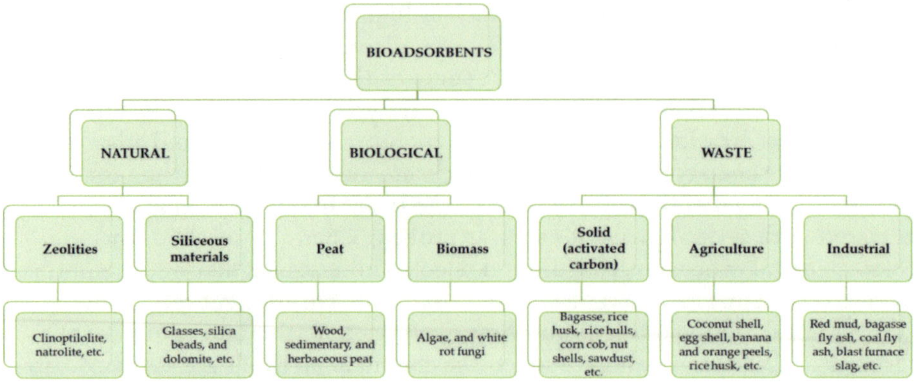

Figure 1.
Classification of bioadsorbents and their examples.

biomass. Waste oriented adsorbents include, bagasse, nutshells, sawdust, sugar beet pulp, bamboo and cassava peel, rice husk, orange and banana peels, egg shell, corn stack, etc. to name a few that come from solid waste, agricultural waste, waste from fruits and vegetables peels, and industrial waste [28]. Some of the examples of natural, biological and waste oriented bioadsorbents are indicated in **Figure 1**.

4. Biosorbents from agro and food waste

Agricultural waste, pulp, peels and seeds of fruits and vegetables are the discarded waste material and due to its several properties it may have a wide range of application in the removal of heavy metal ions. Generally, the agricultural waste having cellulose shows a high metal biosorption capacity. The ingredients of agricultural waste such as proteins, lipids, lignin, hemicelluloses, starch, hydrocarbons, and functional groups facilitate metal complexation which helps in the removal of heavy metals [29]. Several studies reveal [20, 26, 30] that this low cost, readily available, renewable, efficient, and eco-friendly waste material seems to be a feasible option for the removal of toxic metal ions. In the past several years, research has been carried out for the eradication of toxic metal ions using several bioadsorbents that come from agro and food waste such as wheat bran, rice bran, coconut shells, wheat husk, rice husk, saw dust of various plant, maize corn cob, arjun nuts, black gram husk, sugar cane bagasse, coffee beans, apple peels, banana peels, orange peels, sugar beet pulp, and grapes stalks etc. [12, 14, 18, 28, 31]. They have been used either in their natural form or after some modification (may be physical or chemical).

5. Biosorption mechanisms

The mechanism of biosorption is a complex process which involves the binding of sorbate onto the biosorbent [17]. Various materials that are found natural including agricultural wastes can be used as biosorbents which involve the binding of metal ions by several mechanisms including chelation, complexation, ion exchange, chemisorption, reduction, precipitation, and adsorption on the surface and pores [32, 33]. **Figure 2** shows the different mechanisms involved in the biosorption phenomenon. Bisorbents especially made from agro and food waste contain several compounds

Figure 2.
Various mechanisms involved in bioadsorption process.

such as starches, cellulose, simple sugar, hemicellulose, lignin, proteins, hydrocarbons and various functional groups such as carbonyl, amine, amide, sulfonate, carbonyl, phenolic, carboxyl groups, alcohols, and esters that can attract and isolate the metal ions. In the past several years many researchers have confirmed the presence of the aforementioned functional groups in the biosorbents and furthermore, they have also reported their complexation with heavy metals during the biosorption process [34, 35]. Some of the factors that help in controlling and characterizing the mechanism of biosorption are given in **Figure 3**.

5.1 Chelation

It is a mechanism in which an organic complex agent (chelate) binds the metal ions at more than one place at a time in order to form a ring structure. The molecules on an organic compound that form these types of coordination are called as ligands and the ligands—metal association is referred as coordination complex [36]. An increase in the coordination complexes on mineral surfaces weakens the bond of the metal or cation to the crystal lattice resulting in the dissolution. Chelates form several binding with the metal ion at more than one place as compared to the complexes, therefore, chelates are more stable. In the past several years various studies have reported the application of this mechanism for the eradication of toxic metal ions from the waste water that was obtained from different sources [17]. For example, in one study, for the eradication of Cd (II) from the wastewater rice straw has been used successfully as a potential biosorbent [37].

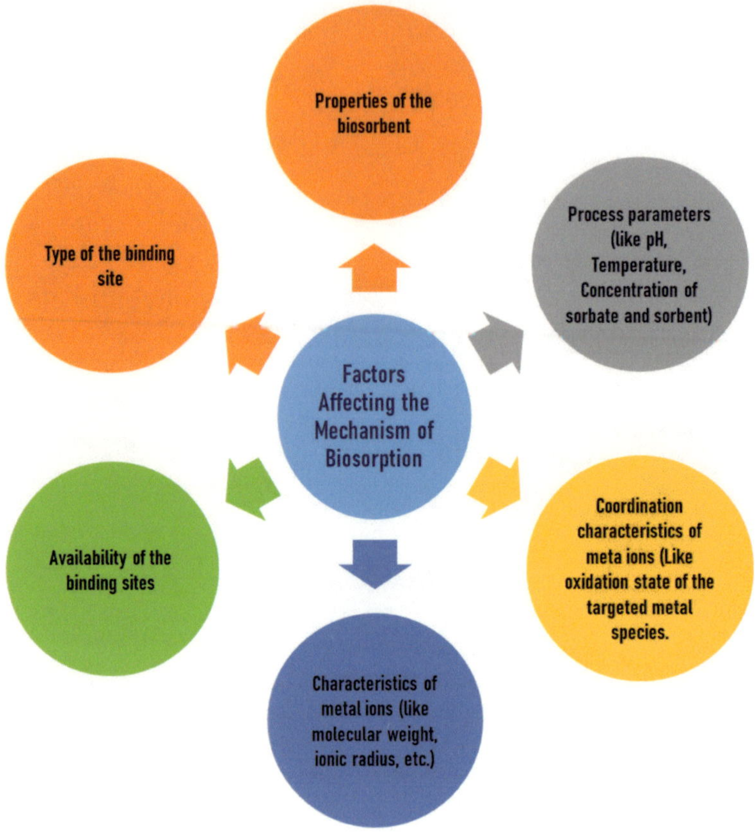

Figure 3.
Factor affecting the mechanism of bioadsorption.

5.2 Complexation

It is a process in which two or more species are associated and they form a complex. When the complexes of metal ions and the ligands are formed in such a way that the single metal atom enjoys the central position then it is called a mononuclear complexes. When more than one metal ion is present in the centre they form a polynuclear complex [38]. During the formation of polynuclear complexes, on the basis of number of binding ligands involved the metal atom may hold a positive, negative, or neutral charge. Several studies have confirmed that the formation of complexes by the mononuclear ligands is more desirable than polynuclear ligands because the latter contains multiple ligands results multiple species binding.

5.3 Chemisorption

Chemical adsorption is also known as chemisorption. It is a process of adsorption which involves a chemical reaction between the surface and the adsorbate. Mainly, it occurs when the adsorbate and adsorbent are attracted via a chemical bond or due to the chemical forces of attraction [39]. During the process of chemisorption only a single layer of adsorbate on adsorbent is formed and this process has high enthalpy of adsorption. With a rise in temperature, the rate of the process of chemisorption first increases and then decreases.

5.4 Ion exchange

This process also plays a vital role in adsorption. During the biosorption process it exchanges the binary metal ions with the counter ions present on the surface of the biosorbent. Various systems that are readily available for the purification of water, work on the ion exchange mechanism. The process of ion exchange generally takes place either by cation or anion exchange. Amino groups are one of the good examples of anion exchangers while carboxyl groups represent the cation exchangers [28]. Several studies have reported the ion exchange mechanism of biosorption using various agro/food waste such as watermelon rind, rice straw etc. for the eradication of toxic metal ions such as chromium, Cu (II), Zn (II), Pb (II), and cadmium [17, 40].

5.5 Surface precipitation

It is a process in which the metal ions present in the aqueous solution form precipitates with the functional groups that are present on the surface of the microbial cells due to which the metal ions remain intact with the microbial cell. Organic and inorganic metal precipitates are generally formed during the process of adsorption. Use of microbial cells forms the organic metal precipitates, that occur due to the excretion of extracellular polymeric substances. In several other cases, insoluble inorganic metal precipitates are also formed. Several studies have reported the involvement of surface precipitation mechanism using the husk of green tomato, soybean meal, and watermelon rind for the eradication of Cu (II), Cu (III), Pb (II), Zn (II), Fe and Mn [41, 42].

5.6 Reduction

Reduction is also an important mechanism of adsorption that plays a vital role in the biosorption of various heavy metals such as gold and palladium. During the process of reduction the metal connects with the functional group, gets reduced, and undergoes the growth of crystals. The metal that is reduced binds the biosorbent at various places. The eradication of numerous heavy metals such as chromium, gold, palladium etc. can be done easily by the process of reduction. For example, by using the process of biosorption, removal of Cr (VI) can be done easily by reducing it into Cr (III) from the aqueous solution [17].

6. Involvement of functional groups in adsorption process

Several types of functional groups are generally involved in the process of adsorption namely, hydroxyl, amino and carboxyl groups [43]. These groups play an influential role in the metal adsorption process. The heavy metals are more efficiently absorbed by the phenolic, lactonic, and oxygen functional groups as compared to the other groups. During the formation of adsorbent the temperature and high degree carbonization majorly affects the mechanism of these functional groups but the other factors such as porosity, surface area, increase in pH, etc. do not alter the mechanism of these groups [28]. Several authors have confirmed the potential role of functional groups in the adsorption process by using the FTIR (Fourier Transform Infrared Spectroscopy) [44]. Low temperature pyrolysis retains the functional groups inside the sample whereas, an increase in temperature during the pyrolysis may lead to the loss of the functional group. For example, when the wood and grass feedstock biomass is heated below 100–200°C then its FTIR spectra shows no prominent change in their functional groups [45]. Some of the external parameters such as pH change the complexity of functional groups when they undergo certain mechanisms. The carboxyl group works efficiently in the adsorption process at the pH range 3–4. Beyond this pH the carboxyl group of biosorbent forms a complex with the positively charged metals after deprotonation. Several authors [17, 33, 35] have observed the changes in the functional groups of adsorbent before and after the adsorption of metals. Various interactions such as complexation, precipitation, cation exchange, electrostatic interaction and chemical reduction have an intrinsic effect to make the process of eradication of heavy metal successful. The functional groups like hydroxyl (-OH) and carboxylate (-COOH) available on the surface of bio char have strong interaction with heavy metals [28].

7. Role of agro/food waste as biosorbents for the eradication of toxic metal ions

Utilization of agro and food waste as biosorbents for the eradication of toxic metal ions from the wastewater and aqueous streams is a promising and innovative technology [35]. In the past several years, this promising technology has gained more attention because these waste materials have shown a higher efficacy towards the removal of heavy metals and the other reasons include low cost, and easy availability. The efficacy of agro/food waste depends upon its phsico-chemical nature, capacity of adsorption and affinity. A lot of research has already been carried out for the removal of heavy metals such as arsenic, lead, mercury, cadmium, chromium, nickel, cobalt, etc. by using various types of biosorbents namely rice husk, wheat bran, peels of apple and banana, etc. [14, 46–48]. They have used biosorbents either in the natural form or in modified form by thermal and chemical treatment to increase their sorption capacities.

7.1 Removal of cadmium

Cadmium metal and its ions are the most severe pollutants because they are highly soluble in water as compared to the other toxic metal ions. Therefore, they are easily mobile in soil and have a higher tendency to bioaccumulate. Some of the basic sources that produce cadmium into the environment include solid waste, sewage irrigation, plastics, application of fertilizer, mining, plating on steel, etc. [49]. If the human body is exposed to cadmium for a long time it may result in severe diseases such as bone damage, kidney and lung cancer [50]. Rice husk, rice bran, wheat bran, black gram husk, rice polish, fig leaves, jack fruit, and orange peels, are some of the waste materials that have shown the excellent efficiency towards the removal of Cd. Several studies have been tried using these materials for the eradication of Cd either in their natural form or in the modified form. In addition, potato peels, olive branches, Musa paradisiacal peels, and coconut waste have also been extensively used for the removal of Cd. The adsorption capacity of coconut waste for the removal of Cd (II) was found to be 285.70 mg/g while, the adsorption capacities of other material such as potato peels, olive branches, and Musa paradisiacal peels, were found to be 125, 38.17, 10.0 mg/g, respectively [15]. Therefore, coconut waste has shown the highest adsorption capacity as compared to the potato peels, olive branches, and Musa paradisiacal peels. Some of the waste materials have shown higher adsorption capacity when they were used at acidic pH namely, fig leaves, medlar peels, beans, and jack fruits [51, 52]. Some of the research work that has already been done for the removal of Cd using various agricultural wastes is presented in **Table 1**.

7.2 Removal of Chromium

Chromium is a naturally occurring heavy metal found in the earth's crust. It can be released into the environment either through the natural process or by manmade industrial activities. Industrial activities include wood preservation, tanning, textile manufacturing, pigments, paints, and dyes manufacturing. Chromium is generally found in a number of oxidation states but Cr (III), and Cr (VI) are the largest threat to the environment [42, 84]. In the last two decades, a lot of researches have been done by using various agricultural wastes such as orange, lemon and banana peels, soybean and rice hulls, hazelnut and peanut shells for the removal of chromium metals and ions from different types of wastewater [26, 85]. These waste materials have shown significant chromium removal efficiency as shown in **Table 1**. Other than these waste materials, neem leaves powder, cactus leaves, coconut shell fiber, and pine needles have also shown a promising efficiency in the range 90–100% for the removal of chromium at optimum pH [71]. Wheat brans have showed the highest adsorption capacity for Cr (VI) at 310.58 mg/g whereas, rice bran is a lesser effective adsorbent for the removal of Cr with only up to 50% efficiency [86]. Saw dust, bagasse, rice husk, and mustard oil cake have also been tried by numerous researchers for the removal of chromium and they have reported significant efficiencies in literature [85].

7.3 Removal of nickel

The removal of nickel is also of great interest due to its high toxicity and its presence in various types of industrial wastewater and effluents. Nickel may enter into the environment by natural and industrial sources. Some of the industrial sources include batteries, coloured ceramics, nickel plating, power plants and trash incinerators [87]. Nickel and its compounds have no taste and odour. Various types of agricultural waste materials such as saw dust of maple, hazelnut and groundnut shell, and waste tea leaves, etc. have been used for the removal of nickel from the effluents either in natural or modified form. These waste materials have shown a very promising and significant efficacy for the removal of nickel. For example, the bagasse of sugarcane showed more than 80% removal efficiency in its natural form for nickel [88]. Similarly, other agro and food wastes such as corncobs, soybeans, cotton seeds, and coir fibers have also been tried by various researchers for the eradication of nickel in their modified form [78, 83]. *Acacia leucocephala*bark has shown the highest adsorption capacity for nickel at 294.10 mg/g and *Cassia fistula* biomass in its natural form has shown the removal efficiency of 99–100% for nickel [15, 73].

7.4 Removal of lead

Various activities such as finishing of tools, steel and cable reclamation, manufacturing of plastics, formation of cathode ray tubes, ceramics and soldering are the major sources of lead discharge into the environment [89]. Large exposure of lead results in various harmful biological effects and it strongly binds itself with the particles of oil, sediments, and sewage sludge due to which its removal has gained much attention. Various types of agro/food waste materials have been used for the removal of lead such as orange peel, chitosan, rice husk, walnut shells, peanut, lemon grass *(Cymbopogoncitratus)*, groundnut (*Arachishypogaea)*,

Heavy Metal Ion	Agro/Food Waste	Efficiency	References
Cd (II)	Wheat bran	>82%	[16]
	Bagasse	90–95%	[53]
	Green coconut shell powder	98%	[54]
	Rice bran	>80%	[55]
	Black gram husk	99%	[56]
	Rice husk	80–97%	[57]
	Papaya wood	98%	[58]
	Wheat bran	87.15%	[59]
	Rice polish	>90%	[59]
	Bagasse fly ash	90%	[60]
Pb (II)	Rice bran	>80%	[55]
	Waste tea leaves	92%	[12]
	Apple residue waste	Nearly 80%	[61]
	Orizasativa husk	98%	[62]
	Black gram husk	Nearly 93%	[63]
	Maple saw dust	80–90%	[64]
	Saw dust of rubber wood	85%	[35]
	Activated carbon of peanut shells	Nearly 75%	[65]
Cr (III)	Oat biomass	>80%	[66]
	Jatropha oil cake, maize corn cob and sugarcane bagasse,	Up to 97%	[9]
Cr (VI)	Wheat bran	>82%	[16]
	Raw rice bran	40–50%	[67]
	Rubber wood saw dust	60–70%	[68]
	Beech saw dust	100%	[69]
	Coconut shell fibers	>80%	[70]
	Neem leaf powder	>96%	[71]
	Bagasse fly ash	96–98%	[72]
Ni (II)	Sugarcane bagasse	>80%	[9]
	Casia fistula biomass	100%	[73]
	Saw dust of oak	70–90%	[74]
	black gram husk	Nearly 93%	[63]
	Tea waste	86%	[75]
	Maple saw dust	75%	[76]
	Waste tea leaves	92%	[12]
	Mustard oil cake	Nearly 94%	[77]
	Defatted rice bran	57%	[78]
Cu (II)	Mango saw dust	60%	[79]
	Wheat shell	99%	[80]
As (III)	Charred saw dust	80%	[81]
	Rice hush	Nearly 71%	[82]

Heavy Metal Ion	Agro/Food Waste	Efficiency	References
Zn (II)	Coir fiber	>70%	[83]
	Black gram husk	Near about 93%	[63]
	Defatted rice bran	87%	[78]
	Rice bran	>80%	[55]
	Bagasse	Near about 95%	[53]
	Papaya wood	67%	[58]

Table 1.
Performance of several agricultural wastes for the removal of toxic metal ions.

Toxic Metal Ions	Agro/Food Waste	Adsorption Capacity (mg/g)	References
Cr (VI)	Almond shell	3.40	[85]
	Banana peel	131.56	[14]
	Walnut shell	1.33, 8.01	[85]
	Hazelnut shell	8.28	[85]
	Coconut waste (puresorbe)	285.70	[30]
	Nut shell	1.47	[92]
	Pea waste	21.2	[93]
	Saw dust	10.01, 16.05	[20]
	Pinusroxburghii bark	4.15	[94]
	Groundnut husk	7.0	[95]
	Bael fruit	17.27	[96]
	Wheat bran (Chemically modified)	93.00	[16]
	Wheat bran	310.58	[28]
	Groundnut husk (Ag coated)	11.40	[95]
	Wheat straw (Chemically modified)	322.58	[97]
	Wheat straw	21.34	[97]
Cd (II)	Rice husk	73.96	[98]
	Banana peel	35.52, 5.71	[99, 100]
	Rice husk (H_3PO_4 treated)	102.00	[47]
	Wheat bran	22.78, 21.00, 15.82	[101, 102]
	Wheat bran (Ultrasonic treated)	51.58	[102]
	Wheat straw	39.22, 21.00,14.56, 11.60	[103, 104]
	Wheat straw (Urea treated)	4.25	[103]
	Orange peel	47.60	[105]
	Mango peel	68.92	[106]
	Tea waste	11.29	[35]
	Rice husk (Alkali treated)	125.94	[98]
	Orange peel (chemically modified)	136.05	[105]
	Raw coffee powder	15.65	[35]

Toxic Metal Ions	Agro/Food Waste	Adsorption Capacity (mg/g)	References
Cu (II)	Wheat bran	8.34	[80]
	Rice husk (acid treated)	29.00	[89]
	Orange peel	50.94	[105]
	Peanut hull pellet	12.00	[107]
	Sago husk ash	12.40	[108]
	Orange peel (chemically modified)	70.67	[105]
	Potato peel ($ZnCl_2$ treated)	74.00	[109]
	Peanut hull	21.25	[110]
	Mango peel	46.09	[106]
	Chestnut shell	12.56	[111]
Ni (II)	Mango peel	39.75	[106]
	Guava seed	18.05	[112]
	Pomegranate peel	52.00	[113]
	Orange peel	158.00	[114]
	Tea waste	73.00	[12]
Pb (II)	Wheat bran (chemically modified)	62.00	[16]
	Hazelnut shell	28.18	[91]
	Mango peel	99.05	[105]
	Coconut (coir pith waste)	263.00	[90]
	Rice husk (acid treated)	108.00	[89]
	Wheat bran	87.00	[115]
	Chitosan	8.30	[91]
	Banana peel	2.18	[100]
Zn (II)	*Cicerarientinum*seed	20.00	[116]
	Wheat bran	16.40	[48]
	Mango peel	28.21	[106]
	Tea waste	8.90	[117]

Table 2.
Biosorption capacities of different agricultural waste and plant products for the eradication of toxic metal ions.

and coir pith waste of coconut [15, 89]. The absorption capacity of coir pith waste of coconut and chitosan is found to be 263.0 mg/g and 8.3 mg/g respectively as shown in **Table 2** [90, 91]. Biosorption capacities of various other agro/food wastes against the removal of toxic heavy metal ions are presented in **Table 2**.

8. Conclusions

The existing physical and chemical methods for the removal of heavy metals have severe limitations such as they require various chemicals, are energy intensive, costly to operate, produce large quantities of sludge whose disposal is a big problem etc. Therefore, there is a need of some alternative method which can attractively remove the toxic metal ions in the best possible way. In this regard biosorption is an attractive alternative method to remove the toxic metal ions effectively from the effluents that are generated from various industrial activities. This method has proven itself to be efficient and environmental friendly. It solves the problems of the health and environmental hazards of heavy metals in our ecosystem and the utilization of huge quantities of agro/food waste. The use of agricultural wastes as bioadsorbent is recommended because of their sustainability, easy availability, low cost, and high affinity towards the removal of heavy metals. Numerous studies have recommended the use of several agro and food wastes such as wheat bran, coconut shell, rice bran, wheat husk, and rice husk, etc. either in their natural form or after some modification for the removal of chromium, cadmium, cobalt, nickel, lead, etc. In literature very few details are available regarding the modification and regeneration of biosorbents, about the pore size distribution of adsorbent, recovery of metal ions, and molecular size of metal ions. Thus, a lot of work is still required to be done in this direction in the coming days for the effective transformation of agro/food waste into the biosorbents.

Author details

Faizan Ahmad and Sadaf Zaidi*
Department of Post Harvest Engineering and Technology, Faculty of Agricultural Sciences, Aligarh Muslim University, Aligarh, India

*Address all correspondence to: sadaf63in@yahoo.com

References

[1] Celik A, Demirbaş A. *Removal of Heavy Metal Ions from Aqueous Solutions via Adsorption onto Modified Lignin from Pulping Wastes*. Energy Sources. 2005;**27**(12):1167-1177

[2] Ahmad, F. and S.T. Khan, *Potential Industrial Use of Compounds from By-Products of Fruits and Vegetables, in Health and Safety Aspects of Food Processing Technologies*, A. Malik, Z. Erginkaya, and H. Erten, Editors. 2019, Springer International Publishing: Cham. p. 273-307.

[3] Stewart M, Hughes EG. *Urinary beta 2 microglobulin in the biological monitoring of cadmium workers*. British Journal of Industrial Medicine. 1981;**38**(2):170-174

[4] Ahmad, F., S. Zaidi, and S. Ahmad, *Role of By-products of Fruits and Vegetables in Functional Foods*, in *Functional Food Products and Sustainable Health*, S. Ahmad and N.A. Al-Shabib, Editors. 2020, Springer Singapore: Singapore. p. 199-218.

[5] Rudra S. et al. *MINE OF NUTRACEUTICALS: FOOD INDUSTRY WASTE*; 2015

[6] Sadh PK, Duhan S, Duhan JS. *Agro-industrial wastes and their utilization using solid state fermentation: a review*. Bioresources and Bioprocessing. 2018;**5**(1):1

[7] Mahmood Q et al. Current Status of Toxic Metals Addition to Environment and Its Consequences. 2012:35-69

[8] Tchounwou PB et al. *Heavy metal toxicity and the environment*. Exp Suppl. 2012;**101**:133-164

[9] Garg UK et al. *Removal of hexavalent chromium from aqueous solution by agricultural waste biomass*. Journal of Hazardous Materials. 2007;**140**(1-2):60-68

[10] Harada M. *Minamata disease: methylmercury poisoning in Japan caused by environmental pollution*. Critical Reviews in Toxicology. 1995;**25**(1):1-24

[11] Abbas M et al. *Vibrio fischeri bioluminescence inhibition assay for ecotoxicity assessment: A review*. Science of the Total Environment. 2018;**626**:1295-1309

[12] Ahluwalia SS, Goyal D. *Removal of Heavy Metals by Waste Tea Leaves from Aqueous Solution*. Engineering in Life Sciences. 2005;**5**(2):158-162

[13] Burakov AE et al. *Adsorption of heavy metals on conventional and nanostructured materials for wastewater treatment purposes: A review*. Ecotoxicology and Environmental Safety. 2018;**148**:702-712

[14] Annadural G, Juang RS, Lee DJ. *Adsorption of heavy metals from water using banana and orange peels*. Water Science and Technology. 2003;**47**(1):185-190

[15] Alalwan HA, Kadhom MA, Alminshid AH. *Removal of heavy metals from wastewater using agricultural byproducts*. Journal of Water Supply: Research and Technology-AQUA. 2020;**69**(2):99-112

[16] Farajzadeh M, Boveiri Monji A. *Adsorption characteristics of wheat bran towards heavy metal cations*. Separation and Purification Technology. 2004;**38**:197-207

[17] Krishna, R., V. Chintalpudi, and S. Muddada, *Application of Biosorption for Removal of Heavy Metals from Wastewater*. 2018.

[18] De Gisi S et al. *Characteristics and adsorption capacities of low-cost sorbents for wastewater treatment: A review*.

Sustainable Materials and Technologies. 2016;**9**:10-40

[19] Luptakova A et al. *Application of physical–chemical and biological–chemical methods for heavy metals removal from acid mine drainage*. Process Biochemistry. 2012;**47**(11):1633-1639

[20] Mathew BB et al. *Role of Bioadsorbents in Reducing Toxic Metals*. Journal of Toxicology. 2016;**2016**:4369604

[21] Azimi, A., et al., *Removal of Heavy Metals from Industrial Wastewaters: A Review.* 2017. **4**: p. 37-59.

[22] Bazrafshan E et al. *Heavy metals removal from aqueous environments by electrocoagulation process– a systematic review*. Journal of Environmental Health Science and Engineering. 2015;**13**(1):74

[23] Khan ST et al. *Metal and Metal Oxide Nanoparticles for Water Decontamination and Purification*. Environmental Nanotechnology for Water Purification. 2020:151-186

[24] Wuana RA, Okieimen FE. *Heavy Metals in Contaminated Soils: A Review of Sources, Chemistry*. *Risks and Best Available Strategies for Remediation.* ISRN Ecology. 2011;**2011**:402647

[25] Mathew BB et al. *Role of Bioadsorbents in Reducing Toxic Metals*. J Toxicol. 2016;**2016**:4369604

[26] Unctad, et al., *Banana Peel: A Green Solution for Metal Removal from Contaminated Waters.* Korean Journal of Environmental Agriculture, 2013. **32**(2): p. 108-116.

[27] McKay G, Otterburn MS, Sweeney AG. *Kinetics of Colour Removal from Effluent Using Activated Carbon*. Journal of the Society of Dyers and Colourists. 1980;**96**(11):576-579

[28] Singh S et al. *Current advancement and future prospect of biosorbents for bioremediation*. Science of the Total Environment. 2020;**709**:135895

[29] Demirbaş A. *Mechanisms of liquefaction and pyrolysis reactions of biomass*. Energy Conversion and Management. 2000;**41**(6):633-646

[30] Pino GH et al. *Biosorption of Heavy Metals by Powder of Green Coconut Shell*. Separation Science and Technology. 2006;**41**(14):3141-3153

[31] Yao Y et al. *Removal of phosphate from aqueous solution by biochar derived from anaerobically digested sugar beet tailings*. Journal of Hazardous Materials. 2011;**190**(1-3):501-507

[32] Fosso-Kankeu E, Mulaba-Bafubiandi A. *Review of challenges in the escalation of metal-biosorbing processes for wastewater treatment: Applied and commercialized technologies*. African Journal of Biotechnology. 2014;**13**:1756-1771

[33] Yang X et al. *Surface functional groups of carbon-based adsorbents and their roles in the removal of heavy metals from aqueous solutions: A critical review*. The Chemical Engineering Journal. 2019;**366**

[34] Gupta VK et al. *Removal of chromium(VI) from electroplating industry wastewater using bagasse fly ash—a sugar industry waste material*. Environmentalist. 1998;**19**(2):129-136

[35] Sud D, Mahajan G, Kaur MP. *Agricultural waste material as potential adsorbent for sequestering heavy metal ions from aqueous solutions - a review*. Bioresource Technology. 2008;**99**(14):6017-6027

[36] Flora SJ, Pachauri V. *Chelation in metal intoxication*. International Journal of Environmental Research and Public Health. 2010;**7**(7):2745-2788

[37] Ding Y et al. *Biosorption of aquatic cadmium(II) by unmodified*

rice straw. Bioresource Technology. 2012;**114**:20-25

[38] Haas KL, Franz KJ. *Application of metal coordination chemistry to explore and manipulate cell biology*. Chemical Reviews. 2009;**109**(10):4921-4960

[39] Deng, F., et al., *5 - Application of Nanomaterials and Nanotechnology in the Reutilization of Metal Ion From Wastewater*, in *Nanomaterials for the Removal of Pollutants and Resource Reutilization*, X. Luo and F. Deng, Editors. 2019, Elsevier. p. 149-178.

[40] Liu C, Ngo H, Guo W. *Watermelon Rind: Agro-waste or Superior Biosorbent?* Applied Biochemistry and Biotechnology. 2012;**167**:1699-1715

[41] García-Mendieta A, Olguín M, Solache M. *Biosorption properties of green tomato husk (Physalis philadelphica Lam) for iron, manganese and iron–manganese from aqueous systems*. Desalination. 2011;**284**

[42] Witek-Krowiak A, Reddy DH. *Removal of microelemental Cr(III) and Cu(II) by using soybean meal waste - Unusual isotherms and insights of binding mechanism*. Bioresource Technology. 2012;**127C**:350-357

[43] Rajapaksha AU et al. *Engineered/ designer biochar for contaminant removal/immobilization from soil and water: Potential and implication of biochar modification*. Chemosphere. 2016;**148**:276-291

[44] Keiluweit M et al. *Dynamic molecular structure of plant biomass-derived black carbon (biochar)*. Environmental Science & Technology. 2010;**44**(4):1247-1253

[45] Yuan JH, Xu RK, Zhang H. *The forms of alkalis in the biochar produced from crop residues at different temperatures*. Bioresource Technology. 2011;**102**(3):3488-3497

[46] Mohanty K et al. *Removal of Cr(VI) from Dilute Aqueous Solutions by Activated Carbon Developed from Terminalia arjuna Nuts Activated with Zinc Chloride*. Chemical Engineering Science. 2005;**60**:3049-3059

[47] Ye H, Zhu Q, Du D. *Adsorptive removal of Cd(II) from aqueous solution using natural and modified rice husk*. Bioresource Technology. 2010;**101**(14):5175-5179

[48] Dupont L et al. *Biosorption of Cu(II) and Zn(II) Onto a Lignocellulosic Substrate Extracted From Wheat Bran*. Environmental Chemistry Letters. 2005;**2**:165-168

[49] Lu, L.-T., et al., *Identification of Pollution Source of Cadmium in Soil. Application of Material Flow Analysis and A Case Study in Taiwan (11 pp)*. Environmental Science and Pollution Research International, 2007. **14**: p. 49-59.

[50] Volesky B, Holan ZR. *Biosorption of heavy metals*. Biotechnology Progress. 1995;**11**(3):235-250

[51] Benaïssa H. *Screening of new sorbent materials for cadmium removal from aqueous solutions*. Journal of Hazardous Materials. 2006;**132**:189-195

[52] Min SH et al. *Improvement of Cadmium Ion Removal by Base Treatment of Juniper Fiber*. Water Research. 2004;**38**:1289-1295

[53] Mohan D, Singh KP. *Single- and multi-component adsorption of cadmium and zinc using activated carbon derived from bagasse--an agricultural waste*. Water Research. 2002;**36**(9):2304-2318

[54] huamán, G., et al., *Biosorption of Heavy Metals by Powder of Green Coconut Shell.* Separation Science and Technology - SEPAR SCI TECHNOL, 2006. **41**: p. 3141-3153.

[55] Montanher SF, Oliveira EA, Rollemberg MC. *Removal of metal*

ions from aqueous solutions by sorption onto rice bran. Journal of Hazardous Materials. 2005;**117**(2-3):207-211

[56] Saeed A, Iqbal M. *Bioremoval of cadmium from aqueous solution by black gram husk (Cicer arientinum)*. Water Research. 2003;**37**(14):3472-3480

[57] Kumar U, Bandyopadhyay M. *Sorption of cadmium from aqueous solution using pretreated rice husk*. Bioresource Technology. 2006;**97**(1):104-109

[58] Saeed A, Akhter M, Iqbal M. *Removal and Recovery of Heavy Metals from Aqueous Solution Using Papaya Wood as a New Biosorbent*. Separation and Purification Technology. 2005;**45**:25-31

[59] Singh KK, Rastogi R, Hasan SH. *Removal of cadmium from wastewater using agricultural waste 'rice polish'*. Journal of Hazardous Materials. 2005;**121**(1-3):51-58

[60] Gupta VK et al. *Removal of cadmium and nickel from wastewater using bagasse fly ash--a sugar industry waste*. Water Research. 2003;**37**(16):4038-4044

[61] Lee SH et al. *Effect of chemical modification of carboxyl groups in apple residues on metal ion binding*. Korean Journal of Chemical Engineering. 1999;**16**(5):576-580

[62] Zulkali MM, Ahmad AL, Norulakmal NH. *Oryza sativa L. husk as heavy metal adsorbent: optimization with lead as model solution*. Bioresource Technology. 2006;**97**(1):21-25

[63] Saeed A, Iqbal M, Akhtar MW. *Removal and recovery of lead(II) from single and multimetal (Cd, Cu, Ni, Zn) solutions by crop milling waste (black gram husk)*. Journal of Hazardous Materials. 2005;**117**(1):65-73

[64] Yu B et al. *The removal of heavy metals from aqueous solutions by sawdust adsorption--removal of lead and comparison of its adsorption with copper*. Journal of Hazardous Materials. 2001;**84**(1):83-94

[65] Wilson K et al. *Select metal adsorption by activated carbon made from peanut shells*. Bioresource Technology. 2006;**97**(18):2266-2270

[66] Gardea-Torresdey JL et al. *Characterization of Cr(VI) binding and reduction to Cr(III) by the agricultural byproducts of Avena monida (oat) biomass*. Journal of Hazardous Materials. 2000;**80**(1-3):175-188

[67] Oliveira EA et al. *Equilibrium studies for the sorption of chromium and nickel from aqueous solutions using raw rice bran*. Process Biochemistry. 2005;**40**(11):3485-3490

[68] Karthikeyan T, Rajgopal S, Miranda LR. *Chromium(VI) adsorption from aqueous solution by Hevea Brasilinesis sawdust activated carbon*. Journal of Hazardous Materials. 2005;**124**(1-3):192-199

[69] Acar FN, Malkoc E. *The removal of chromium(VI) from aqueous solutions by Fagus orientalis L*. Bioresource Technology. 2004;**94**(1):13-15

[70] Mohan D, Singh KP, Singh VK. *Trivalent chromium removal from wastewater using low cost activated carbon derived from agricultural waste material and activated carbon fabric cloth*. Journal of Hazardous Materials. 2006;**135**(1-3):280-295

[71] Venkateswarlu P et al. *Removal of chromium from an aqueous solution using Azadirachta indica (neem) leaf powder as an adsorbent*. Int. J. Phys. Sci. 2007;**2**

[72] Gupta VK, Ali I. *Removal of lead and chromium from wastewater using bagasse fly ash--a sugar industry waste*. Journal of Colloid and Interface Science. 2004;**271**(2):321-328

[73] Hanif MA et al. *Ni(II) biosorption by Cassia fistula (Golden Shower) biomass*. Journal of Hazardous Materials. 2007;**139**(2):345-355

[74] Sciban M, Klasnja M, Skrbic B. *Modified Hardwood Sawdust as Adsorbent of Heavy Metal Ions from Water*. Wood Science and Technology. 2006;**40**:217-227

[75] Malkoc E, Nuhoglu Y. *Investigations of nickel(II) removal from aqueous solutions using tea factory waste*. Journal of Hazardous Materials. 2005;**127**(1-3):120-128

[76] Shukla SS et al. *Removal of nickel from aqueous solutions by sawdust*. Journal of Hazardous Materials. 2005;**121**(1-3):243-246

[77] Ajmal M, Rao RAK, Khan MA. *Adsorption of copper from aqueous solution on Brassica cumpestris (mustard oil cake)*. Journal of Hazardous Materials. 2005;**122**(1-2):177-183

[78] Marshall WE, Johns MM. *Agricultural by-products as metal adsorbents: Sorption properties and resistance to mechanical abrasion*. Journal of Chemical Technology & Biotechnology. 1996;**66**(2):192-198

[79] Ajmal M et al. *Role of sawdust in the removal of copper(II) from industrial wastes*. Water Research. 1998;**32**(10):3085-3091

[80] Basci N, Kocadagistan E, Kocadagistan B. *Biosorption of Copper(II) from Aqueous Solutions by Wheat Shell*. Desalination. 2004;**164**:135-140

[81] Gupta S, Babu DBV. *Removal of Toxic Metal Cr(VI) from Aqueous Solutions Using Sawdust as Adsorbent: Equilibrium, Kinetics and Regeneration Studies*. Chemical Engineering Journal. 2009;**150**:352-365

[82] Mohan D, Pittman CU. *Arsenic removal from water/wastewater using adsorbents—A critical review*. Journal of Hazardous Materials. 2007;**142**(1):1-53

[83] Shukla SR, Pai RS. *Adsorption of Cu(II), Ni(II) and Zn(II) on modified jute fibres*. Bioresource Technology. 2005;**96**(13):1430-1438

[84] Yu B et al. *The removal of heavy metal from aqueous solutions by sawdust adsorption - removal of copper*. Journal of Hazardous Materials. 2000;**80**(1-3):33-42

[85] Pehlivan E, Altun T. *Biosorption of chromium(VI) ion from aqueous solutions using walnut, hazelnut and almond shell*. Journal of Hazardous Materials. 2008;**155**(1-2):378-384

[86] Singh K et al. *Removal of Cr (VI) from aqueous solutions using wheat bran*. Chemical Engineering Journal. 2009;**151**:113-121

[87] Akhtar N, Iqbal J, Iqbal M. *Removal and recovery of nickel(II) from aqueous solution by loofa sponge-immobilized biomass of Chlorella sorokiniana: characterization studies*. Journal of Hazardous Materials. 2004;**108**(1-2):85-94

[88] Garg U et al. *Removal of Nickel(II) from Aqueous Solution by Adsorption on Agricultural Waste Biomass Using a Response Surface Methodological Approach*. Bioresource Technology. 2008;**99**:1325-1331

[89] Wong KK et al. *Removal of Cu and Pb by tartaric acid modified rice husk from aqueous solutions*. Chemosphere. 2003;**50**(1):23-28

[90] Kadirvelu K, Namasivayam C. *Agricutural By-Product as Metal Adsorbent: Sorption of Lead(II) from Aqueous Solution onto Coirpith Carbon*. Environmental Technology. 2000;**21**(10):1091-1097

[91] Pehlivan E et al. *Lead sorption by waste biomass of hazelnut and almond*

shell. Journal of Hazardous Materials. 2009;**167**(1-3):1203-1208

[92] Orhan Y, Büyükgüngör H. *The Removal of Heavy Metals by Using Agricultural Waste*. Water Science and Technology. 1993;**28**:247-255

[93] Anwar J et al. *Removal of chromium from water using pea waste – a green approach*. Green Chemistry Letters and Reviews. 2010;**3**(3):239-243

[94] Sarin V, Pant KK. *Removal of chromium from industrial waste by using eucalyptus bark*. Bioresource Technology. 2006;**97**(1):15-20

[95] Dubey SP, Gopal K. *Adsorption of chromium(VI) on low cost adsorbents derived from agricultural waste material: a comparative study*. Journal of Hazardous Materials. 2007;**145**(3):465-470

[96] Anandkumar J, Mandal B. *Removal of Cr(VI) from aqueous solution using Bael fruit (Aegle marmelos correa) shell as an adsorbent*. Journal of Hazardous Materials. 2009;**168**(2-3):633-640

[97] Wang XS et al. *Removal of Cr (VI) with wheat-residue derived black carbon: reaction mechanism and adsorption performance*. Journal of Hazardous Materials. 2010;**175**(1-3):816-822

[98] Akhtar M et al. *An economically viable method for the removal of selected divalent metal ions from aqueous solutions using activated rice husk*. Colloids and Surfaces. B, Biointerfaces. 2010;**75**(1):149-155

[99] Memon JR et al. *Characterization of banana peel by scanning electron microscopy and FT-IR spectroscopy and its use for cadmium removal*. Colloids and Surfaces. B, Biointerfaces. 2008;**66**(2):260-265

[100] Anwar J et al. *Removal of Pb(II) and Cd(II) from water by adsorption on peels of banana*. Bioresource Technology. 2010;**101**(6):1752-1755

[101] Nouri L et al. *Batch sorption dynamics and equilibrium for the removal of cadmium ions from aqueous phase using wheat bran*. Journal of Hazardous Materials. 2007;**149**(1):115-125

[102] Nouri L, Hamdaoui O. *Ultrasonication-assisted sorption of cadmium from aqueous phase by wheat bran*. The Journal of Physical Chemistry. A. 2007;**111**(34):8456-8463

[103] Farooq U et al. *Effect of modification of environmentally friendly biosorbent wheat (Triticum aestivum) on the biosorptive removal of cadmium(II) ions from aqueous solution*. Chemical Engineering Journal - CHEM ENG J. 2011;**171**:400-410

[104] Tan G, Xiao D. *Adsorption of cadmium ion from aqueous solution by ground wheat stems*. Journal of Hazardous Materials. 2009;**164**(2-3):1359-1363

[105] Liang, S., et al., *Adsorption of Cu2+ and Cd2+ from aqueous solution by mercapto-acetic acid modified orange peel.* Colloids and surfaces. B, Biointerfaces, 2009. **73**: p. 10-4.

[106] Iqbal M, Saeed A, Kalim I. *Characterization of Adsorptive Capacity and Investigation of Mechanism of Cu2+, Ni2+ and Zn2+ Adsorption on Mango Peel Waste from Constituted Metal Solution and Genuine Electroplating Effluent*. Separation Science and Technology. 2009;**44**(15):3770-3791

[107] Johnson PD et al. *Peanut hull pellets as a single use sorbent for the capture of Cu(II) from wastewater*. Waste Management. 2002;**22**(5):471-480

[108] Quek SY, Wase DAJ, Forster CF. *The Use of Sago Waste for the Sorption of Lead and Copper*. Water SA. 1998;**24**:251-256

[109] Moreno-Piraján J, Giraldo L. *Activated carbon obtained by pyrolysis of potato peel for the removal of heavy metal copper (II) from aqueous solutions*. Journal of Analytical and Applied Pyrolysis. 2011;**90**:42-47

[110] Zhu C-S, Wang L-P, Chen W-B. *Removal of Cu(II) from aqueous solution by agricultural by-product: peanut hull*. Journal of Hazardous Materials. 2009;**168**(2-3):739-746

[111] Yao ZY, Qi JH, Wang LH. *Equilibrium, kinetic and thermodynamic studies on the biosorption of Cu(II) onto chestnut shell*. Journal of Hazardous Materials. 2010;**174**(1-3):137-143

[112] Ramana DKV et al. *Removal of heavy metals from aqueous solutions using activated carbon prepared from Cicer arietinum*. Toxicological & Environmental Chemistry. 2010;**92**(8):1447-1460

[113] Bhatnagar A, Minocha AK. *Biosorption optimization of nickel removal from water using Punica granatum peel waste*. Colloids and Surfaces. B, Biointerfaces. 2010;**76**(2):544-548

[114] Ajmal M et al. *Adsorption studies on Citrus reticulata (fruit peel of orange): removal and recovery of Ni(II) from electroplating wastewater*. Journal of Hazardous Materials. 2000;**79**(1-2):117-131

[115] Naiya TK et al. *The sorption of lead(II) ions on rice husk ash*. Journal of Hazardous Materials. 2009;**163**(2-3):1254-1264

[116] Mohammad M et al. *Metal ion removal from aqueous solution using physic seed hull*. Journal of Hazardous Materials. 2010;**179**(1-3):363-372

[117] Choi SB, Yun YS. *Lead biosorption by waste biomass of Corynebacterium glutamicum generated from lysine fermentation process*. Biotechnology Letters. 2004;**26**(4):331-336

Chapter 9

Impact of Emerging Agricultural Contaminants on Global Warming

Nosiri Chidi, Anyanwu Chukwuma and Nwaogwugwu Joel

Abstract

There are many definitions of emerging contaminants (ECs). They are not usually new chemicals and could be substances that have stayed long in the environment with their presence and importance being recognized now. They may be chemicals or microorganisms which are not usually monitored in the environment but with known or suspected potential to cause ecological damage or adverse human effects. Some natural products and those transformed through biochemical processes from synthetic chemicals may be formed in the environment as ECs. Emerging Agricultural Contaminants are released to the environment or enter indirectly to the soil during the application of manure, fertilizers, biosolids or other solid waste materials. Once they enter the soil, they may be transported by leaching, runoff and drainage processes to water bodies. The extent of the transport is dependent on the persistence of the EC and on how it interacts with soil and air. These ECs contribute to global warming through the emissions of Greenhouse gases. The largest source of GHG emission from Agriculture is Nitrous oxide (N_2O) and it accounts for 38% of the total global emission through the process of nitrification and denitrification, anthropogenic activities (use of nitrogen fertilizer, production of nitrogen-fixing crops and forages, retention of crop residues, application of managed livestock manure) which are either through direct additions and/or through indirect additions (atmospheric deposition of applied nitrogen). The natural digestive processes in ruminants otherwise known as enteric fermentation account for the key source of methane production under livestock production hence the second largest source of total agricultural emission with 34% global share and rice cultivation being the third with 11%. The three important greenhouse gases (Methane, Carbon dioxide and Nitrous oxide) are not harmful in naturally occurring quantities for their atmospheric presence helps in sustaining life on the planet when they trap heat energy near the surface of the earth. Concentration of greenhouse gases from both the natural and human factors have been increasing and contributing to Global Warming and Climate Change. Increase in greenhouse gases may cause tremendous changes to our civilization positively or negatively but the total impact is uncertain. Climate change comes as a result of a warming planet which can affect the weather adversely in many ways. So, as climate changes, extreme weather activities release severe threats on human society. Indicators of global warming include sea surface temperature, temperature over land, snow cover on hills, temperature over land and humidity. It is expected that climate change may cause more floods, storms, droughts, heatwaves and other extreme weathers activities. IPCC estimated that temp may rise from 2 to 6°C within 2021. Mitigation of greenhouse effect could be achieved through Biochemical methods on enteric fermentation, development of good environmental policies even Methanotrophs also aid in recycling the atmospheric Methane.

Keywords: emerging contaminants, pesticides, fertilizers, livestock, manure, global warming

1. Introduction

Global warming: The temperature of the earth since 1950 has been increasing above average temperature. This can also be referred to as climate change. Both climate change and Global warming refer to increase in average global temperature. Climate change involves a change in climate measured using statistical properties such as mean surface temperature which can range from months to millions of years. One of the facets of climate change is global warming as it has been described as an average increase in global surface temperature. The causes of Global warming are both from natural and human events. One of the natural factors is Greenhouse effect. This effect is a blanketing one by which greenhouse gases keep the surface of the earth warm [1]. The human influenced greenhouse gases (CO_2, CH_4, and N_2O) are also essential in climate change [2]. The energy that gets to the earth is absorbed by earth's surface and later re-radiated back as heat energy towards the space. Greenhouse gases equally and importantly trap this re-radiated energy within the atmosphere hence increase the temperature of earth's surface [3].

Climate change occurs as a result of high concentration of greenhouse gases (GHG) like carbon dioxide (CO_2), nitrous oxide (N_2O) and methane (CH_4), sulfur dioxide (SO_2) etc. [4]. The increase in high GHG emissions are associated with economic activity especially as related to industry, energy, transport and methods of land use (Agricultural production and deforestation [5]. Greenhouse gases are still building up in the atmosphere and prompting shifts across the globe in the climate.

Agriculture is a well-known source of global emission of greenhouse gases. Rearing of livestock and application of fertilizers and other related land clearing affects the levels of both GHG in the atmosphere and the possibilities of carbon sequestration and storage. Agricultural pollutants are the contaminants that are released into the environment as a by-product from growing and raising of food crops and livestock. However, these pollutants are biotic and abiotic by-products of farming practices (pesticides, fertilizers and livestock manure) that result in degradation or contamination of the environment and ecosystem hence leading to injury in humans and their economic interests. There has been an increasing concern over the emerging and growing increase in emission of Greenhouse gases such as CH_4, CO_2 and N_2O, etc. as they have detrimental impact on the environment leading to global warming which results in climate change and environmental degradation [6, 7].

1.1 Sources of emerging contaminants

Different routes of release of emerging contaminants can be summed-up into 3 major categories as follows:

1. Use of Livestock: This includes manure and flatulent gas (methane) being released directly from animals and their compost-releasing Methane gas into the atmosphere, artificial fertilizers, hormones used in livestock and all other veterinary composts.

2. Plant Protection Activities: These include mainly nanomaterials designed as "smart" chemicals or "smart" pesticides that have the ability for selective toxicities. Unfortunately, these nanomaterials enter the ecological system virtually undetected. Pesticides which include Rodenticides, Fungicides, bactericides, weed toxins, zootoxins which kill small animals like rodents, phycotoxins which inhibit the growth of algae, different personal and household use products, all that have the ability to produce other compounds that are consequently released directly or indirectly into the environment especially through air and wastewater as emerging contaminants.

3. Human Use Activities: These include chemicals such as new chemical compounds and pharmaceuticals synthesized by humans that are released directly and indirectly. These compounds usually pass into wastewater treatments to land resulting in sludge and biosolids, and or from irrigation to waste-water effluent. Pesticides which may be released as nanoparticles are also part of this group with increased absorption ability, increased solubilities, or increased toxicities.

2. Emerging contaminants from agrochemicals

Contaminants refer to biological, chemical, physical or radiological substances found in considerable concentrations that can adversely affect human beings through soil, food contamination and air. For a long time now Contaminants from Agricultural practices have been known to be as a result of human activities both man-made and natural. Their increasing effect on the ecological system including resultant Green-House effects poses a disturbing concern.

What are the emerging agricultural contaminants? Agricultural contaminants can generally be grouped into some major classifications, but a major source of concern for the environment is the emergence of animal waste and its release of Methane gas (CH4) into the atmosphere. Another growing source of concern is the emergence of pharmaceutical wastes as contaminants which manifests typically in run-off water in the agricultural watersheds. Although there are many definitions of what constitutes "emerging contaminants" it is of paramount importance to explain the word "emerging" attached to the contaminants as chemical compounds which have not been known to be in existence or have not been extensively studied. Again, emerging contaminants can be classified into chemical compounds which have been known previously but whose impacts on health and environment have not been fully comprehended. The definitions of emerging contaminants are numerous and vary from region to region of the world. They may be chemicals that have been shown to be threatening to the environment or public health through different routes without enough data to assess their risk levels. Those with potential health and environmental threat without regulatory standards also fall into emerging contaminants just as chemicals or substances that penetrate the environment through different routes but lack not only risk stratification, but inadequate detection capabilities. Although detecting emerging contaminants can be more challenging, they can penetrate the environment causing adverse biological and ecological effects [8]. The United States Geological Survey describes emerging contaminants as chemicals or microorganisms which are not usually monitored in the environment but with known or suspected potential to cause ecological damage or adverse human effects. New Synthetic chemicals or changes in use and disposal of existing ones can also be referred to as a means of creating emerging contaminants. Emerging contaminants in another way is defined as chemical substances or compounds that are characterized by a perceived challenge to human health or the environment lacking published health criteria. They may also be identified from sources that are unknown, new exposure to humans, or a new detection approach or technology [9]. Emerging contaminants (ECs) are not usually new chemicals. They could be substances or chemicals that have stayed long in the environment with their presence and importance being recognized now. There are many proposed and contradictory definitions of "Emerging Contaminants". Boxall [10] defines it as a contaminant from a chemical class that has not been studied extensively where either there is a concern from scientists, Regulators, NGOs or other Stakeholders that the contaminant class may be having an impact on human health or environment or where there is a concern that an existing environmental assessment paradigms are not appropriate for the contaminant class.

Recently, there has been a great concern on the environmental challenges caused by Emerging contaminants. These ECs emanate from by-products of a variety of materials such as paints and coatings, pharmaceuticals, personal care products, agrochemicals, veterinary medicines etc. Some natural products and those transformed through biochemical processes from synthetic chemicals may be formed in the environment as ECs.

2.1 Emerging contaminants from agricultural activities

Emerging Contaminants are released to the agricultural environment via different pathways. They may be released to the environment directly (veterinary medicines that are used in the treatment of animals at pasture) or enter indirectly to the soil during the application of manure, biosolids or other solid waste materials. Once they enter the soil, they may be transported by leaching, runoff and drainage processes to water bodies. The extent of the transport is dependent on the persistence of the EC and on how it interacts with soil and air.

2.1.1 Agricultural soils

The largest source of GHG emission from Agriculture is Nitrous oxide (N_2O) and it accounts for 38% of the total global emission. Through the process of nitrification and denitrification, it is produced naturally from soil. Agricultural activity which is anthropogenic may add nitrogen directly or indirectly to soils. The use of nitrogen fertilizer, production of nitrogen-fixing crops and forages, retention of crop residues, application of managed livestock manure, sewage sludge and cultivation of soils with high organic matter content are the direct additions. Through indirect additions are surface runoff and leaching of applied nitrogen into surface and ground water, volatilization and atmospheric deposition of applied nitrogen [11]. Other anthropogenic sources of GHG emissions from agriculture include manufacture of equipment, pesticides, fertilizers, on-farm use of fuels and the transport of agricultural products [12]. Emissions of GHG from agricultural soils (N_2O) and enteric fermentation and management of manure (CH_4) linked with livestock production contribute to the largest share in agricultural sector. So, from agricultural sector, sources of GHG emissions are accounted for as follows: Residue burning/forest cleaning (13%), Methane and Nitrous oxide from Management of manure (7%), Methane from Rice cultivation (11%), Nitrous oxide from fertilizers (37%) and Methane from Livestock (32%) [11]. These emissions are expected to increase in future due to high demand for agricultural products and enhanced nutrition [13]. The direct addition of both synthetic and organic nitrogen containing fertilizers will possibly be a key source of rise in emission of N_2O with the highest coming from developing countries that are seen using 36 million tons more than the developed nations [14].

2.1.2 Livestock

Emissions from Livestock constitute 9% of CO2 equivalent obtained from all human related activities and produces 65% of human-related nitrous oxide, and 37 percent of human induced methane, and 64% ammonia especially from manure. Considering agriculture sector, livestock account for nearly 80% of all emissions [15]. All these emissions from livestock make it a major target for mitigation options. Feed produced locally such as grass and other roughage, together with wastes that are rich in nutrient from farms and households when replaced with feed manufactured from feed mills containing some secondary plant metabolites able to reduce the emission of methane gas could lead to negative climate impact.

The natural digestive processes in ruminants otherwise known as enteric fermentation in sheep and cattle account for the key source of methane production under this category hence the second largest source of total agricultural emission with 34% global share and rice cultivation being the third with 11%. Other domesticated animals that also emit methane (methanogenesis) as a by- product of enteric fermentation include Horses, swine and poultry. The estimates of enteric methane from ruminants produced globally are about 80,000 Gg [16].

2.1.3 Management of manure

Handling, treatment and storage (Management of manure) of manure account for 7% of emissions from agriculture. Anaerobic breakdown of manure produces methane (methanogenesis) while aerobic handling of manure produces nitrous oxide (Nitrification) and then anaerobically (denitrification) hence often increased when the available nitrogen exceeds that of plant requirements [12]. Due to high demand expected for beef and dairy products globally especially from the developing world, methane emissions from enteric fermentation are expected to increase by 32% [11].

3. Impact of emerging agricultural contaminants on global warming

The three important greenhouse gases (Methane, Carbon dioxide and Nitrous oxide) are not harmful in naturally occurring quantities for their atmospheric presence helps in sustaining life on the planet when they trap heat energy near the surface of the earth. Concentration of greenhouse gases from both the natural and human factors have been increasing and contributing to Global Warming and Climate Change.

Increase in greenhouse gases may cause tremendous changes to our civilization positively or negatively but the total impact is uncertain. Climate change comes as a result of a warming planet which can affect the weather adversely in many ways. So, as climate changes, extreme weather activities release severe threats on human society. Indicators of global warming include ocean heat content, sea ice, sea level, sea surface temperature, temperature over ocean, temperature over land, tropospheric temperature, snow cover on hills, temperature over land and humidity. It is expected that climate change may cause more floods, storms, droughts, heatwaves and other extreme weathers activities. IPCC estimated that temp may rise from 2 to 6°C within 2021 [4].

Industrial Agriculture is one of the causes of climate change with the practices having impact on it. The currently practiced industrial agriculture with Agrochemicals and monocultures globally is a key contributor to climate change for it causes the emission of Green House Gases (GHG) via changes in land use and soil degradation or losses, via technologies in agriculture and from livestock [17].

According to Food and Agriculture Organization (FAO), Livestock production sector as the major emitter of all the greenhouse gases is responsible for about 18% on one fifth of human-induced greenhouse gas emissions [18]. For instance, in almost every step of egg, meat and milk production in agriculture, greenhouse gases that cause climate change are released into the atmosphere disturbing temperature, weather and health of ecosystem. Mitigating these problems will require changes in agricultural practices and livestock consumption.

Livestock Production as one of the Agricultural food-based industries has been described as leading or major contributor of the anthropogenic source of CH_4 and CO_2 greenhouse gases [19]. They contribute immensely to global warming, pollution and environmental degradation because of the large amount of greenhouse gases produced during the ruminal fermentation of feeds. For instance, anytime a cow burps or passes out gas, it puffs out a little methane which wafts into the atmosphere. Each of these puffs from the cow put together can result to a big effect on climate because methane as a potent greenhouse gas is about 25 times more powerful in trapping heat and increasing earth's global warming than Carbon dioxide on equal mass basis. The livestock sector therefore accounts for about 18% of CH_4 and 9% of CO_2 of all the greenhouse gases emissions [20] hence methane accounts for 50–60% of emitted gases during livestock production [21]. A lot of proposals have been focused on soil and plant production for the reduction of emission from agriculture leaving livestock production which accounts for major amounts of CH_4, N_2O, CO_2 and NH_3 emissions.

Several gases cause chemical reactions that can result in producing other greenhouse gases among other effects and some trap heat energy better than others. Considering the lifetime of greenhouse gases, some persist in the atmosphere for different durations of time and some contribute to global warming due to their long duration [22].

4. Mitigation

There are differences in Global Warming Potential (GWP) and lifetime of greenhouse gases in the atmosphere. One GWP has been assigned to CO_2 and the warming potential of other gases are based and expressed relative to CO_2 GWP [23]. For instance, 1 tonne of CH_4 has a warming effect of about 25 and 72 tons of CO_2 over 100- and 20 year periods respectively according to IPCC. Studies have shown that CH_4 is more potent than CO_2, so reducing CH_4 emission will show more immediate and significant impact on mitigating changes in climate than reducing CO_2 emissions [24] Another potent greenhouse gas is N_2O for it remains in the atmosphere for 114 years [25, 26]) and 298 times as potent as CO_2 over 100 years [27] (see **Table 1** below).

Increased Industrial agriculture is proposed to be a solution to the problems of climate change which is a contributor without its impact being addressed. In the

GHG	Atmospheric lifetime	GWP	GWP (100 years)
CO_2	Centuries to Millenia	1	1
CH_4	About 10 years	72	25
N_2O	114 years	289	298

Source: [27].

Table 1.
Global Warming Potential (GWP) and lifetime of greenhouse gases in the atmosphere.

negotiation for a new climate treaty that has been negotiated to follow the Kyoto protocol in 2012, agricultural practices have been proposed to be a means for climate change mitigation and partly for carbon trading [17].

Livestock farming in Agriculture is one of the largest sources of water contaminant and this contamination maybe the easiest route of emerging agricultural contamination in getting to humans and the eco-system. Animal dung production helps in releasing massive amount of Methane into the environment hence posing challenging situations as intensive farming serves in meeting global food challenges. Also, Algae and Fungi bloom intensifies methane release into the environment, thus having adverse effects on Green-house gases. To mitigate emerging contaminants the following can be employed:

1. Development of good environmental policies to mitigate unwarranted release of emerging contaminants such as standards for water policy, improved testing on pollution and emission, assessments on environmental impact of farms and irrigation systems.

2. Enhancing the intake of digestible forage will mitigate GHG emissions from rumen fermentation and stored manure (For example, to reduce enteric methane emission replace corn silage with grass silage in diet, apply dietary lipids, use legume over grass as it contains more fiber in addition to replacing inorganic fertilizer) [28].

3. Release of pollutants at the source of production should be reduced because of its easy entrance to vulnerable environment.

4. Reduction of the application of fertilizers and pesticides and introduction of natural pest-control methods should be encouraged.

5. Restoration of degraded pastureland by utilization of traditional approaches or techniques.

6. Methanotrophs otherwise known as the methane-eating bacteria found in the ecosystem where methane is produced ensures the conversion of methane into substances such as sugars before being released into the atmosphere. In this way, the methanotrophs help in recycling methane all over the globe.

7. Biochemical method of mitigation of Greenhouse gases: This is the employment of feed supplements that have the potential of reducing methane emission. The bulk of complex molecules contained in ruminant feeds are broken down during fermentation by microbes releasing energy and other chemical compounds [7]. Plant Natural Products (PNP) are engaged to aid in the challenges observed in livestock production where there is wastage of nutrient caused by excess excretion and poor or inefficient digestibility of feeds, emission of CO_2 and CH_4 that results to 2–12% loss of feed energy [29]. These PNPs include Saponins, terpenoids, tannins, flavonoids, phenols, glycosides, alkaloids and essential oils These challenges brought the attention of Biochemists, Nutritionists and Microbiologists to look into the PNP application [30], yeasts [31] and enzymes as additives to feed. This incorporation as additives is meant to modify fermentation kinetics in ruminants and manipulate microbial ecosystems. The primary purpose of these modifications is to improve the utilization of animal feed and digestibility of fibrous feeds, decrease degradability of protein [32], inhibit the proliferation of pathogenic bacteria in the gastrointestinal tract [33] enhance

performance of animals, reduce loss of dietary energy during fermentation in rumen and mitigate CH_4 and CO_2 productions. PNPs have been found to be natural and safer feed additives as they demonstrate vital role in mitigating methane and carbon dioxide productions without affecting fermentation in rumen during their production [6].

5. Conclusion

Emerging Agricultural Contaminants play both positive and negative roles in global warming which leads to the change in climate. The impact can be reduced through mitigation of the underlying causes such as the events that result in high emissions of the greenhouse gases. Climate change is serious and may likely worsen leading to land loss and unpredictable changes that will emanate from natural growing conditions.

Author details

Nosiri Chidi[1*], Anyanwu Chukwuma[2] and Nwaogwugwu Joel[1]

1 Department of Biochemistry, Faculty of Biological Sciences, Abia State University, Uturu, Nigeria

2 Informatics Div, Department of Health, Bureau of Epidemiology, Center for Disease Control and Prevention, Houston, TX, USA

*Address all correspondence to: nosiric@yahoo.com

References

[1] Le Treut H., Somerville R, Cubasch U, et al. 2007. Historical overview of climate change. In: Solomon S, Qin D, Manning M, et al (eds.), Climate change 2007: the physical science basis. Contribution of Working Group I to the Fourth Assessment Report of the Intergovernmental Panel on Climate Change (Cambridge, United Kingdom and New York, NY, USA: Cambridge University Press, p. 96).

[2] Forster P, Ramaswamy V, Artaxo P, et al. 2007. Changes in atmospheric constituents and in radiative forcing. In: Solomon S, Qin D, Manning M, et al (eds.), Climate change 2007: the physical science basis. Contribution of Working Group I to the Fourth Assessment Report of the Intergovernmental Panel on Climate Change (Cambridge, United Kingdom and New York, NY, USA: Cambridge University Press, p. 135-136, FAQ 2.1).

[3] Le Treut H., Somerville R, Cubasch U, et al. 2007. Historical overview of climate change. In: Solomon S, Qin D, Manning M, et al (eds.), Climate change 2007: the physical science basis. Contribution of Working Group I to the Fourth Assessment Report of the Intergovernmental Panel on Climate Change (Cambridge, United Kingdom and New York, NY, USA: Cambridge University Press, p. 115 FAQ 1.3).

[4] Singh, B. R and Singh, O. (2012). Study of Impacts of Global Warming on Climate Change: Rise in Sea Level and Disaster Frequency. In Global Warming-Impacts and Future Perspective. Edited by Bharat Raj Singh Doi:10.5772/50464

[5] World Resources Institute (WRI) (2008) Climate Analysis Indicators Toolkit (CAIT) [January 2008, available online at: http://cait.wri.org/].

[6] Patra, A and Saxena,J (2010). ChemInform Abstract: A New Perspective on the Use of Plant Secondary Metabolites to Inhibit Methanogenesis in the Rumen Phytochemistry 71 (11-12):1198-222 DOI:10.1016/j.phytochem.2010.05.010

[7] IPCC (2007) Summary for policy makers. Climate Change 2007: Synthesis Report. Fourth Assessment Report of the Intergovernmental Panel for Climate Change. [Available online at http://www.ipcc.ch/pdf/assessment-report/ar4/ syr/ar4_syr_spm.pdf]

[8] Snow, DD; Bartelt-Hunt, SL; Saunders, SE and Cassada, DA. (2007) Detection, Occurrence, and Fate of Emerging Contaminants in Agricultural Environments Water. Environment Research Vol. 79, No. 10, Literature Reviews [CD-ROM content] (2007), pp. 1061-1084 (24 pages). Published By: Wiley. DOI: 10.2307/29763261 https://www.jstor.com/stable/29763261

[9] Murnyak, G,. Vandenberg, J., Yaroschak, J., Williams, L., Prabhakaran, K. and J. Hinz, J. (2009)"Emerging contaminants: presentations at the 2009 Toxicology and Risk Assessment Conference," Toxicology and Applied Pharmacology, vol. 254, no. 2, pp. 167-169, 2011.

[10] Boxall, A.B. A. (2012). New and Emerging Water Pollutants arising from Agriculture, in Water Quality and Agriculture: Meeting the Policy Challenge. (OECD report, 2012). Downloaded at *www.oecd.org/agriculture/water*

[11] U.S. Environmental Protection Agency. Frequently asked questions about global warming and climate change: back to basics, p. 3. http://www.epa.gov/climatechange/downloads/Climate_Basics.pdf.

[12] Rosegrant, M. W., Ewing, M., Yohe, G., Burton, I., Huq, S., Valmonte-Santos, R. (2008). Climate Change and Agriculture Threats and Opportunities pp 1-36 http://www.gtz.de/climate

[13] Delgado, C., M. Rosegrant, H. Steinfeld, S. Ehui, C. Courbois (1999) Livestock to 2020: The Next Food Revolution. Food, Agriculture, and Environment. Discussion Paper No.28, International Food Policy Research Institute, Washington D.C.

[14] Bumb, B. and C. Baanante (1996) World Trends in Fertilizer Use and Projections to 2020. 2020 Brief #38, International Food Policy Research Institute, Washington, D.C.

[15] Steinfeld H., Gerber P., Wassenaar T., Castel V., Rosales M. & de Haan C. (2006): Lifestock's long shadow. Environmental issues and options. FAO, Rome

[16] Ku-Vera Juan Carlos, Rafael Jiménez-Ocampo, Sara Stephanie Valencia-Salazar, María Denisse Montoya-Flores , Isabel Cristina Molina-Botero , Jacobo Arango , Carlos Alfredo Gómez-Bravo , Carlos Fernando Aguilar-Pérez and Francisco Javier Solorio-Sánchez. (2020) Role of Secondary Plant Metabolites on Enteric Methane Mitigation in Ruminants. Front. Vet. Sci., https://doi.org/10.3389/fvets.2020.00584

[17] Paul, H., Semino, S., Lorch, A., Anderson, B.H., Gura, S., Ernsting., A (2009). (2009) Agriculture and climate change: Real problems, false solutions Preliminary report by Grupo de Reflexion Rural, Biofuelwatch, EcoNexus and NOAH - Friends of the Earth Denmark pp 1-32 http://www.econexus.info/pdf/agriculture-climate-change-june-2009.pdf

[18] Steinfeld H, Gerber P, Wassenaar T, Castel V, Rosales M, and de Haan C. 2006. Livestock's long shadow: environmental issues and options. Food and Agriculture Organization of the United Nations, p. xxi. 65 Steinfeld H, Gerber P, Wassenaar T, Castel V, Rosales M, and de Haan C. 2006. Livestock's long

[19] Audsley, E and Wilkinson, M. (2014). What is the potential for reducing national greenhouse gas emissions from crop and livestock production systems? Journal of Cleaner Production 73:263-268 DOI: 10.1016/j.jclepro.2014.01.066

[20] FAO (Food and Agricultural Organization). (2013). Tackling climate change through livestock: A global assessment of emissions and mitigation opportunities. Food and Agricultural Organization of the United Nations (FAO) Rome

[21] Mirzaei-Ag, Syed Alireza Sy, Hasan Fathi , Sohrab Rasouli, Mohammad Sadaghian, Mohamad Tarahomi. (2012). Garlic in Ruminants Feeding. Asian Journal of Biological Sciences 5(7):328-340. DOI: 10.3923/ajbs.2012.328.340

[22] McDonald, J. (2018) How Potent Is Methane? FactCheck.org® *A Project of The Annenberg Public Policy Center* , SCICHECK

[23] Paustian K, Antle JM, Sheehan J, and Paul EA. 2006. Agriculture's role in greenhouse gas mitigation. Pew Center on Global Climate Change, p. 3.

[24] Moore FC and MacCracken MC. 2009. Lifetime-leveraging: an approach to achieving international agreement and effective climate protection using mitigation of short-lived greenhouse gases. International Journal of Climate Change Strategies and Management 1(1):42-62, pp. 46-47, 49.

[25] Solomon S, Qin D, Manning M, et al. 2007. Technical summary. In: Solomon S, Qin D, Manning M, et al (eds.), Climate change

2007: the physical science basis. Contribution of Working Group I to the Fourth Assessment Report of the Intergovernmental Panel on Climate Change (New York, NY: Cambridge University Press, p. 33 Table TS.2). http://www.ipcc.ch/pdf/assessment-report/ar4/wg1/ar4-wg1-ts.pdf.

[26] Steinfeld H, Gerber P, Wassenaar T, Castel V, Rosales M, and de Haan C. 2006. Livestock's long shadow: environmental issues and options. Food and Agriculture Organization of the United Nations, p. 82. http://www.fao.org/docrep/010/a0701e/a0701e00.HTM.

[27] Forster P, Ramaswamy V, Artaxo P, et al. 2007. Changes in atmospheric constituents and in radiative forcing. In: Solomon S, Qin D, Manning M, et al (eds.), Climate change 2007: the physical science basis. Contribution of Working Group I to the Fourth Assessment Report of the Intergovernmental Panel on Climate Change (Cambridge, United Kingdom and New York, NY, USA: Cambridge University Press, p. 212 Table 2.14).

[28] Hristov, A.N., Oh, J., Lee, C., Meinen, R., Montes, F., Ott, T., Firkins, J., Rotz, A., Dell, C., Adesogan, A., Yang, W., Tricarico, J., Kebreab, E., Waghorn, G., Dijkstra, J. & Oosting, S. 2013. Mitigation of greenhouse gas emissions in livestock production – A review of technical options for non-CO2 emissions. Edited by Pierre J. Gerber, Benjamin Henderson and Harinder P.S. Makkar. FAO Animal Production and Health Paper No. 177. FAO, Rome, Italy.

[29] Hristov, A.N, Joonpyo Oh , Fabio Giallongo, Tyler W. Frederick, Michael T. Harper, Holley L. Weeks, Antonio F. Branco, Peter J. Moate, Matthew H. Deighton, S. Richard O. Williams, Maik Kindermann, and Stephane Duval (2015). An inhibitor persistently decreased enteric methane emission from dairy cows with no negative effect on milk production. PNAS, 112(34) 10663-10668. https://doi.org/10.1073/pnas.1504124112

[30] Elghandour M.Y., , Poonooru R. K., Salem, A.Z., Pandu, R., Iqbal, H., Alberto, B. and Duvvuru, Y. (2018). Plant Bioactives and Extracts as Feed Additives in Horse Nutrition. Journal of Equine Veterinary Science 69(2018): 66-77. DOI: 10.1016/j.jevs.2018.06.004

[31] Shurson, G.C. 2017. Review - Yeast and yeast derivatives in feed additives and ingredients: Sources, characteristics, animal responses, and quantification methods. Anim. Feed Sci. Technol. 235:60-76.

[32] Salem Mohammed, Leonardo Souza Ahmed Serag, Fernie R. Alisdair, Ferag,A. Mohammed,Ezzat M. Shahira and Saleh Alseekh (2012) Metabolomics in the Context of Plant Natural Products Research: From Sample Preparation to Metabolite Analysis. Metabolites 10(1):37. DOI: 10.3390/metabo10010037

[33] Arowolo Muhammed and He Jianhua. (2018). Use of probiotics and botanical extracts to improve ruminant production in the tropics: A review. Animal Nutrition. 4(3): 241-249. Doi: 10.1016/j.aninu.2018.04.010

Section 3

Pharmaceuticals in the Environment

Chapter 10

Pharmaceutical Antibiotics at a Significant Level in Nature: From Hospitals, Livestock, and Plants to Soil, Water, and Sea

Mauricio Corredor and Amalia Muñoz-Gómez

Abstract

Antibiotics were the world's great therapeutic hope after the Second World War, but today, unmonitored use has become one of the greatest risks for humanity. Without overestimation, one of the last scientific books on antibiotics was entitled: Antibiotics, the perfect storm. Before to environmental contamination by antibiotics, the pathogens got resistant to them. Because of the radical changes that antibiotics have brought about, they can generate new resistant bacteria in the environment that were previously harmless. These microorganisms will be exposed to concentrations of antibiotics never reached or will be exposed to unknown molecules that, for many of them, in certain environments, have never been exposed before. Initially, many of these antibiotics did not penetrate soils with high agricultural production, but in the following decades, they were even interspersed into crops. Nowadays, hundreds of tons of antibiotics are dumped into rivers and the sea. Many hospitals have water treatment facilities to prevent significant contamination, but not all companies, farms, and hospitals in developed, emerging, or poor countries apply wastewater treatment. Antibiotics are incorporated into wild microorganisms and plants, triggering a broad "unnatural" resistance, which will rapidly incorporate this information into the genome of other pathogenic microorganisms by horizontal transfer. On the other hand, antibiotics could be incorporated into drinking water and water intended for human or agricultural consumption that travels without being detected or monitored. This review covers the most important aspects of environmental pollution by antibiotics.

Keywords: antibiotics, water, hospitals, farms, livestock, soil, sea

1. Introduction

Antibiotics exist in normal (low) concentration in nature [1–3]. Bacteria, fungi in general all organisms develop and metabolize these molecules to survive [4]. The discovery of penicillin split the history of humanity in two and that magnificent discovery is only comparable to the handling of electricity, atomic energy, space travel, and anesthesia, among others [5]. The penicillin and antibiotics brought about the saving of millions of lives, but their use and application were exaggerated and misguided to the point that antibiotics are used as food supplements in the livestock, poultry, swine, and crop industries [6–9]. However, pharmaceutical antibiotics at a significant level in nature and different environments has been detected in previous studies and monitoring in plains, valleys, coast, and mountains, when were monitored in hospital wastewater, farms, rivers, and coastal water [10–13]. Antibiotics emerge as new environment contaminants as plastics, pesticides, among others. The main risk and concern as the pesticides are microbial antibiotics and multidrug resistance. The human being is developing littles Frankenstein, likely by carelessness and naivety, contrary to genetically modified organisms (GMOs) opinion, the scare of gene manipulation, and its ethical reflection by the Scientifics and public [14, 15]. The consumer has scary to GMOs, but without laboratory use, the human being is making GMOs resistant to antibiotics. The controversy is rising, why the public is concerned by artificial experiments but carefree by antibiotics pollutants? This environmental risk goes in the same way as global warming.

Probably some natural antibiotic-resistant bacteria in soil gain multidrug resistance consequence by human practices as livestock or water activities. However, antibiotics come to the soil by wastewater from human or animal feces together with the selection of antibiotic-resistant microbiome [16]. Antibiotics and bacteria remain in the soil until water carries them to stream or rivers or are transferred by

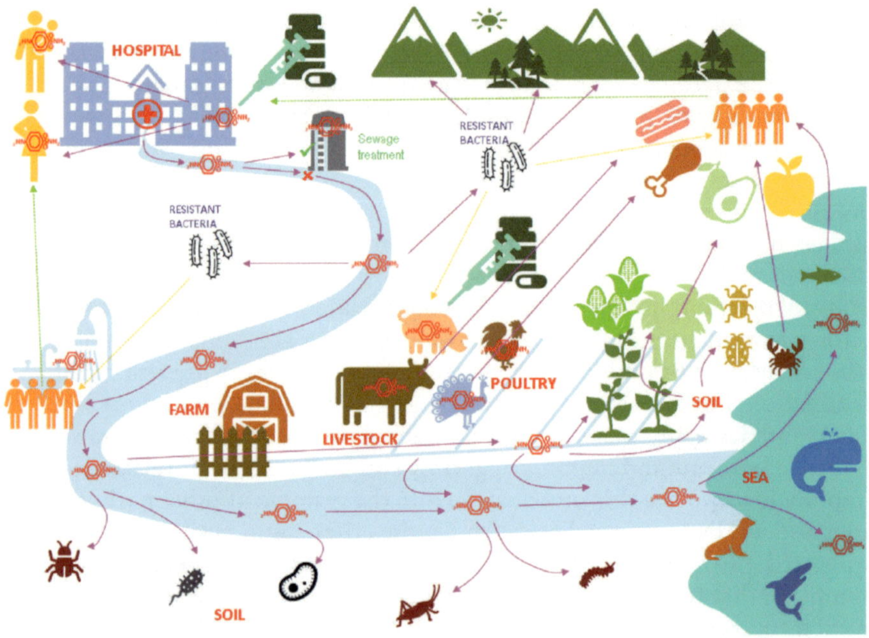

Figure 1.
Water carries on antibiotics. *Water from hospitals and farms could pass through sewage treatment, but in many countries, those health and environmental rules are not applied or respected, especially in developing countries by poor investment and low budget. Water is used for human consummation, livestock, poultry, swine, and soil irrigation to crop grass and agriculture plants. But lockdown animals, cattle, and others get sick and must be under antibiotic treatment. Then, this sanitary water passes directly to the sewer rarely treated and directly deposed in the streams and rivers. Antibiotics are used in some countries to yield more muscular biomass of livestock, poultry, and swine. This practice increases the level of antibiotics in soil, humans, flora, and fauna. Likely, some soils intersperse antibiotics in roots, stems, leaves, and fruits for human consummation. Even, antibiotics in the soil are absorbed by bacteria, fungi, protists, and invertebrates. Soils stock, pick up, and pour antibiotics in streams, then to rivers and finally to the sea. The pharmaceutical antibiotics were poured into the sea for 70 years. Surely the antibiotics levels in nature and water were not the same before the pharmaceutical antibiotic' revolution and production. This image is an actual case in South America, where some hospitals lack the budget to invest in the sewage treatment plant. Near this real example, there are other little villages where its hospital has two water treatment plants, one for human consummation and the other for sewage. In that case, contaminated hospital water never goes back to flow, however, many farms invest weakly in the water treatment process, ejecting polluted water to the environment with antibiotics and pesticides.*

roots in plants reaching leaves, which will be consumed by insects, incorporating new molecular information to wild species (**Figure 1**).

2. Antibiotics real state consummation and biodegradability

Tons of pharmaceutical antibiotics have been produced by the industry. The worldwide production of antibiotics is estimated to be in the order of 100.000 tons per year [17], and 73% in the world and 80% in the United States is used for livestock consummation [18]. Different from the European Union, where the amount of antibiotic used is 60% for humans and 40% for animals and crops [19]. The great threat as warned by the WHO is to increase the resistance to antibiotics not only by pathogens but also by environmental organisms, which can lead to possible environmental damage (https://www.who.int/news/item/07-11-2017-stop-using-antibiotics-in-healthy-animals-to-prevent-the-spread-of-antibiotic-resistance). Now WHO is recommending avoiding the use of antibiotics in healthy animals.

Antibiotics were approved in the USA in the next order: sulfonamides in 1935, penicillin in 1941, aminoglycosides in 1944, cephalosporins in 1945, chloramphenicol in 1949, tetracyclines in 1950, macrolides/lincosamides/streptogramins in 1952, glycopeptides in 1956, rifamycins in 1957, nitroimidazoles in 1959, quinolones in 1962, trimethoprim in 1968, oxazolidinones in 2000, lipopeptides in 2003 [20]. **Table 1** relates the antibiotics with their molecular functional group and in the last column their resistance model in microorganisms. Another problem is that antibiotics are not used alone and come with stimulators and enhancers, such as clavulanic acid which helps overcome β-lactam resistance.

Class	Example(s)	Functional group or whole molecule	Model(s) of resistance
β-Lactams	Penicillins (ampicillin), cephalosporins (cephamycin), penems (meropenem), monobactams (aztreonam)	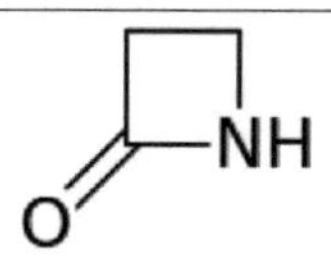	Hydrolysis, efflux, altered target
Aminoglycosides	Gentamicin, streptomycin, spectinomycin	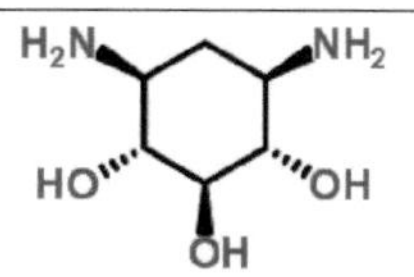	Phosphorylation, acetylation, nucleotidylation, efflux, altered target
Glycopeptides	Vancomycin, teicoplanin	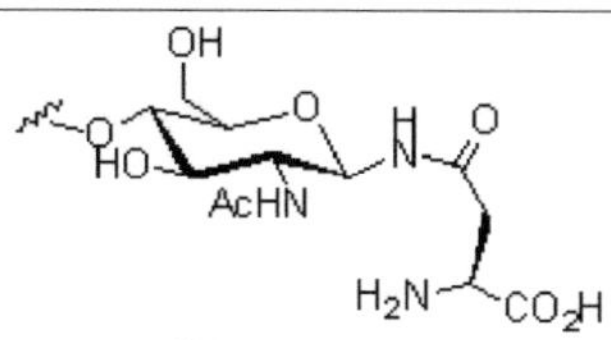 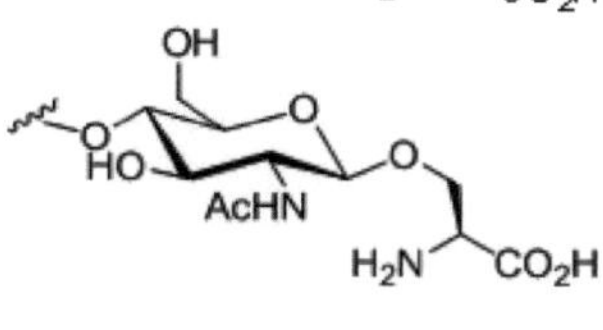	Reprogramming peptidoglycan biosynthesis

Class	Example(s)	Functional group or whole molecule	Model(s) of resistance
Tetracyclines	Minocycline, tigecycline	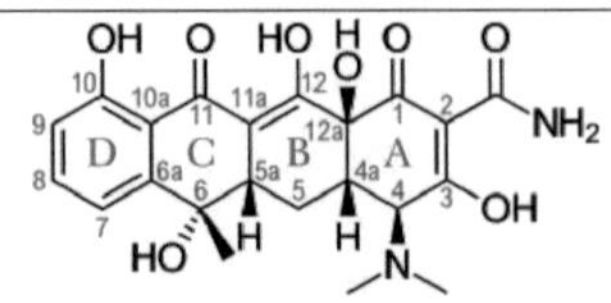	Monooxygenation, efflux, altered target
Macrolides	Erythromycin, azithromicin	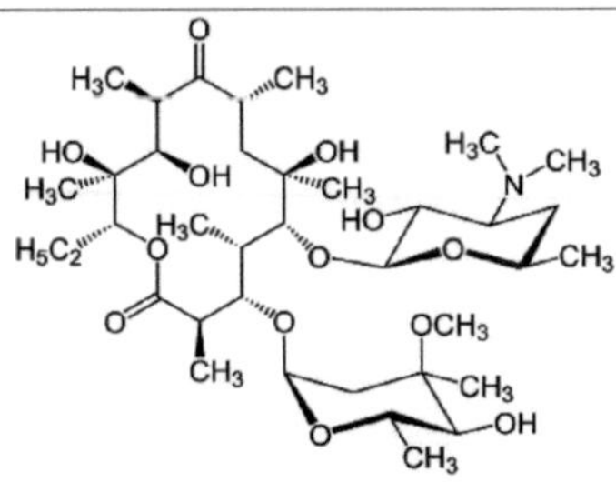	Hydrolysis, glycosylation, phosphorylation, efflux, altered target
Lincosamides	Clindamycin	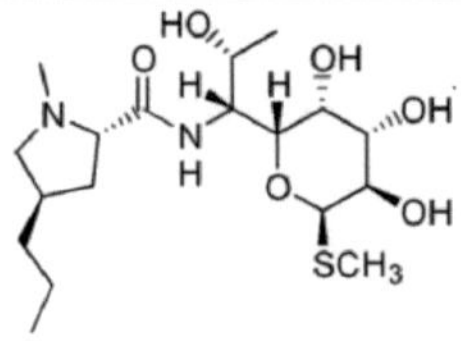	Nucleotidylation, efflux, altered target
Streptogramins	Synercid	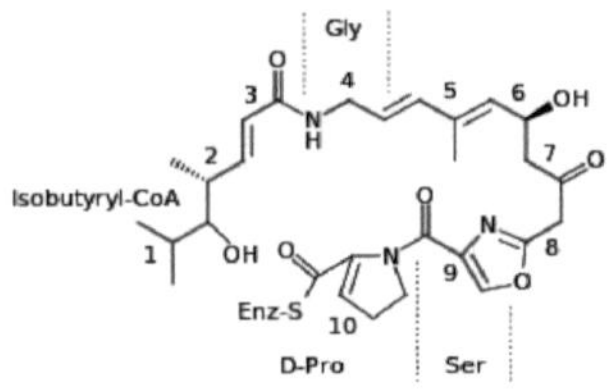	C-O lyase (type B streptogramins), acetylation (type A streptogramins), efflux, altered target
Oxazolidinones	Linezolid	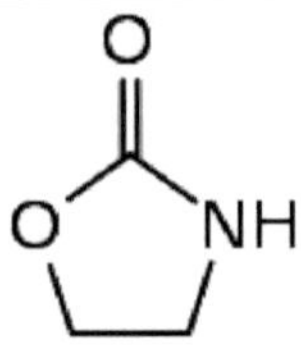	Efflux, altered target
Phenicols	Chloramphenicol	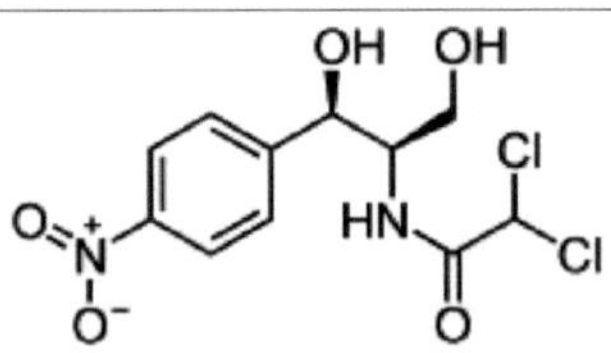	Acetylation, efflux, altered target
Quinolones	Ciprofloxacin	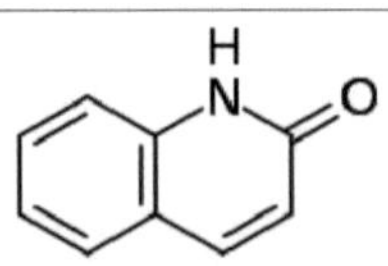	Acetylation, efflux, altered target
Pyrimidines	Trimethoprim	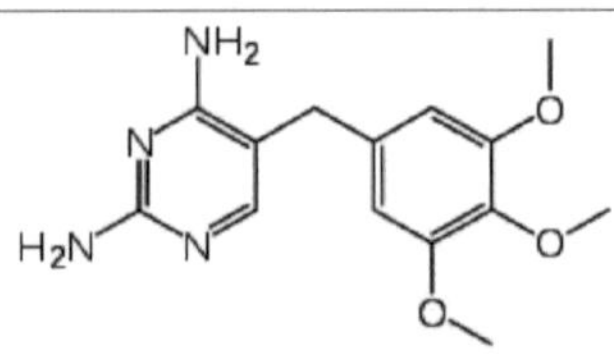	Efflux, altered target

Class	Example(s)	Functional group or whole molecule	Model(s) of resistance
Sulfonamides	Sulfamethoxazole	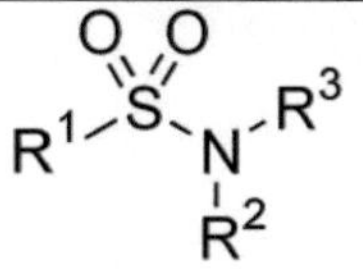	Efflux, altered target
Rifamycins	Rifampin	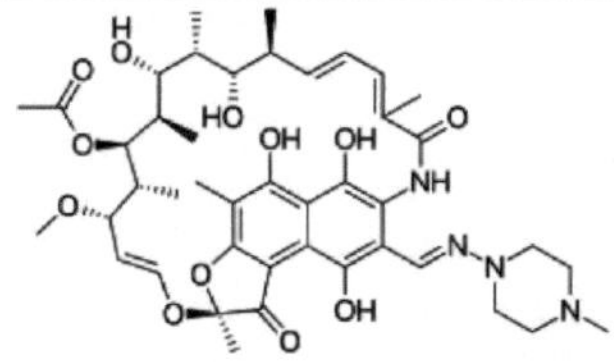	ADP-ribosylation, efflux, altered target
Lipopeptides	Daptomycin	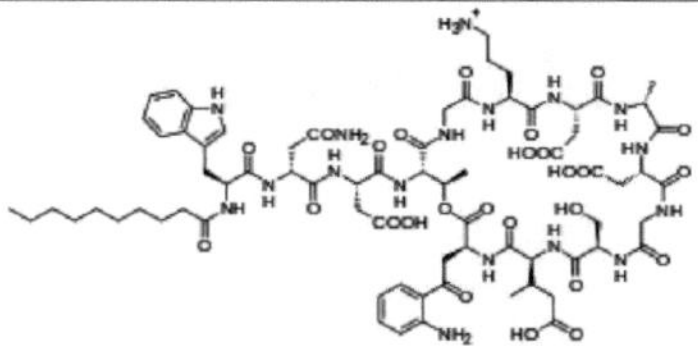	Altered target
Cationic	Colistin	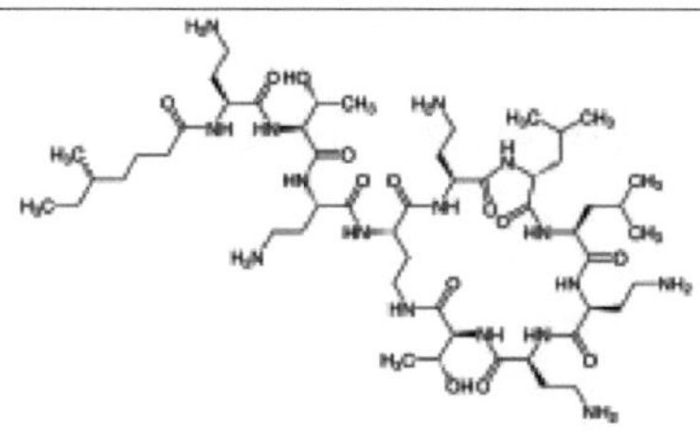	Altered target, efflux
Nystatin	Mycostatin	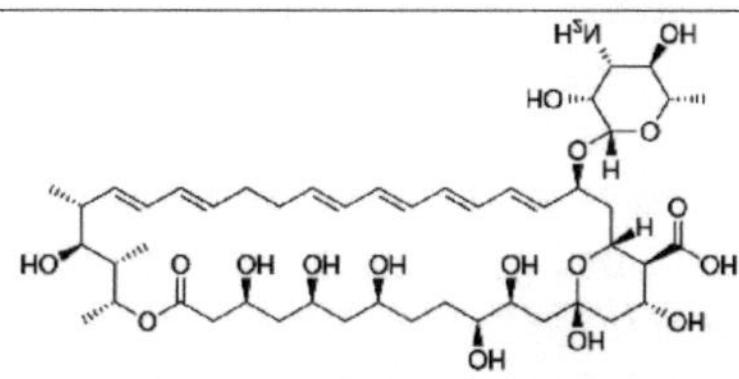	Ergosterol pathway Efflux

The most used pharmaceutical antibiotics, showing class, examples, functional group or whole molecule and models of resistance. Base on from the Gartiser et al. [21], Davies and Davies [22], Li & Zhang [23] and Zhi et al. [24].

Table 1.
Biodegradability and resistance.

Currently, animal production practices are linked with the routine use of antibiotics, and the selection pressure on bacteria is increasing the potential resistance [18]. For example, at the United States, the antibiotics use is broadly applied for animals and crops. This has been a controversial topic for over 30 years. A 1998 report from the prestigious Institute of Medicine of the National Academy of Sciences noted that about 4 million pounds (2000 tons) of antibiotics were used to treat sick farm animals and another 16 million pounds (8000 tons) were used as growth promotions (low doses of antibiotics usually included in animal feed) for animals every year [3]. Worldwide, in 2010, livestock consumed at least 63,200 tons of antibiotics, more than all human consumption [18].

Many investigations pointed out to de biodegradation of antibiotics, but some ones are biodegradable and other ones are accumulative in soil and water [21, 23, 24]. However, this concern is poorly studied today, for example Erythromycin, nystatin and sulfomethoxazole,

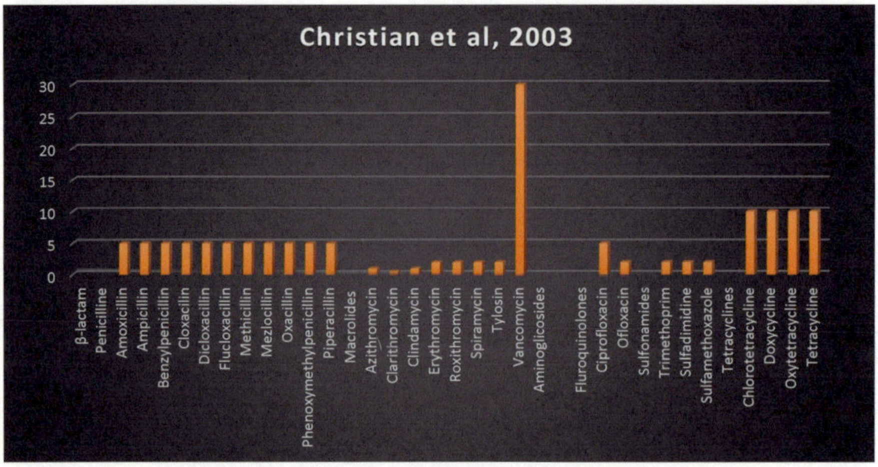

Figure 2.
Level of some antibiotic's detection. *The detection level of certain antibiotics in ng/L evaluated in soil by [19] work. The data from [29] report is not diagrammed, only penicillin and macrolides were in the highest concentration for the detection, which would leave [19]'s data almost at baseline.*

3. Level detection in water and soil and new antibiotics

Various studies have evaluated the presence of antibiotics in soil and water [21, 25–27], as it will see below. Some authors have measured bacterial activity, through the bacterial resistance causes by them or evaluating its concentration in fruits or leaves caused by plants translocation [25, 28]. Besides, others have used direct methods to assess their presence in the environment, using techniques such as HPLC, LCMS, or gas chromatography. As a reference work, we will mention the research of Christian et al., (2003), who evaluated many antibiotics, among which the following stand out 11 β-lactams, 8 macrolides, 3 Sulfonamides, 2 fluoroquinolones, and 4 tetracyclines.

A remarkable review of the last decade with data from the last century is the work of Thiele-Bruhn, [29], but Cristian et al. [19] publish laboratory data in the same year. However, [29], summarized chemistry data such as solubility, molecular weight, polarity or not, etc., as follows. We plotted the data from both European outstanding publish, see **Figure 2.**

4. New antibiotics synthetic molecules

In the race for the discovery of antibiotics some molecules paid a high price since they never reached the consumer, others on the contrary were the panacea to fight hundreds of pathogens [30]. However, the development of resistance from pathogens to antibiotics and their irresponsible use have led to the discouragement of production by pharmaceutical companies [31], while bioaccumulation and biodegradation are the new challenges added to new antibiotics [32]. In addition to fighting the same pathogens, new antibiotics shall aim to preserve soil and water quality and will prevent their accumulation and bioaccumulation. It is likely that the same active principle is used [33], perhaps better than synthesizing new molecules that are more complex to degrade. Anyway, new antibiotics must also fight against the same pathogens; nonetheless, they must also preserve soil and water quality, avoiding its accumulation and bioaccumulation, and perhaps that will be the hardest task, but it is worth it. "Even with more appropriate prescribing, it seems likely that

DOI: http://dx.doi.org/10.5772/intechopen.95368

antibacterial resistance will continue to accumulate in many pathogens and settings, especially in hospitals" [34].

5. Antibiotics in wastewater from hospital

As emphasized in **Figure 1**, some hospitals around the world have their wastewater treatment plant (WWTP). However, the vast majority in the world in both developed [35] and developing countries [36] or in poor countries [37], hospitals water goes directly into the sewers, which in many cases reach the rivers and the sea carrying a huge load of antibiotics. Even, the larger hospitals in many towns dumped antibiotics are into the sewers. Recently, Bansal [38] published an excellent review about hospital effluents, proposing finally bioremediation for control and lowing the multidrug-resistant bacteria, and author concludes with the sentence: "Antibiotic compounds have a suspicious reputation among the class of pollutants referred to as 'emerging contaminants' as the biological activity of antibiotics is an intrinsic characteristic of their functional design". Then actually, there are two concerns: first the resistant bacteria [39] and second the antibiotic pollutant. Consequently, both problems would be repaired by not throwing more antibiotics into the environment.

The Antibiotics' ecotoxicological risks over the effluent discharge on the aquatic ecosystem is the main concern. Aydin et al. [36] found azithromycin, clarithromycin, and ciprofloxacin as most abundant in WWTP with seasonal differences ranged from 21.2 ± 0.13 to 4886 ± 3.80 ng/L in summer and from 497 ± 3.66 to 322,735 ± 4.58 ng/L in winter analyzing WWTP from some hospitals in Turkey. The total antibiotic load to the influent in those WWTP was 3.46 g/day in summer and 303.2 g/day in winter.

Hocquet et al. [40] say that in France and worldwide a few countries demand or recommend previous treatment of hospital effluents before discharge into the main wastewater stream for treatment at municipal WWTP. There are well-known cases such as those of the city of Montreal related at the St. Lawrence River [41]. In Norway Langford & Thomas [42] study by LCMS two antibiotics (clotrimazole and other not specified) among 38 pharmaceutical compounds, showing that perhaps the use of antibiotics in that country has already begun to be discouraged or used with responsibly in hospitals, not being the case for most of the remaining compounds. Santos et al. [43] made a biggest elaborated work where they found in WWTP 4 antibiotics as follow: fluoroquinolone, macrolide, sulfamethoxazole, trimethoprim calculated by mg per day per 1000 inhabitant were ranged between 19 and 1337,24–53, 75–199, and n.d.–43 respectively. The total antibiotics calculated by mg per day per 1000 inhabitant was ranged between 174 and 1612 and other antibiotics were ranged between 2 and 67.

6. Antibiotics in drinking water, farm water, and aquifers

Perhaps one of the most worrisome issues with non-treatment of hospital water is that water with antibiotics can reach farms and/or water for human or animal consumption and even plant irrigation. In Kumasi, a big city in Ghana, the hospital wastewater and effluents from waste stabilization ponds, are directly discharged as low-quality water into nearby streams which are eventually used to irrigate vegetables. The presence of 12 commonly used antibiotics was detected: metronidazole, ciprofloxacin, erythromycin, trimethoprim, ampicillin, cefuroxime, sulfamethoxazole, amoxicillin, tetracycline, oxytetracycline, chlortetracycline, and doxycycline [44]. The authors reported 15 μg/L for ciprofloxacin in hospital wastewater, and after the antibiotic concentration in irrigation water was up from 0.2 μg/L to 200 ng/L; these quantities lead us to think or almost elucubrating a refined method to purify and reuse those antibiotic as the recycling urine to purify drink water at International Station at the space. Seriously, we imagine that astronauts did not consume antibiotics before (https://asgardia.space/en/news/Drugs-in-Space-What-Can-Get-an-Astronaut-High) [45]. Peculiar because everybody on the Earth consumes antibiotics without prescription, for instance in one preschool in Hong Kong, 31 children were detected veterinary antibiotics in urine, not human antibiotics. Norfloxacin and penicillin were the main antibiotics detected (48.4% and 35.5%, respectively), with median concentrations of 0.037 and 0.13 ng/mL, respectively [46]. In the same study enrofloxacin, penicillin, and erythromycin were the most detected veterinary antibiotics in raw and cooked food.

In many countries, farm irrigation water is already a vehicle for various antibiotics causing bioaccumulation in crop plants such as tomatoes and wheat crops [25, 47]. However, the bioaccumulation of antibiotics on tomatoes irrigated with wastewater is no consistent and depends on the high concentration of antibiotics on soil [25]. But researches reveal that hazardous quotidian (HQ) values show that the consumption of fruits harvested from tomato plants irrigated for long period with the wastewater applied for irrigation under field conditions in this study represents a low risk to human health [25]. On the other hand, in wheat plants that were spray-irrigated with wastewater treatment plant effluent, ofloxacin was found throughout the plant, with higher concentrations in the straw (10.2 ± 7.05 ng/g) and lower concentrations in the grain (2.28 ± 0.89 ng/g). Trimethoprim was found only on grain or straw surfaces, whereas carbamazepine and sulfamethoxazole were concentrated within the grain (1.88 ± 2.11 and 0.64 ± 0.37 ng/g, respectively). These findings demonstrate that pharmaceuticals and personal care products (PPCPs) can be taken up into wheat plants and adhere to plant surfaces when WWTP effluent is spray-irrigated [47].

Identical to what happened with pesticides, cyanide, and mercury, one of the biggest pollution problems today is that of aquifers. Wang et al. [48] evaluated the presence of antibiotics in groundwater up to 50 meters of aquifers, finding seasonal differences in spring, summer, and winter an average value of 1.60 μg/L, 0.772 μg/L, and 0.546 μg/L respectively. The predominant antibiotics were fluoroquinolones and tetracyclines, but the highest risk probably will be erythromycin for algae in surface waters and in deep waters, where ciprofloxacin would be the most concentrated among the 14 antibiotics.

7. Antibiotics in agricultural practices in farms

A disturbing issue today in farms and ranches as industry type is that in these places the handling of antibiotics is on a large scale even in higher quantities than in hospitals. There is overestimation since we mention before that in the United States the 80% of antibiotics are used for livestock consummation [18], which must be very similar in many countries in which these levels are allowed, or where there is not even a single rule about it. We previously mentioned the irrigation of tomato and wheat with water contaminated with antibiotics [25, 47], that is, in an indirect relationship to antibiotic management. We will now talk about the direct handling of antibiotics by the farms and what consequences this has brought to their soil and water. No, the issue is another, for example, many farms have their aqueducts or crop irrigation systems. There are very few studies on pesticide and antibiotic composition, and except for supported studies, where health and environment are the concerns and priority. From those studies is possible to collect valuable information. It is important to emphasize that some veterinary antibiotics have been handled as dietary supplements or as a supplement for spraying crops. This is the case of Chung et al. [49], who investigates veterinary antibiotics in soil experimentally contaminated by chlortetracycline, enrofloxacin, and sulphathiazole and translocated and bioaccumulates in roots and leaves of radish. They established that the concentration of chlortetracycline was lower than 2.73%, for enrofloxacin ranged 0.08–3.90%, and sulphathiazole lower than 1.64%. In another study in north China [50], the authors look at antibiotics in manure, soil, vegetables, and groundwater. This interesting work mentioned the complex antibiotic translocation from manure to the soil, establishing residual rate (RR) of antibiotics (mg/kg ha yr). The antibiotics used were sulfamethoxazole (SMZ), sulfadoxine (SDO), sulfachlorpyridazine (SCP) and chloramphenicol (CAP), oxytetracycline (OTC), tetracycline (TC), chlortetracycline (CTC), lincomycin (LIN), ofloxacin (OFL), ciprofloxacin (CIP). They conclude that with the application of manure containing antibiotics to organic vegetable bases, the residues of antibiotics in soil, vegetables, and groundwater were widely detected, mainly OTC, TC, CTC, and SCP with RR from 1,6 to 43% and the range of antibiotics in vegetable was 0.1-532 mg/kg using radish, rape, celery, coriander, which not all antibiotics were detectable. An interesting large study in northeastern China [51] using feces from chicken, pig, and dairy cow, 14 veterinary antibiotics such as tetracycline, oxytetracycline, chlortetracycline, ciprofloxacin, sulfaguanidine (SG), sulfanilamide(SA), sulfamethoxazole, sulfamonomethoxine (SMM), sulfamerazine (SMR), sulfachlorpyridazine (SCP), norfloxacin (NOR), enrofloxacin (ENR), difloxacin (DIF), and tylosin (TYL) were detected by HPLC, which six antibiotics (OTC, CTC, TC, TYL, SCP, and SMZ) were assessed using the hazard quotient (HQ). The 14 veterinary antibiotics detected in three types of animal feces the authors found that antibiotics occurred 7,41 to 57,41% inside 54 samples, and the levels ranged from 0,08 to 56.81 mg/kg. TCs were predominant with a maximum level of 56.81 mg/kg mostly detected in pig feces. SAs were common and detected with the highest concentration of 7.11 mg/kg. Fluoroquinolones were more widely detected in chicken feces rather than in pig or dairy cow manure, which contained the dominant antibiotic ENR [51].

Other interesting studies in Waseca and Staples, Minnesota USA, used chlortetracycline, monensin, sulfamethazine, tylosin, and virginiamycin evaluated in 11 vegetable crops in two different soils fertilized with raw versus composted turkey and hog manure or inorganic fertilizer [52], the authors conclude that all antibiotics in vegetable tissues were generally less than the limits of quantification were less than 10 μg/kg, and radish peel and spinach leaf were the vegetables that bioaccumulated more antibiotics than others ranging from 4 to 6 μg/kg. In other research, the authors evaluated the transfer of five different antibiotics such as tetracycline, sulfamethazine, norfloxacin, erythromycin, and chloramphenicol (CAP) evaluated in carrot, tomato, and lettuce under two levels of antibiotic-contaminated wastewater irrigation and animal manure fertilization [53]. TC, NOR, and CAP were accumulated at higher levels in the shoots/leaves than the other two compounds, except for CAP in carrot. Likewise, this work concludes that the levels of antibiotics are in acceptable daily intake (ADI), indicating that the main problem is not bioaccumulation in plants or animals, but bioaccumulation in multi-drug resistant bacteria related to environmental contamination.

As it was previously mentioned, Sulfonamides (SAs) are one of the most persistent antibiotics in soil and water, but especially bioaccumulate, and more than three tons of SM are used annually in Europe for swine production [54]. Li et al. [54] evaluated SAs in pakchoi cabbage such as sulfadiazine (SDZ), sulfamethazine (SM_2), and SMZ. The three antibiotics are spiked in soil could be taken up by pakchoi cabbage. SM_2 and SMZ were accumulated more easily by pakchoi cabbage than SDZ. The dissipation half-lives of SMZ (16.8 d) and SM_2 (16.7 d) in soil were significantly longer than SDZ (10.8 d).

In a study in an agricultural-livestock region of Japan, loads of veterinary antibiotic oxytetracycline (OTC) in stream waters were investigated [55]. OTC was detected in the concentration range of 2 ng/L to 68 μg/L. An increase in daily OTC loads was observed during the winter as a result of the common veterinary practice of using higher doses of antibiotics as prophylaxis in the prevention of winter diseases. Also, OTC concentrations were observed in stream water near one of the cattle farms. According to the authors, the daily OTC load rate was in a reasonable range to the number of animals, but it was not looked at the antibiotic loads in soil or the associated resistant bacteria.

As we mentioned in the last section the antibiotic accumulation is the second main concern after natural pathogen bacteria. Osterman et al. [56] elaborate one investigation about veterinary antibiotics leached in calcareous Chinese croplands. They found that in daily farm practice, sulfamethazine was constantly detected in the leachate up to 120 ng/L, and conclude that the overall substances were still detected in the soil after 53 days, suggesting that there was no overall rapid and complete dissipation, indicating that strong lixiviation under rainfall does not eliminate the totality of antibiotics and its incomplete dissipation increases risks of their accumulation in soil. Other Chinese work developed antibiotics detection in the agricultural soils from the Yangtze River Delta [57], among which five antibiotics were prevalent in 241 soil samples with a 100% detection rate and total concertation ranging from 4.55 to 2010 ng/g dry weight. The concentrations were as follow: quinolones with mean 48.8 ng/g, tetracyclines with mean 34.9 ng/g, sulfonamides with mean 2.35 ng/g, ciprofloxacin with mean 27.7 ng/g, and oxytetracyclines with a mean of 18.9 ng/g, being the most prevalent the two last ones.

Sarkar et al. [58] did an excellent review of antibiotics in the farm, livestock, aquaculture, and plants, and they concluded that the indiscriminate use of veterinary and human antibiotics not only created multidrug resistant human or animal bacteria, but plant resistant bacteria. Complex and jeopardy manipulation because the human being was transformed the overall bacteria species multidrug-resistant without suspecting. On the other hand, it is not to make the studies unsettling or reassuring that human or livestock antibiotics concentration is an acceptable level when we have to transform rapidly the species and filling the Earth with antibiotics. The most important now is to make clever decisions such as those discussed by Kuppusamy et al. [59]. They comment in their critical review that precisely the veterinary antibiotics are usually poorly sorbed in the animal gut, and the majority is excreted unchanged or as their recalcitrant metabolites in feces and urine.

8. Antibiotics in livestock, poultry, and swine

Antibiotics are very important in the prescription and treatment, to recover the health of the animals. However, there is no logical explanation for antibiotics as a supplementary feed, except to clarify that the animal biomass increases. There is no scientific explanation about the use and human health of the consumption of beef, chicken or pork fed with no antibiotic supplement. There is a valuable review in these topics especially antibiotics in livestock [58, 60, 61].

There are three main concerns about antibiotics in livestock, which could be in poultry and swine: 1- the high antibiotic concentration in treatment [62], 2- the high concentration in supplementary foods [63], and 3- the extra antibiotics pick out by livestock from water or grass [64]. And a final concern will be, what is the final concentration of antibiotics in the beef after cooking or treatment in the sausage or deli process. However, most of the antibiotics consumed by livestock are found in the feces [65], indicating poor assimilation, overuse, and uselessness, or little control over their management. If the farmer gives to the animal too much or little antibiotics, it will heal and fatten [61]. Therefore, the excessive cost of antibiotics and their great management on farms is almost unjustifiable, saving a lot of money in the decrease of management that could preserve the health of the troop and the environment by eliminating multi-resistant bacteria from its environment decontaminated [66].

There is no difference in antibiotics panorama in poultry and swine [67]. For example, Zheng et al. [68] determined the presence of 15 sulfonamides in livestock and poultry manure standardizing a new technique with UPLC-MS/MS plus Fe_3O_4-MWCNTs. The SAs recovered oscillated from 89 to 119% and the limit of detection of the method was reached 0,5 to 1 μg/kg dry weight, and the limit of quantification was between 1 and 3 μg/kg dry weight. Swine manure contains up to 12 antibiotics, was detected by Rasschaert et al. [69] work. The most frequent antibiotic detected were doxycycline, sulfadiazine, and lincomycin. Doxycycline was found in the highest concentration in manure with a mean of 1476 μg/kg from 8 to 13632 μg/kg, tylosin was found in manure with mean concentrations of 784 μg/kg from 17 to 5599 μg/kg, oxytetracycline was found in manure with mean concentrations of 482 μg/kg from 11 to 3865 μg/kg, lincomycin, was found in manure with mean concentrations of 177 μg/kg manure from 9 to 3154 μg/kg. The remained 18 antibiotics were found in manure with mean concentrations of less than 100 μg/kg.

9. Antibiotics in aquaculture or fish-farming

Aquaculture is another prominent fish and meat business and veterinary antibiotics are also used in aquaculture for the preservation of Salmonidae, and other large-scale cultured species. Currently, aquaculture almost reaches the same level of production as fishing, and likely it will exceed the production of world fisheries in the coming years (FAO) (http://www.fao.org/faostat/en/#data/CL). Chen et al. [70], investigate the bioaccumulation of ciprofloxacin and enrofloxacin in some tissues of culture grass carp such as plasma, bile, liver, and muscle tissues, They calculated the mean values of log bioaccumulation factors (Log BAFs) for these antibiotics and main results were for plasma in the range of 0,43-3,70, for bile in the range 0,36-4,75, for liver in the range − 0,31-4,48, and for muscle in the range 0,23-4,33. This work also calculated hazard quotients, human health risk evaluation, however, the authors state that those grass carp do not pose risk for human health and clarifying that very little has been published in this regard.

Sankar et al. [58] reviewed other previous articles reporting that Kurunasagar et al. [59], published the use and application of cotrimoxazole, chloramphenicol, streptomycin, erythromycin being already a habit in fish aquaculture in India. But as declared Burridge et al. [71], "the use and potential effects of these compounds are herein summarized for the four major salmon producing nations: Norway, Chile, UK and Canada" and state that around 75 percent of the antibiotics fed to fish are excreted into the water. On the other hand, Romero et al. [72], looked the effect of oxytetracycline by HSP70:GFP expression in fish larvae after 48 hours, and how oxytetracycline-triggered the stress and the immune response. In consequence, they believe that is possible new alternative practices for the prevention in aquaculture due to the use, overuse, or abuse of antibiotics, which promote the contamination of the environment and rise resistant bacteria. They consider alternative solutions such as developing strict regulations controlling the use of antibiotics and having led to only a few antibiotics being licensed for use in aquaculture and preventing the high proportions of antibiotic-resistant bacteria which still persist in sediments and farm surroundings and suggesting the implementation of rearing practices that reduce the level of stress from fish larvae to adult, which could reduce the likelihood of infections requiring antibiotic treatment.

10. Antibiotics in soil

Probably one of the main concerns about antibiotic contamination is their accumulation in soil, by the effect on resistant bacteria and direct consequence in human and animal health. Likely the easier indirect antibiotic detection is to calculate the concentration in crops [73, 74], as was mentioned by Christou et al. [25] reporting antibiotics in tomatoes from the soil. This work elaborated during three years shows that the accumulation of DCF, SMX, and TMP in the soil detected by MeOH-NaOH solution and Chromatography MS/MS, The results show an imminent physical translocation as we shown in **Figure 1**, and show great variability in the accumulation of the three antibiotics during those 3 years, with increases or decreases depending on the irrigation using wastewater and depending on the duration of irrigation and the origin of the wastewater applied. Other authors monitoring sulfonamides and tetracyclines over 18 years from an agricultural field site in Lower Saxony, Germany [75]. SAs and TCs are largely fixed in the upper soil layer. The analysis showed a strong decrease below the sampling depth of 30 cm which is the depth of the plow and below other antibiotics such as SG and SMZ were detected up to 90 cm, TC was shown to persist >100 μg/kg in topsoil, however this antibiotic no leachate in deeper soil segments or groundwater. They conclude that some SAs disappeared, but other veterinary antibiotics are even more persistent than expected.

In Kenia, Africa Yang et al. [76] analyzed four soils in the suburban area of Narok. Over 12 antibiotics analyzed they found that SMX, SMZ, OTC, and ENR were the major antibiotics that polluted those soils among 12 antibiotics monitored and the highest average value was for Narok 43,64 μg/kg dry weight, and the other three average values were Mai Mahiu 26.70 μg/kg dry weight, Juja 24.41 μg/kg dry weight, and Mount Suswa Conservancy 12,21 μg/kg dry weight. They advise more attention to reduce the misuse of SAs. Pan and Chu [27], investigated TC, NOR, SM2, CAP, Erythromycin (EM), antibiotics in agricultural soils, and was probably the first study. The Persistence and environmental risk profile of the five antibiotics were TC > NOR>EM > CAP>SM2, but "the study suggests that the adsorption of antibiotics in soil depends on the physicochemical properties of both antibiotics and soil". Probably this study was the first to develop a model for predict antibiotic persistence in soil.

11. Antibiotics in crops plants

Antibiotics for plants were used since the 1950s to control certain bacterial diseases the most common used are oxytetracycline and streptomycin. In the United State antibiotics for plants represent 0,5% of total antibiotics used in this country. However, the emergence of antibiotic resistance of *Xanthomonas campestris*, *Erwinia amylovora*, and *Pseudomonas* spp., and has blocked the control of their diseases [77]. Then plants could pick up three kinds of antibiotics: antibiotics plants, human antibiotics [44], and veterinary antibiotics [78].

Some, but not many countries have rules for quality plant irrigation, but that concern is complex because WWPTs have some low load of antibiotics. The other side of the problem is irrigating plants directly with WW [25, 28]. As mentioned above, different antibiotics were detected in crops plant [58, 79, 80], those research emphasize the low risk for human health as daily intake, forgetting the drug charge in the environment and the resistant bacteria even now for control pathogen of the plant.

12. Antibiotics in rivers and sea

Finally, we arrive at the climax where all the antibiotics will finally be deposited if their degradation does not occur or if the increase of antibiotics continues its course as if such substances do not represent a risk for all the species in nature that has learned to fight diseases with evolution as antibiotic resistance [81]. Rivers and the sea are the final deposits of antibiotics and or following studies and many others that we cannot all cite shown that biodegradation, degradation, and bioaccumulation can once again put the health of the planet at risk.

The previous data show that antibiotics degradation is not totally in agreement with antibiotics in soils, farms, and WWTPs. We know today the antibiotic loads in pristine rivers as Amazona, Niger, Congo, Zambezi, Lena, Amour, and Yenisei carrying low concentrations of antibiotics, but the remaining rivers of the worldwide receive greater loads of antibiotics [41, 73, 82]. We could again make the list of antibiotics and see that they follow the same comparison of antibiotics and all effluents from WWTPs [81], the main technical solution will be the degradation of antibiotics before reach the river stream. About this concern, some researchers begin to develop this challenge [48, 83, 84].

As Zang et al., [85], mentioned," the ocean is an important sink of land-based pollutants", and exactly as crop plants allow detection of antibiotics, marine food will be the best way to follows the study of bioaccumulation of antibiotics [86, 87]. Liu et al. [86], found 9 SAs, TMP, 5 FQ, and 4 ML, which were in gill, muscle, kidney, and liver tissues of seven wild fish species collected from Laizhou Bay, North China. As previous famous works with mercury accumulation of marine fish, now there are antibiotics reports in tissue fish in higher concentrations than aquaculture fish.

Alga will be another antibiotic "*bio-accumulator*" to detect antibiotics in rivers and sea. Seoane et al. [88] developed an experimental evaluation of chloramphenicol (CHL), florfenicol (FLO), and oxytetracycline (OTC) in marine microalga *Tetraselmis suecica*, showing that three antibiotics inhibited the growth after 96 h with IC_{50}. Finally, we think that in the future it will be necessary to look for antibiotics in river and sea sediments. But these first works are already beginning to fill the research data worldwide and to conclude without doubts that antibiotics are the "new emergent pollutants of Earth". Pioneer work from Fernandes et al., [89] in Portugal found azithromycin in Leça river at 2819 ng/g in water but in sediments at 43,2 ng/g.

13. Strategies to change all previous practices, challenges, and hopes

Table 2 is not a summary of this review of all articles. It is a simple exercise to verify that perhaps no antibiotic has disappeared from industrial production and it generates a contradictory taste associated with the fact that pharmaceutical companies are discouraged from manufacturing new antibiotics, but the sale and

Type	Antibiotic	References
Agricultural soil		
Tetracyclines	tetracycline, doxycycline, oxytetracycline	[27]
Fluoroquinolones	ofloxacin, ciprofloxacin, norfloxacin, enrofloxacin	[57]
Sulfonamides	sulfadimethoxine, sulfamethazine, sulfamethizole, sulfamerazine, sulfamethoxazole, sulfadiazine	[74]
Macrolides	erythromycin	[76]
Amphenicols	chloramphenicol	[79]
Aquaculture		
Sulfonamides	sulfamethoxazole, sulfaquinoxaline	[58]
Tetracyclines	oxytetracycline	[70]
Fluoroquinolones	norfloxacin, ofloxacin, ciprofloxacin, enrofloxacin, pefloxacin	[72]
Macrolides	*erythromycin*-H_2O	
Amphenicols	chloramphenicol, florphenicol,	
Aminoglycosides	streptomycin	
Combination of trimethoprim and sulfamethoxazole	cotrimoxazole	
Sulfonamide/Diaminopyridine	sulfadimethoxine/ormetoprim	
Crops		
Tetracyclines	Tetracycline, oxytetracycline, chlortetracycline, doxycycline	[28]
Sulfonamides	sulfamethoxazole	[58]
Diaminopyrimidines	trimethoprim	[74]
Macrolides	erythromycin	[79]
Azoles	metronidazole	[80]
Cephalosporins	Cefuroxime, second generation	
β-lactams	ampicillin, amoxicillin	
Fluoroquinolones	ciprofloxacin (CIP); norfloxacin (NOR)	
Aminoglycosides	blasticidin S, validamycin	
antibiotics of fungal origin	strobilurin, odemansins.	

Type	Antibiotic	References
In agricultural practices in farms		
Azoles	metronidazole	[25]
Macrolides	erythromycin, norfloxacin, tylosin, spiramycin	[44]
Sulfonamides	sulfamethoxazole, Sulfadiazine: sulfamethazine, sulfamethoxazol sulphathiazole sulfaguanidine, sulfanilamide, sulfamonomethoxine, sulfamerazine, sulfachlorpyridazine, sulphadimethoxine, sulphadimidine, sulfapyridine	[49–54] [60]
Diaminopyrimidines	trimethoprim	[78]
Tetracyclines	tetracycline, oxytetracycline, chlortetracycline, doxycycline	
Cephalosporins (second generation)	cefuroxime	
β-lactams	ampicillin, amoxicillin, benzylpenicillin, cloxacilin, dicloxacillin, oxacillin	
Amphenicols	chloramphenicol	
Lincosamides	lincomycin	
Quinolones	ofloxacin, pefloxacin, ciprofloxacin, enrofloxacin, difloxacin, marbofloxacin	
Aminoglycosides	Streptomycin, apramycin, kanamycin, spectinomycin	
Peptides	virginiamycin	
Ionophores	monensin	
Peptidomimetics	bacitracin	
Diaminopyrimidines	trimethoprim	
Cephalosporins	ceftiofur, cefquinom	
Hospital		
β-lactams	ampicillin amoxicillin	[39]
Sulfonamides	streptomycin; sulfamethoxazole	[44]
Tetracyclines	tetracycline, oxytetracycline chlortetracycline and doxycycline	
Aminoglycosides	streptomycin	
Fluoroquinolones	ciprofloxacin;	
Cephalosporins	cefotaxime;	
Cephamycin (second-generation)	cefoxitin	
Azoles	metronidazole	
Macrolides	erythromycin	
Diaminopyrimidines	trimethoprim	
Cephalosporins (second generation)	cefuroxime	

Type	Antibiotic	References
Livestock, poultry and swine		
Tetracyclines:	Tetracycline, oxytetracycline, chlortetracycline doxycycline	[51]
Sulfonamide Aminoglycoside:	sulfamethazine neomycin	[56]
Ionophore:	monensin	[67]
Macrolide:	tylosin	[69]
β-lactams	amoxicillin	[78]
Cephalosporines	ceftiofur	[90]
Polypeptide	colistin	[91]
Diaminopyrimidines	trimethoprim	
Lincosamides	lincomycin	
Sulfonamides	sulfachlorpyridazine, sulfadiazine, sulfadimethoxine, sulfamerazine, sulfamethazine, sulfamethoxazole, sulfamethoxypyridazine, sulfapyridine, sulfathiazole, sulfaguanidine, sulfanilamide, sulfamonomethoxine, sulfachlorpyridazine	
Quinolones	norfloxacin, ciprofloxacin, enrofloxacin, difloxacin	
Rivers		
β-lactams	amoxicillin	[73]
Sulfonamides	sulfamethoxazole	[79]
Tetracyclines	tetracycline oxytetracycline	[81]
Quinolones	nalidixic acid, fluoroquinolones	
Macrolides		
Diaminopyrimidines	trimethoprime	
Sea		
Sulfonamides		[86]
fluoroquinolones		[88]
Macrolides		
Diaminopyrimidines	trimethoprim	
Amphenicols	chloramphenicol, florphenicol	

DOI: http://dx.doi.org/10.5772/intechopen.95368

Type	Antibiotic	References
Wastewater treatment plants (WWTPs)		
Quinolones	ofloxacin, ciprofloxacin	[25]
Sulfonamides	sulfamethoxazol	[44]
Diaminopyrimidines	trimethoprim	[47]
Tetracyclines	tetracycline, oxytetracycline, chlortetracycline, doxycycline	
β-lactams	ampicillin, amoxicillin	
Cephalosporins (second generation)	cefuroxime	
Macrolides	erythromycin	
Azoles	metronidazole	
Veterinary and human antibiotics		
B-lactams	emoxycilin	[58]
Macrolides	tylosin, azithromycin	
Aminoglycosides	neomycin, gentamycin	
Fluoroquinolones	enrofloxacin, ciprofloxacin, levofloxacin, ofloxacin	
Tetracyclines	oxytetracycline, chlortetracycline	
Streptogramins	virginiamycin	
Glycopeptide	avoparcin	
Phenicol	florphenicol	
Cephalosporins	cephalexin	
Polypeptides	enramycin	
Pleuromultilin	tiamulin	
Foods		
Fluoroquinolones	Enrofloxacin,	[46]
β-lactams	penicillin	[92]
Macrolides	Erythromycin, tilmicosin	

This table is not a summary of the articles reviewed, and the goal was compared to all human practices with antibiotics used, which it is clear that antibiotics production and release in nature was risen in the last years and the discouragement to discover new antibiotics are not in a relationship with production.

Table 2.
Some selected antibiotics in the practices were reviewed

large scale production is not a compromise. This alone is one of the most profitable businesses in the pharmaceutical industry with a high cost of sacrifice from nature to our food. The bioaccumulation and accumulation of antibiotics in all our activities is already a risk rather than a fact. We did not do the review of antibiotics in food, but several authors who have reviewed other practices independent of human health have realized that there will be no health if the whole environment is contaminated to our food. We did not do the review of antibiotics in food, but several authors who have reviewed other practices independent of human health have realized that there will be no health if the whole environment is contaminated to our food.

A prime candidate as a bioaccumulator of antibiotics and other contaminants is milk and is perhaps one of the best-monitored foods in the world. In a pioneer study in the north of Italy Chiesa et al. [93] developed the detection by LC-HRMS veterinary antibiotics in milk. The researcher detected Lincomycin 30 samples (17,29 ppm the mean) from 254 raw milk samples and oxytetracycline in three samples and two samples with cefapirin and spiramycin. Part per million is the of an antibiotic is a negligible amount when absorbed by humans, however, the remaining concern

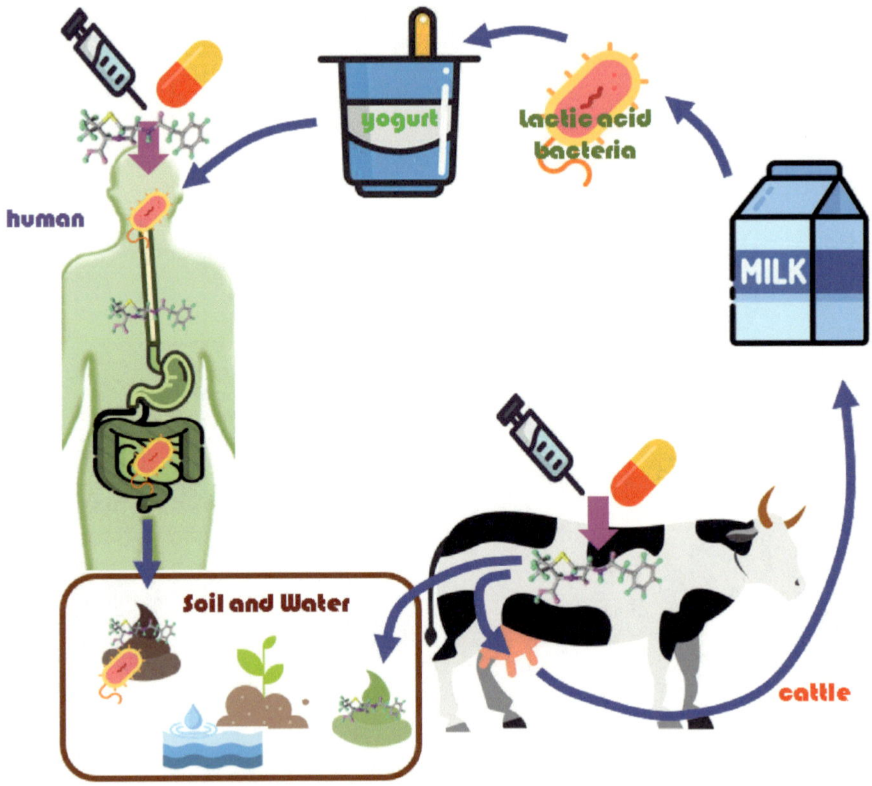

Figure 3.
Practice and management of antibiotics, milk, and lactic acid bacteria. *The practice of antibiotics (penicillin molecule example on diagram) treatment of mastitis in cattle has generated problems for the milk elaboration and consumer. Antibiotics were found in the milk in enough concentrations to inhibit dairy starter microorganisms and cause economic losses to the cheese and fermented milk industries from the sixties years [94]. Today this practice is monitored in many farms and countries, however, is not extensively applied in every country or farm, therefore, dairy industries are obliged and encouraged by national health and food institutions to monitoring antibiotics levels in milk. Other interesting studies with lactic acid bacteria [95], show that bacteria had developed important levels of antibiotic resistance. Confinement of cattle and water treatment will prevent new emergent antibiotic-resistant bacteria in soil and water and that will be the first step to lower the level of pharmaceutical antibiotics in nature. Programs for monitoring antibiotic-resistant wild bacteria must rise in each country in the next years.*

is whether the trace amount is a stimulator of resistance in lactic acid bacteria. An excellent measure will be to continue monitoring if lactic acid bacteria in milk are developing resistance to antibiotics. Assimilating antibiotic-resistant bacteria from our microbiome may have future consequences that we cannot yet measure. Therefore, the work of Chiesa et al., [93] serves us to predict the possible exposure of lactic acid bacteria to veterinary antibiotics. See **Figure 3**.

14. Conclusions

Perhaps we never imagined that after successful treatment with antibiotics that allowed us to leave the hospital, our true and lasting health was associated with them. The hospital's water will end up watering our food, which will end up on our plates as we have already seen to the satiety. The institutes and ministries of health, agriculture, and the environment, now have the same role and associated with the ministries of economy, education. If we do not realize the damage caused, will continue to save publications to demonstrate that there is no risk with antibiotics, and hospitals and farms will never change their irrational practices of consuming antibiotics as food.

Antibiotics monitoring will be a challenge in the future, and elimination will be a sustainable economic future. We cannot advise: "stopping the use of antibiotics for human or animal disease treatment", never, we do not have a better option. But we are sure that the rational application of good practices goes on to help the rapid elimination or accumulation of antibiotics in every environment as soil or sea. Biodegradation and bioremediation are an excellent opportunity. Firstly, antibiotics are metabolized rapidly by millions of microorganism, while as long as the antibiotic load in nature is reasonable, in other words, an antibiotic load low or close to the natural concentration; secondly the chemical, microbiological and mechanical techniques to eliminate antibiotics in nature will contribute to decontaminate the environments and to lower their indiscriminate use.

Several authors are assuring the risk if we do not stop [96, 97]. New alternatives and innovations will be promising new research since demonstrating a low risk of bioaccumulation will not eliminate practices that have led to the filling of soil and water in the rivers and sea with antibiotics. Many of us believe that detection techniques and new practices will change the future, such as those suggested by Behera et al. [98]. Decreasing antibiotic use will be the only way to decontaminate soil and water and reduce antibiotic-resistant bacteria.

Acknowledgements

We thank the University of Antioquia for their financial support to the project CODI 2017-15753.

Author details

Mauricio Corredor and Amalia Muñoz-Gómez*
Gebiomic Group, Faculty of Exact and Natural Sciences, University of Antioquia, Medellín, Colombia

*Address all correspondence to: mauricio.corredor@udea.edu.co

References

[1] Davies, J. (2006). Are antibiotics naturally antibiotics? Journal of Industrial Microbiology and Biotechnology, 33(7), 496-499.

[2] Martínez, J. L. (2008). Antibiotics and antibiotic resistance genes in natural environments. Science, *321*(5887), 365-367.

[3] Shlaes, D. M. (2010). Resistance Chapter. The Perfect Storm. In *Antibiotics* (pp. 16). Springer, Dordrecht.

[4] Aminov, R. I. (2009). The role of antibiotics and antibiotic resistance in nature. Environmental Microbiology, *11*(12), 2970-2988.

[5] Rubin, R. P. (2007). A brief history of great discoveries in pharmacology: in celebration of the centennial anniversary of the founding of the American Society of Pharmacology and Experimental Therapeutics. Pharmacological Reviews, 59(4), 289-359.

[6] Cromwell, G. L. (2002). Why and how antibiotics are used in swine production. Animal Biotechnology, 13(1), 7-27.

[7] Katakweba, A. A. S., Mtambo, M. M. A., Olsen, J. E., & Muhairwa, A. P. (2012). Awareness of human health risks associated with the use of antibiotics among livestock keepers and factors that contribute to selection of antibiotic resistance bacteria within livestock in Tanzania. Livestock Research for Rural Development, 24(10), 170.

[8] Manikandan, M., Chun, S., Kazibwe, Z., Gopal, J., Singh, U. B., & Oh, J. W. (2020). Phenomenal bombardment of antibiotic in poultry: Contemplating the environmental repercussions. International Journal of Environmental Research and Public Health, 17(14), 5053.

[9] Ramessar, K., Peremarti, A., Gómez-Galera, S., Naqvi, S., Moralejo, M., Munoz, P., ... & Christou, P. (2007). Biosafety and risk assessment framework for selectable marker genes in transgenic crop plants: A case of the science not supporting the politics. Transgenic Research, 16(3), 261-280.

[10] Díaz-Quiroz, C. A., Hernandez-Chavez, J. F., Ulloa-Mercado, G., Gortáres-Moroyoqui, P., Martínez-Macías, R., Meza-Escalante, E., & Serrano-Palacios, D. (2018). Simultaneous quantification of antibiotics in wastewater from pig farms by capillary electrophoresis. Journal of Chromatography B, 1092, 386-393.

[11] Johnson, A. C., Keller, V., Dumont, E., & Sumpter, J. P. (2015). Assessing the concentrations and risks of toxicity from the antibiotics ciprofloxacin, sulfamethoxazole, trimethoprim and erythromycin in European rivers. Science of the Total Environment, 511, 747-755.

[12] Lu, J., Wu, J., Zhang, C., Zhang, Y., Lin, Y., & Luo, Y. (2018). Occurrence, distribution, and ecological-health risks of selected antibiotics in coastal waters along the coastline of China. Science of the Total Environment, 644, 1469-1476.

[13] Paulus, G. K., Hornstra, L. M., Alygizakis, N., Slobodnik, J., Thomaidis, N., & Medema, G. (2019). The impact of on-site hospital wastewater treatment on the downstream communal wastewater system in terms of antibiotics and antibiotic resistance genes. International Journal of Hygiene and Environmental Health, 222(4), 635-644.

[14] Davison, J., & Ammann, K. (2017). New GMO regulations for old:

Determining a new future for EU crop biotechnology. GM crops & food, 8(1), 13-34.

[15] Pedro, J. S. (2017). Gene Editing. Time to Reflection. *Rev. Bioetica & Derecho*, 40, 157.

[16] Witte, W. (2000). Selective pressure by antibiotic use in livestock. International Journal of Antimicrobial Agents, *16*, 19-24

[17] Bbosa, G. S., Mwebaza, N., Odda, J., Kyegombe, D. B., & Ntale, M. (2014). Antibiotics/antibacterial drug use, their marketing and promotion during the post-antibiotic golden age and their role in emergence of bacterial resistance. Health, 2014.

[18] Van Boeckel, T. P., Brower, C., Gilbert, M., Grenfell, B. T., Levin, S. A., Robinson, T. P., ... & Laxminarayan, R. (2015). Global trends in antimicrobial use in food animals. Proceedings of the National Academy of Sciences, *112*(18), 5649-5654.

[19] Christian, T., Schneider, R. J., Färber, H. A., Skutlarek, D., Meyer, M. T., & Goldbach, H. E. (2003). Determination of antibiotic residues in manure, soil, and surface waters. Acta Hydrochimica et Hydrobiologica, 31(1), 36-44.

[20] Shlaes, D. M. (2010b). The Industry Chapter. The Perfect Storm. In *Antibiotics* (pp. 58). Springer, Dordrecht.

[21] Gartiser, S., Urich, E., Alexy, R., & Kümmerer, K. (2007). Anaerobic inhibition and biodegradation of antibiotics in ISO test schemes. Chemosphere, 66(10), 1839-1848.

[22] Davies, J., & Davies, D. (2010). Origins and evolution of antibiotic resistance. Microbiology and Molecular Biology Reviews, 74(3), 417-433.

[23] Li, B., & Zhang, T. (2010). Biodegradation and adsorption of antibiotics in the activated sludge process. Environmental Science & Technology, 44(9), 3468-3473.

[24] Zhi, D., Yang, D., Zheng, Y., Yang, Y., He, Y., Luo, L., & Zhou, Y. (2019). Current progress in the adsorption, transport and biodegradation of antibiotics in soil. Journal of Environmental Management, 251, 109598.

[25] Christou, A., Karaolia, P., Hapeshi, E., Michael, C., & Fatta-Kassinos, D. (2017). Long-term wastewater irrigation of vegetables in real agricultural systems: Concentration of pharmaceuticals in soil, uptake and bioaccumulation in tomato fruits and human health risk assessment. Water Research, 109, 24-34.

[26] Marti, R., Scott, A., Tien, Y. C., Murray, R., Sabourin, L., Zhang, Y., & Topp, E. (2013). Impact of manure fertilization on the abundance of antibiotic-resistant bacteria and frequency of detection of antibiotic resistance genes in soil and on vegetables at harvest. Applied and Environmental Microbiology, 79(18), 5701-5709.

[27] Pan, M., & Chu, L. M. (2016). Adsorption and degradation of five selected antibiotics in agricultural soil. Science of the Total Environment, 545, 48-56.

[28] Azanu, D., Mortey, C., Darko, G., Weisser, J. J., Styrishave, B., & Abaidoo, R. C. (2016). Uptake of antibiotics from irrigation water by plants. Chemosphere, 157, 107-114.

[29] Thiele-Bruhn, S. (2003). Pharmaceutical antibiotic compounds in soils–a review. Journal of Plant Nutrition and Soil Science, 166(2), 145-167.

[30] Chen, W. (2019). Rational Use of Antibiotics under the Context of China's Health System Reform.

[31] Alanis, A. J. (2005). Resistance to antibiotics: Are we in the post-antibiotic era? Archives of Medical Research, 36(6), 697-705.

[32] Puckowski, A., Mioduszewska, K., Łukaszewicz, P., Borecka, M., Caban, M., Maszkowska, J., & Stepnowski, P. (2016). Bioaccumulation and analytics of pharmaceutical residues in the environment: A review. Journal of Pharmaceutical and Biomedical Analysis, 127, 232-255.

[33] Clardy, J., Fischbach, M. A., & Walsh, C. T. (2006). New antibiotics from bacterial natural products. Nature Biotechnology, 24(12), 1541-1550.

[34] Livermore, D. M. (2004). The need for new antibiotics. Clinical Microbiology and Infection, 10, 1-9.

[35] Pena, A., Paulo, M., Silva, L. J. G., Seifrtová, M., Lino, C. M., & Solich, P. (2010). Tetracycline antibiotics in hospital and municipal wastewaters: A pilot study in Portugal. Analytical and Bioanalytical Chemistry, 396(8), 2929-2936.

[36] Aydin, S., Aydin, M. E., Ulvi, A., & Kilic, H. (2019). Antibiotics in hospital effluents: Occurrence, contribution to urban wastewater, removal in a wastewater treatment plant, and environmental risk assessment. Environmental Science and Pollution Research, 26(1), 544-558.

[37] Laffite, A., Kilunga, P. I., Kayembe, J. M., Devarajan, N., Mulaji, C. K., Giuliani, G., ... & Pote, J. (2016). Hospital effluents are one of several sources of metal, antibiotic resistance genes, and bacterial markers disseminated in sub-Saharan urban rivers. Frontiers in Microbiology, *7*, 1128.

[38] Bansal, O. P. (2019). Antibiotics in hospital effluents and their impact on the antibiotics resistant bacteria and remediation of the antibiotics: A review. Network Pharmacology, 4(3-4), 6-30.

[39] Galvin, S., Boyle, F., Hickey, P., Vellinga, A., Morris, D., & Cormican, M. (2010). Enumeration and characterization of antimicrobial-resistant Escherichia coli bacteria in effluent from municipal, hospital, and secondary treatment facility sources. Applied and Environmental Microbiology, *76*(14), 4772-4779.

[40] Hocquet, D., Muller, A., & Bertrand, X. (2016). What happens in hospitals does not stay in hospitals: Antibiotic-resistant bacteria in hospital wastewater systems. Journal of Hospital Infection, 93(4), 395-402.

[41] Marcogliese, D. J., Blaise, C., Cyr, D., de Lafontaine, Y., Fournier, M., Gagné, F., ... & Hudon, C. (2015). Effects of a major municipal effluent on the St. Lawrence River: A case study. Ambio, 44(4), 257-274.

[42] Langford, K. H., & Thomas, K. V. (2009). Determination of pharmaceutical compounds in hospital effluents and their contribution to wastewater treatment works. Environment International, 35(5), 766-770.

[43] Santos, L. H., Gros, M., Rodriguez-Mozaz, S., Delerue-Matos, C., Pena, A., Barceló, D., & Montenegro, M. C. B. (2013). Contribution of hospital effluents to the load of pharmaceuticals in urban wastewaters: Identification of ecologically relevant pharmaceuticals. Science of the Total Environment, 461, 302-316.

[44] Azanu, D., Styrishave, B., Darko, G., Weisser, J. J., & Abaidoo, R. C. (2018). Occurrence and risk assessment of antibiotics in water and lettuce in Ghana. Science of the Total Environment, 622, 293-305.

[45] Prescott, J. F. (1997). Antibiotics: Miracle drugs or pig food?. The Canadian Veterinary Journal, *38*(12), 763.

[46] Li, N., Ho, K. W., Ying, G. G., & Deng, W. J. (2017). Veterinary antibiotics in food, drinking water, and the urine of preschool children in Hong Kong. Environment International, 108, 246-252.

[47] Franklin, A. M., Williams, C. F., Andrews, D. M., Woodward, E. E., & Watson, J. E. (2016). Uptake of three antibiotics and an antiepileptic drug by wheat crops spray irrigated with wastewater treatment plant effluent. Journal of Environmental Quality, 45(2), 546-554.

[48] Wang, C., Yin, L., Xu, Z., Niu, J., & Hou, L. A. (2017). Electrochemical degradation of enrofloxacin by lead dioxide anode: Kinetics, mechanism and toxicity evaluation. Chemical Engineering Journal, 326, 911-920.

[49] Chung, H. S., Lee, Y. J., Rahman, M. M., Abd El-Aty, A. M., Lee, H. S., Kabir, M. H., ... & Nahar, N. (2017). Uptake of the veterinary antibiotics chlortetracycline, enrofloxacin, and sulphathiazole from soil by radish. Science of the Total Environment, 605, 322-331.

[50] Hu, X., Zhou, Q., & Luo, Y. (2010). Occurrence and source analysis of typical veterinary antibiotics in manure, soil, vegetables and groundwater from organic vegetable bases, northern China. Environmental Pollution, 158(9), 2992-2998.

[51] Li, Y. X., Zhang, X. L., Li, W., Lu, X. F., Liu, B., & Wang, J. (2013a). The residues and environmental risks of multiple veterinary antibiotics in animal faeces. Environmental Monitoring and Assessment, 185(3), 2211-2220.

[52] Kang, D. H., Gupta, S., Rosen, C., Fritz, V., Singh, A., Chander, Y., ... & Rohwer, C. (2013). Antibiotic uptake by vegetable crops from manure-applied soils. Journal of Agricultural and Food Chemistry, 61(42), 9992-10001.

[53] Pan, M., & Chu, L. M. (2017). Transfer of antibiotics from wastewater or animal manure to soil and edible crops. Environmental Pollution, 231, 829-836.

[54] Li, X., Yu, H., Xu, S., & Hua, R. (2013b). Uptake of three sulfonamides from contaminated soil by pakchoi cabbage. Ecotoxicology and Environmental Safety, 92, 297-302.

[55] Matsui, Y., Ozu, T., Inoue, T., & Matsushita, T. (2008). Occurrence of a veterinary antibiotic in streams in a small catchment area with livestock farms. Desalination, 226(1-3), 215-221.

[56] Ostermann, A., Siemens, J., Welp, G., Xue, Q., Lin, X., Liu, X., & Amelung, W. (2013). Leaching of veterinary antibiotics in calcareous Chinese croplands. Chemosphere, 91(7), 928-934.

[57] Sun, J., Zeng, Q., Tsang, D. C., Zhu, L. Z., & Li, X. D. (2017). Antibiotics in the agricultural soils from the Yangtze River Delta, China. Chemosphere, 189, 301-308.

[58] Sarkar, D. J., Mukherjee, I., Shakil, N. A., Rana, V. S., Kaushik, P., & Debnath, S. (2018). Antibiotics in agriculture: Use and impact. Indian Journal of Ethnophytopharmaceuticals, 4 (1), 4-19.

[59] Karunasagar, I., Pai, R., Malathi, G. R., & Karunasagar, I. (1994). Mass mortality of Penaeus monodon larvae due to antibiotic-resistant Vibrio harveyi infection. Aquaculture, 128(3-4), 203-209

[60] Kuppusamy, S., Kakarla, D., Venkateswarlu, K., Megharaj, M., Yoon, Y. E., & Lee, Y. B. (2018). Veterinary

antibiotics (VAs) contamination as a global agro-ecological issue: A critical view. Agriculture, Ecosystems & Environment, 257, 47-59.

[61] Massé, D. I., Saady, N. M. C., & Gilbert, Y. (2014). Potential of biological processes to eliminate antibiotics in livestock manure: An overview. Animals, 4(2), 146-163.

[62] Holman, D. B., Yang, W., & Alexander, T. W. (2019). Antibiotic treatment in feedlot cattle: A longitudinal study of the effect of oxytetracycline and tulathromycin on the fecal and nasopharyngeal microbiota. Microbiome, 7(1), 1-14.

[63] Alexander, T. W., Yanke, L. J., Topp, E., Olson, M. E., Read, R. R., Morck, D. W., & McAllister, T. A. (2008). Effect of subtherapeutic administration of antibiotics on the prevalence of antibiotic-resistant Escherichia coli bacteria in feedlot cattle. Applied and Environmental Microbiology, 74(14), 4405-4416.

[64] Almeida, C. M. R., Santos, F., Ferreira, A. C. F., Lourinha, I., Basto, M. C. P., & Mucha, A. P. (2017). Can veterinary antibiotics affect constructed wetlands performance during treatment of livestock wastewater? Ecological Engineering, 102, 583-588.

[65] Gros, M., Marti, E., Balcázar, J. L., Boy-Roura, M., Busquets, A., Colón, J., ... & Petrovic, M. (2019). Fate of pharmaceuticals and antibiotic resistance genes in a full-scale on-farm livestock waste treatment plant. Journal of Hazardous Materials, 378, 120716.

[66] Menz, J., Olsson, O., & Kümmerer, K. (2019). Antibiotic residues in livestock manure: Does the EU risk assessment sufficiently protect against microbial toxicity and selection of resistant bacteria in the environment? Journal of Hazardous Materials, 379, 120807.

[67] Zhao, L., Dong, Y. H., & Wang, H. (2010). Residues of veterinary antibiotics in manures from feedlot livestock in eight provinces of China. Science of the Total Environment, 408(5), 1069-1075.

[68] Zheng, Y., Fan, L., Dong, Y., Li, D., Zhao, L., Yuan, X., ... & Zhao, S. (2020). Determination of Sulfonamide Residues in Livestock and Poultry Manure Using Carbon Nanotube Extraction Combined with UPLC-MS/MS. *Food Analytical Methods*, 1-12.

[69] Rasschaert, G., Elst, D. V., Colson, L., Herman, L., Ferreira, H. C. D. C., Dewulf, J., ... & Heyndrickx, M. (2020). Antibiotic residues and antibiotic-resistant bacteria in pig slurry used to fertilize agricultural fields. Antibiotics, 9(1), 34.

[70] Chen, H., Liu, S., Xu, X. R., Diao, Z. H., Sun, K. F., Hao, Q. W., ... & Ying, G. G. (2018). Tissue distribution, bioaccumulation characteristics and health risk of antibiotics in cultured fish from a typical aquaculture area. Journal of Hazardous Materials, 343, 140-148.

[71] Burridge, L., Weis, J. S., Cabello, F., Pizarro, J., & Bostick, K. (2010). Chemical use in salmon aquaculture: A review of current practices and possible environmental effects. Aquaculture, 306(1-4), 7-23.

[72] Romero, J., Feijoó, C. G., & Navarrete, P. (2012). Antibiotics in aquaculture–use, abuse and alternatives. Health and environment in aquaculture, 159.

[73] Li, X., Shi, H., Li, K., Zhang, L., & Gan, Y. (2014b). Occurrence and fate of antibiotics in advanced wastewater treatment facilities and receiving rivers in Beijing, China. Frontiers of *Environmental Science & Engineering*, 8(6), 888-894.

[74] Li, C., Chen, J., Wang, J., Ma, Z., Han, P., Luan, Y., & Lu, A. (2015).

Occurrence of antibiotics in soils and manures from greenhouse vegetable production bases of Beijing, China and an associated risk assessment. Science of the Total Environment, 521, 101-107.

[75] Spielmeyer, A., Petri, M. S., Höper, H., & Hamscher, G. (2020). Long-term monitoring of sulfonamides and tetracyclines in manure amended soils and leachate samples-a follow-up study. Heliyon, 6(8), e04656.

[76] Yang, Y., Owino, A. A., Gao, Y., Yan, X., Xu, C., & Wang, J. (2016). Occurrence, composition and risk assessment of antibiotics in soils from Kenya, Africa. Ecotoxicology, 25(6), 1194-1201.

[77] McManus, P. S., Stockwell, V. O., Sundin, G. W., & Jones, A. L. (2002). Antibiotic use in plant agriculture. Annual Review of Phytopathology, 40(1), 443-465.

[78] Tasho, R. P., & Cho, J. Y. (2016). Veterinary antibiotics in animal waste, its distribution in soil and uptake by plants: A review. Science of the Total Environment, 563, 366-376.

[79] Li, X. W., Xie, Y. F., Li, C. L., Zhao, H. N., Zhao, H., Wang, N., & Wang, J. F. (2014a). Investigation of residual fluoroquinolones in a soil–vegetable system in an intensive vegetable cultivation area in northern China. Science of the Total Environment, 468, 258-264.

[80] Zhao, F., Yang, L., Chen, L., Li, S., & Sun, L. (2019). Bioaccumulation of antibiotics in crops under long-term manure application: Occurrence, biomass response and human exposure. Chemosphere, 219, 882-895.

[81] Proia, L., Anzil, A., Subirats, J., Borrego, C., Farrè, M., Llorca, M., ... & Servais, P. (2018). Antibiotic resistance along an urban river impacted by treated wastewaters. Science of the Total Environment, 628, 453-466.

[82] Alygizakis, N. A., Besselink, H., Paulus, G. K., Oswald, P., Hornstra, L. M., Oswaldova, M., ... & Slobodnik, J. (2019). Characterization of wastewater effluents in the Danube River basin with chemical screening, in vitro bioassays and antibiotic resistant genes analysis. Environment International, 127, 420-429

[83] Song, X., Huang, D., Zhang, L., Wang, H., Wang, L., & Bian, Z. (2020). Electrochemical degradation of the antibiotic chloramphenicol via the combined reduction-oxidation process with Cu-Ni/graphene cathode. Electrochimica Acta, 330, 135187

[84] Xia, Y., & Dai, Q. (2018). Electrochemical degradation of antibiotic levofloxacin by PbO2 electrode: Kinetics, energy demands and reaction pathways. Chemosphere, 205, 215-222.

[85] Zhang, Ruijie, Jianhui Tang, Jun Li, Qian Zheng, Di Liu, Yingjun Chen, Yongde Zou, Xiaoxiang Chen, Chunling Luo, and Gan Zhang. "Antibiotics in the offshore waters of the Bohai Sea and the Yellow Sea in China: occurrence, distribution and ecological risks." Environmental Pollution 174 (2013): 71-77.

[86] Liu, S., Zhao, H., Lehmler, H. J., Cai, X., & Chen, J. (2017). Antibiotic pollution in marine food webs in Laizhou Bay, North China: Trophodynamics and human exposure implication. Environmental Science & Technology, 51(4), 2392-2400.

[87] Liu, S., Bekele, T. G., Zhao, H., Cai, X., & Chen, J. (2018). Bioaccumulation and tissue distribution of antibiotics in wild marine fish from Laizhou Bay, North China. Science of the *Total Environment,* 631, 1398-1405

[88] Seoane, M., Rioboo, C., Herrero, C., & Cid, Á. (2014). Toxicity induced by three antibiotics commonly used in aquaculture on the marine microalga Tetraselmis suecica (Kylin) butch. Marine Environmental Research, 101, 1-7.

[89] Fernandes, M. J., Paíga, P., Silva, A., Llaguno, C. P., Carvalho, M., Vázquez, F. M., & Delerue-Matos, C. (2020). Antibiotics and antidepressants occurrence in surface waters and sediments collected in the north of Portugal. Chemosphere, 239, 124729.

[90] Mobasseri, G., Teh, C. S. J., Ooi, P. T., & Thong, K. L. (2019). The emergence of colistin-resistant Klebsiella pneumoniae strains from swine in Malaysia. Journal of global antimicrobial resistance, 17, 227-232.

[91] Spielmeyer, A., Höper, H., & Hamscher, G. (2017). Long-term monitoring of sulfonamide leaching from manure amended soil into groundwater. Chemosphere, 177, 232-238.

[92] Yamaguchi T, Okihashi M, Harada K, Konishi Y, Uchida K, Hoang Ngoc Do M, Thi Bui L, Nguyen TD, Phan HB, Dang Thien Bui H, Nguyen PD. Detection of antibiotics in chicken eggs obtained from supermarkets in Ho Chi Minh City, Vietnam. Journal of Environmental Science and Health, Part B 52, no. 6 (2017): 430-433.

[93] Chiesa, L. M., DeCastelli, L., Nobile, M., Martucci, F., Mosconi, G., Fontana, M., ... & Panseri, S. (2020). Analysis of antibiotic residues in raw bovine milk and their impact toward food safety and on milk starter cultures in cheese-making process. LWT, 131, 109783.

[94] Kantiani, L., Farré, M., & Barceló, D. (2009). Analytical methodologies for the detection of β-lactam antibiotics in milk and feed samples. TrAC *Trends in Analytical Chemistry*, 28(6), 729-744.

[95] Zhou, N., Zhang, J. X., Fan, M. T., Wang, J., Guo, G., & Wei, X. Y. (2012). Antibiotic resistance of lactic acid bacteria isolated from Chinese yogurts. Journal of Dairy Science, *95*(9), 4775-4783.

[96] Alduina, R. (2020). Antibiotics and environment. Antibiotics, 202.

[97] Wang, B., Sun, C., & Hu, G. J. (2007). Residue antibiotics in environment: Potential risks and relevant studies [J]. Environmental Science & Technology, 3

[98] Behera, B. K., Rout, P. K., & Behera, S. (2019). Sustainable Livestock Farming for Zero Hunger. In Move Towards Zero Hunger (pp. 141-159). Springer, Singapore.

Chapter 11

Effects of Antibiotics on Impacted Aquatic Environment Microorganisms

Lívia Caroline Alexandre de Araújo, Sivoneide Maria da Silva, Rafael Artur de Queiroz Cavalcanti de Sá, Ana Vitoria Araujo Lima, Amanda Virginia Barbosa, Jaqueline dos Santos Silva, Kaleen Massari Leite, Wellenilton Jose do Nascimento Júnior, Vladimir da Mota Silveira-Filho, Carina Lucena Mendes-Marques, Francisco Henrique Da Silva and Maria Betânia Melo de Oliveira

Abstract

Due to their mass production and intense consumption in human medicine, veterinary, and aquaculture, antibiotics have been widely detected in different ecosystems, leading to a growing worldwide concern. These and their byproducts are being continuously discarded in natural ecosystems via excretion of human and animal urine and feces, also domestic and hospital effluents. Residues of these drugs can persist in natural environments through bioaccumulation due to their difficult biodegradation. Also, they have a gradual deposition in sediments, aquatic surfaces, and groundwater. Studies have shown the presence of these drugs in aquatic environments, which can trigger severe changes in the composition and structure of the bacterial community, such as the ability to develop and propagate genes resistant to these pollutants. In this context, this review aims to address the effects of the antibiotics on microorganisms present in impacted aquatic environments.

Keywords: bacteria, resistance genes, pollutants

1. Introduction

Water, as an essential resource for the maintenance of life, must be ingested in significant amounts daily. However, when this substance does not correspond to levels suitable for human consumption, it can represent a means of transmitting diseases [1]. According to the World Health Organization – WHO (2016) [2], approximately half of the developing world population will be affected by infections directly related to water outside the quality standards either-or with lack of adequate or even non-existent sanitation.

The inappropriate disposal of residential waste, industrial, and agricultural effluents cause an imbalance in natural aquatic ecosystems. The amount of pollutants interferes with the proper functioning of the fauna and flora of these environments and, consequently, cause damage to the man himself [3]. Among the polluting agents of these ecosystems are pharmaceutical products, which are called emerging contaminants, of which antibiotics have been the subject of studies [4, 5].

Antibiotics are natural, synthetic, or semi-synthetic compounds developed and widely prescribed for therapeutic and prophylactic use against microbial infections in humans and animals [6, 7]. Besides, they have been widely used to stimulate the growth and production of animal [8–10]. The consumption of these drugs has increased considerably in recent decades, which is considered an emerging contaminant increasingly present in several environments, such as water [11, 12].

Despite being found in low concentrations in aquatic environments, data from the literature show that the presence of these contaminants causes negative impacts on several non-target organisms, such as microbial populations [13–15]. Changes in these populations can affect the maintenance of biological processes and change the structure of the entire ecosystem [16].

These drugs are considered the main source of changes in natural environments and influence a diversity of processes, including the development of antibiotic resistance and the spread of genes through horizontal gene transfer (HGT) [12, 17, 18]. Thus, the aquatic environment is considered efficient for the selection of bacterial populations resistant to antimicrobials, through mobile genetic elements [8]. Increased resistance to antimicrobials is often associated with a high adaptive capacity of microorganisms, with bacteria being more susceptible to changes and gene acquisition [19, 20].

Constant monitoring of bacterial communities is a tool that can lead to an understanding of the complexity of this phenomenon and the impact of human activities on natural environments [21]. In this context, this review aims to address the effects of antibiotics on microorganisms present in impacted aquatic environments.

2. Antibiotics

Antibiotics can be either from chemical compositions, derived from natural sources, or synthetic or semi-synthetic formulations. Compounds capable of inhibiting the growth of the invading microorganism, are classified as bacteriostatic. However, those that promote their microbial destruction are bactericidal [22]. The most appropriate antibiotics for the treatment of infections, by any infectious agents, are those that exert their respective mechanisms of action, directly on the pathogen without causing damage to the host, in other words, they present selective toxicity. However, all of these medications can have side effects [23].

They are usually classified according to their respective spectra and mechanisms of action. Through the description and synthesis of new compounds, it was necessary to organize them into the following groups: inhibitors of cell wall synthesis, inhibitors of cytoplasmic membrane synthesis, protein synthesis inhibitors in ribosomes, alteration of nucleic acid synthesis, and alteration of cellular metabolism [24]. The main targets of antibiotic action can be seen in **Figure 1**.

Its prescription, often, considers the symptoms presented by the patients and has no basis in more accurate clinical diagnoses that indicate the need for certain antibiotics or the appropriate dosages for each case [25]. Generally, there are several

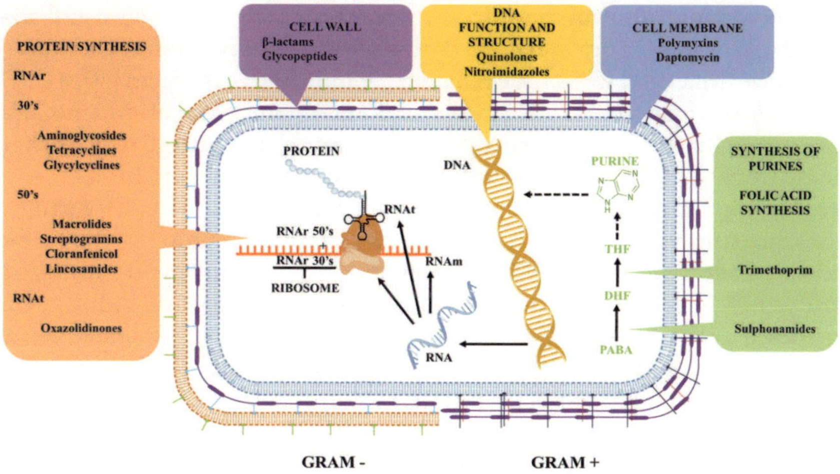

Figure 1.
Classifying scheme of the action mechanisms of antibiotics.

independent factors associated with non-compliance with the guidelines for antimicrobial prophylaxis, such as the lack of well-defined protocols, the lack of knowledge and updating, the lack of communication or disagreement between the prescribing professionals and the absence of tax audits [26]. For this reason, all health institutions and clinical analysis laboratories must use the same procedures and interpretive criteria, especially in the case of bacteria producing carbapenemases [27].

The discovery of antibiotics, initially penicillin by Alexander Fleming in 1928, was a milestone in the history of medicine and represented one of the most important therapeutic interventions for the control of diseases caused by bacteria. Thus, until the early 1960s, more than 20 classes of antibiotics were marketed worldwide. However, since that period, only two new classes were added to this group of drugs [28].

Thus, the obtention of new effective active ingredients capable of reaching different bacterial targets has been significantly reduced, so that new therapeutic approaches are essential to minimize the damage generated to human health [29, 30]. Also, The search for new molecules that have at least one synergistic mechanism in association with drugs already known, thus enhancing their effects, have been highlighted and may help the therapy of infections caused by microorganisms [31, 32].

The diversity of resistance genes identified in pathogenic bacteria, it is pointed out as one of the main causes of the growing increase in antibiotic resistance. However, the prevalence of bacterial resistance both regionally and on a global scale is highly attributed to the mechanisms of dissemination of virulent bacteria, which can colonize and adapt in various environments, including the aquatic [33, 34].

3. Antibiotics in aquatic environments

The extensive and continuous consumption of antibiotics is directly related to their presence in the environment, and for this reason, they have aroused a growing worldwide concern [12, 35]. Also, it estimated that a few thousand tons of these drugs are produced annually, and part of them is introduced into various ecosystems, including aquatic ones [36–38].

It is known that antibiotics can occur naturally in the environment. However, anthropogenic activities are considered the major source of these pollutants. Among these activities, the use in human and veterinary medicine for the treatment and prevention of bacterial infections stands out [7, 39].

After the administration of these drugs, they are not fully metabolized and absorbed by the body of the human or animal, there about 90% of the residues of each dose is excreted in the urine, and 75% of the remainder eliminated in the feces [40, 41]. These non-metabolized compounds are discarded in domestic and hospital effluents, and they can be partially disposed of in water treatment plants, following until they reach natural aquatic environments, such as rivers, lakes, seas, and groundwater [42, 43].

Another important source of environmental contamination by these drugs is through aquaculture, which also contributes to the contamination of the surface, groundwater, and sediment due to the use of antibiotics as fish food additives to promote growth and prevent disease [44, 45]. The pharmaceutical industry can also be considered a relevant source due to the disposal of effluents generated during the synthesis, processing, and filing of these compounds [46–48].

The dynamics of these compounds in the environment is determined by various physical-chemical and biological processes in soil-water systems, being dependent on structural characteristics of antibiotics, properties of soil and water, as well as environmental parameters, such as geographic location, temperature, and pH [10, 49–52]. In aquatic environments, antibiotics are being detected in low concentrations (ng/L e μg/L) [53–55]. However, some of these drugs can persist in the environment due to their low biodegradation. Consequently, they trigger their bioaccumulation and biomagnification [46, 56].

The presence of antibiotics in the natural environment, even in low concentrations, can affect the survival, reproduction, metabolism of populations of organisms, in addition to changing the structure of communities and ecological functions of ecosystems [57–59]. In this context, several studies have shown that these drugs trigger toxic, acute, either-or chronic effects in organisms of different trophic levels, such as bacteria, fungi, algae, invertebrates, and fish [60–62].

4. Effects of antibiotics on microorganisms in aquatic environments

Antibiotics were strategically developed to act against pathogenic microorganisms, and their advantages for the control and treatment of infectious diseases are indisputable [62]. However, its bioactive properties, the increase in its presence in the environment, and its toxicity on non-target organisms have aroused growing concern worldwide [15, 63].

Among the different groups of organisms, microorganisms are considered the most sensitive and most affected by the action of these drugs [61, 64]. Their bactericidal and bacteriostatic effects can trigger changes in the structure of microbial populations and their ecological functions in the environment. Microbial biodiversity is essential for the maintenance of biological processes in aquatic and terrestrial environments, including biogeochemical cycles [16].

Studies have shown that the presence of antibiotics, even at low concentrations, can cause changes in essential microbial processes for the ecosystems, such as nitrogen transformation, methanogenesis, sulfate reduction, nutrient cycle, and degradation of organic matter [16, 59, 65].

Another widely discussed effect of this class of drugs on the environment, it is the contribution to the emergence of antibiotic-resistant bacteria (ARB), and genes of resistance to antibiotics (GRA) [66]. Antibiotic resistance may occur naturally in the environment, but the increase has been reported due to the overuse and inappropriate disposal of these drugs. According to WHO, the occurrence of BRA and GRA is recognized as one of the most worrying public health problems of the century, with GRA being considered an emerging environmental pollutant [12, 67].

Among the different ecological niches, the aquatic environment is considered the most efficient for the selection of resistant bacterial populations. This environment is prone to constant changes that can apply selective pressure, favoring the evolution of these microorganisms regarding the adaptation process to various antimicrobial agents [68].

Literature data have been reporting the presence of antibiotic-resistant opportunistic pathogens such as *Escherichia coli*, *Klebsiella pneumoniae*, *Acinetobacter* spp., *Pseudomonas* spp., *Shigella* spp. [69–71], in several urban rivers, lakes, and streams, which receive an untreated effluent discharge domestic, hospital, and industrial, or inefficient water treatment plants [72–74].

Studies have shown that ARBs are introduced in these environments mainly by contamination by enteric bacteria that are exposed to high concentrations of antibiotics in the gastrointestinal tract of humans and animals [75]. Among various sources of these bacteria, urban water treatment stations stand out due to their high microbial density and the high presence of contaminants such as antibiotic residues, pharmaceuticals, heavy metals, among others. The combination of these substances puts selective pressure on these microorganisms, amplifying the development of resistance and its dissemination [76–78].

These bacteria act as a source of resistance in natural environments and can spread GRAs present in the environment [79]. Literature data suggest that the increase in the frequency of resistant bacteria in the water is influenced by the spread of GRAs among bacterial populations through vertical gene transfer (VGT) and horizontal gene transfer (HGT). VGT is parental genetic transmission via reproduction, and mutations in bacterial DNA can occur. On the other hand, HGT consists of the exchange of genes between microorganisms, intra or inter-specific, through mobile genetic elements, such as plasmids and transposons encoding antimicrobial resistance [80–82].

As mentioned above, microorganisms present in these impacted aquatic environments can develop different resistance mechanisms to adapt to the stress caused by various contaminants, such as heavy metals. Studies suggest that contamination by metals in the natural environment may play an important role in maintaining and proliferating resistance to antimicrobials [83–85]. In these environments, the selective pressure exerted by metals can select resistant isolates in a similar way to antibiotics, since both resistance genes are often located in the same moving elements [86, 87].

Martins et al. [88] observed that *P. aeruginosa* isolate, obtained from a contaminated river in southeastern Brazil, presented a conjugative plasmid with co-resistance to tetracycline and copper, reinforcing that resistance to antibiotics can be induced by the selective pressure of heavy metals in the environment. Rasmussen and Sorensen [89] demonstrated an increase in the occurrence of conjugative plasmids in contaminated sites. They observed that the mercury and tetracycline resistance genes were located on the same plasmid. This data that corroborates with the study carried out by Araújo et al. [90], who verified genes for resistance to mercury and tetracycline in isolates of *K. pneumoniae* obtained from an impacted urban stream in Pernambuco, Brazil.

In this way, not only the indiscriminate use of antibiotics but also environmental contamination by other substances can pose risks and damage to the environment and human health, since resistance genes can be transferred horizontally from environmental microorganisms to human commensals [91, 92]. Given this problem, there is a clear need to develop research and technologies for monitoring and removing these substances in the aquatic environment, in order to reduce their impact on the environment and public health.

5. Conclusion

The presence of antibiotics in aquatic ecosystems, although considered less than other contaminants, can alter the microbiota of these environments and favor the emergence of multi-resistant bacterial populations. Considering the importance of water bodies for human subsistence and due to the risk of bacterial spread, strategies for controlling the disposal of these contaminants must be evaluated, as well as the other agents that impact aquatic environments.

Author details

Lívia Caroline Alexandre de Araújo[1], Sivoneide Maria da Silva[1], Rafael Artur de Queiroz Cavalcanti de Sá[1], Ana Vitoria Araujo Lima[1], Amanda Virginia Barbosa[1], Jaqueline dos Santos Silva[1], Kaleen Massari Leite[1], Wellenilton Jose do Nascimento Júnior[2], Vladimir da Mota Silveira-Filho[3], Carina Lucena Mendes-Marques[1], Francisco Henrique Da Silva[1] and Maria Betânia Melo de Oliveira[1*]

1 Laboratory of Molecular Biology (BioMol), Biochemistry Department, Federal University of Pernambuco, Recife, Pernambuco, Brazil

2 Laboratory of Food Engineering, Federal University of Pernambuco, Recife, Pernambuco, Brazil

3 Biology Department, University of Pernambuco, Recife, Pernambuco, Brazil

*Address all correspondence to: maria.bmoliveira@ufpe.br

References

[1] Mouly D, Goria S, Mounié M, Beaudeau, Galey C, Anne Gallay A, et al. Waterborne Disease Outbreak Detection: A Simulation-Based Study. International Journal of Environmental Research and Public Health. 2018; 157: 1505.

[2] WHO. Sanitation safety planning: manual for safe use and disposal of wastewater, greywater and excreta. World Health Organization, 20 Avenue Appia, 1211 Geneva 27. 2016.

[3] Bashir I, Lone FA, Bhat RA, Mir AA, Dar ZA, Dar AS. Concerns and Threats of Contamination on Aquatic Ecosystems. Springer Nature Switzerland AG. 2020; 1-26.

[4] Gaffney V, Cardoso V, Benoliel MJ, Almeida C. Emerging contaminants - pharmaceuticals: monitoring, environmental and human health risk assessment. Água & Resíduos - APESB. 2016; 1: 1-27.

[5] Kraemer SA, Ramachandran A, Perron GG. Antibiotic Pollution in the Environment: From Microbial Ecology to Public Policy. Microorganisms. 2019; 7: 1-24.

[6] Kümmerer K. Antibiotics in the aquatic environment—a Review-Part 1. Chemosphere. 2009a 75, 417-434.

[7] Carvalho IT, Santos L. Antibiotics in the aquatic environments: a review of the European scenario. Environment International. 2016; 94, 736-757.

[8] Von Wintersdorff CJH, Penders J, Van Niekerk JM, Mills ND, Majumder S, Van Alphen LB, et al. Dissemination of Antimicrobial Resistance in Microbial Ecosystems through Horizontal Gene Transfer. Frontiers in Microbiology. 2016; 7: 173.

[9] Cycoń M, Mrozik A, Piotrowska-Seget Z. Antibiotics in the soil environment—degradation and their impact on microbial activity and diversity. Frontiers Microbiology. 2019; 8(10):338.

[10] Zhao RX, Feng J, Liu J, Fu WJ, Li XY, Li B. Deciphering of microbial community and antibiotic resistance genes in activated sludge reactors under high selective pressure of diferente antibiotics. Water Research. 2019; 151, 388-402.

[11] Liu JL, Wong MH. Pharmaceuticals and personal care products (PPCPs): a review on environmental contamination in China. Environment International. 2013;59:208-224.

[12] Ben YJ, Fu CX, Hu M, Liu L, Whong MH, Zheng CM. Human health risk assessment of antibiotic resistance associated with antibiotic residues in the environment: a review. Environmental Research. 2019; 168, 483-493.

[13] Minguez L, Pedelucq J, Farcy E, Ballandonne C, Budzinski H, Halm-Lemeille MP. Toxicities of 48 pharmaceuticals and their freshwater and marine environmental assessment in northwestern France. Environmental Science and Pollution Research. 2016; 23:4992-5001.

[14] Straub JO, Gysel D, Kastl U, Klemmer J, Sonderegger M, Studer M. Environmental risk assessment for ancillary substances in biotechnological production of pharmaceuticals. Environmental Toxicology and Chemistry. 2012; 31:681-687.

[15] Kumar M., Jaiswal S, Sodhi KK, Shree P, Singh DK, Agrawal PK, Shukla P. Antibiotics bioremediation: perspectives on its ecotoxicity and resistance. Environment International. 2019; 124, 448-461.

[16] Grenni P, Ancona V, Caracciolo A. Ecological effects of

antibiotics on natural ecosystems: a review. Microchemical Journal. 2018; 136, 25-39.

[17] Qiao M, Ying GG, Singer AC, Zhu YG. Review of antibiotic resistance in China and its environment. Environment. International. 2018; 110:160-172.

[18] Zhang T, Hu Y, Jiang, L, Yao S, Lin, K, Zhou Y, Cui C. Removal of antibiotic resistance genes and control of horizontal transfer risk by UV, chlorination and UV/ chlorination treatments of drinking water. Chemical Engineering Journal. 2019; 358:589-597.

[19] Cho S, Nguyen HAT, McDonald JM, Woodley TA, Hiott LM, Barrett JB, et al. Genetic Characterization of Antimicrobial-Resistant *Escherichia coli* Isolated from a Mixed-Use Watershed in Northeast Georgia, USA. International Journal of Environmental Research and Public Health. 2019; 16: 3761.

[20] Dai Z, Sevillano-Rivera MC, Calus ST, Santos QMB, Eren AM, Wielen PWJJ, et al. Disinfection exhibits systematic impacts on the drinking water microbiome. Microbiome. 2020; 8: 42.

[21] Jutkina J, Rutgersson C, Flach CF, Larsson DG. An assay for determining minimal concentrations of antibiotics that drive horizontal transfer of resistance. Science of the Total Environment. 2016; 548-549.

[22] Guimarães DO, Momesso LS, Pupo MT. Antibióticos: importância terapêutica e perspectivas para a descoberta e desenvolvimento de novos agentes. Química. Nova. 2010; 33: 667-679.

[23] Fernandes R, Amador P, Prudêncio C. β-Lactams: chemical structure, mode of action and mechanisms of resistance. Reviews in Medical Microbiology. 2013; 24: 7-17.

[24] Madigan MT, Martinko JM, Bender KS, Buckley DH, Stahl DA. Microbiologia de Brock. 14 ed. Porto Alegre: ArtMed. 2016.

[25] Leekha S, Terrell CL, Edson RS. General Principles of Antimicrobial Therapy. Mayo Clinic Proceedings. 2011; 86: 156-167.

[26] Calbo E, Álvarez-Rocha L, Gudiold F, Pasquau J. A review of the factors influencing antimicrobial prescribing. Enfermedades Infecciosas y Microbiología Clínica. 2013; 31: 12-15.

[27] Eichenberger EM, Thaden JT. Epidemiology and Mechanisms of Resistance of Extensively Drug Resistant Gram-Negative Bacteria. Antibiotics. 2019; 8: 37.

[28] Coates A, Halls G, Hu Y. Novel classes of antibiotics or more of the same? British Journal of Pharmacology. 2011; 163: 184-194.

[29] Nogueira HS, Xavier AREO, Xavier MAS, Carvalho AA, Monção GA, Barreto NAP. Antibacterial: main classes, mechanisms of action and resistance. Revista Unimontes Cientifica. 2016; 18: 96-108.

[30] Santos DVA, Oliveira GA, Pacheco LG, Faria LMO, Cunha JC, Mello TM. Antibiotics through the approach of mechanism of bacterial resistance. Ciência Atual. 2018; 11: 4-14.

[31] Mabou FD, Tamokou JD, Ngnokam D, Nazabadioko LV, Kuiate JR, Bag PK. Complex secondary metabolites from *Ludwigia leptocarpa* with potent antibacterial and antioxidant activities. Drug Discoveries & Therapeutics. 2016; 10: 141-149.

[32] Kobayashi J. Search for New Bioactive Marine Natural Products and Application to Drug Development. Chemical and Pharmaceutical Bulletin. 2016; 64: 1079-1083.

[33] Rojas GC, Ulate L. A. Resistencia antimicrobiana: microorganismos más resistentes y antibióticos con menor actividad. Revista Medica de Costa Rica y Centroamerica. 2016; 621: 757-763.

[34] Motlagh AM, Yang Z. Detection and occurrence of indicator organisms and pathogens. Water Environment Research. 2019; 91: 1402-1408.

[35] Kümmerer K, Alexy R, Hüttig J, Schöll A. Standardized tests fail to assess the effects of antibiotics on environmental bacteria. Water Research. 2004; 38, 2111-2116.

[36] Ji, X, Shen Q, Liu F, Ma J, Xu G, Wang Y, Wu M. Antibiotic resistance gene abundances associated with antibiotics and heavy metals in animal manures and agricultural soils adjacent to feedlots in Shanghai; China. Journal of Hazardous Materials. 2012; 235, 178-185.

[37] Harnisz M, Korzeniewska E, Gołaś I. The impact of a freshwater fish farm on the community of tetracycline-resistant bacteria and the structure of tetracycline resistance genes in river water. Chemosphere. 2015a; 128, 134-141.

[38] Al Aukidy M, Verlicchi P, Jelic A, Petrovic M, Barcelò D. Monitoring release of pharmaceutical compounds: occurrence and environmental risk assessment of two WWTP effluents and their receiving bodies in the Po Valley. Italy. Science of the Total Environment. 2012; 438, 15-25.

[39] Berkner S, Konradi S, Schönfeld J. Antibiotic resistance and the environment—there and back again. EMBO Reports. 2014; 15, 740-1744.

[40] Sarmah AK, Meyer MT, Boxall ABA. A global perspective on the use, sales, exposure pathways, occurrence, fate and effects of veterinary antibiotics (VAs) in the environment. Chemosphere. 2006; 65:725-759.

[41] Zuccato E, Castiglioni S, Bagnati R, Melis M, Fanelli R. Source, occurrence and fate of antibiotics in the Italian aquatic environment. Journal of Hazardous Materials. 2010; 179 (1), 1042-1048.

[42] Moldovan Z. Occurrences of pharmaceutical and personal care products as micropollutants in rivers from Romania. Chemosphere. 2006; 1808-1817.

[43] Lapworth DJ, Baran N, Stuart ME, Ward RS. Emerging organic contaminants in groundwater: a review of sources, fate and occurrence. Environmental Pollution. 2012; 163, 287-303.

[44] Pham-Duc P, Cook MA, Cong-Hong H, Nguyen-Thuy H, Padungtod P, Nguyen-Thi H, et al. Knowledge, attitudes and practices of livestock and aquaculture producers regarding antimicrobial use and resistance in Vietnam. PLoS ONE. 2019; 14(9): e0223115.

[45] Hossain A, Nakamichi S, Habibullah-Al-Mamun M, Tani K, Masunaga S, Matsuda H. Occurrence, distribution, ecological and resistance risk of antibiotics in surface water of finfish and shellfish aquaculture in Bangladesh. Chemosphere. 2017;188:329-336.

[46] Carvalho, E.V. et al. Aspectos legais e toxicológicos do descarte de medicamentos. Revista Brasileira de Toxicologia. 2009; 22 (1-2):1-8.

[47] Américo, J.H.P. et al. Ocorrência, destino e potenciais impactos dos fármacos no ambiente. Saúde e Biologia. 2013; 8 (2):59-72.

[48] Giri P, Pal C. Ecotoxicological aspects of pharmaceuticals on aquatic Environment. American Journal of Drug Discovery. 2014; 1:10-24.

[49] Andreozzi R, Caprio V, Ciniglia C, de Champdoré M, Lo Giudice R, Marotta R, ZuccatoE. Antibioticsintheenvironment: occurrence in Italian STPs fate, and preliminary assessment on algal toxicity of amoxicillin. Environmental. Science. Technology. 2004; 38 (24), 6832-6838.

[50] Lee YJ, Lee SE, Lee DS, Kim YH. Risk assessment of human antibiotics in Korean aquatic environment. Environmental Toxicology and Pharmacology. 2008; 26 (2), 216-221.

[51] Brausch JM, Connors KA, Brooks BW, Rand GM. Human pharmaceuticals in the aquatic environment: a review of recent toxicological studies and considerations for toxicity testing. Reviews of Environmental Contamination and Toxicology. 2012; 218, 1-99.

[52] Mitchell SM, Ullman JL, Teel AL, Watts RJ. pH and temperature effects on the hydrolysis of three -lactam antibiotics: ampicillin, cefalotin and cefoxitin. Science of the Total Environment. 2014; 466, 547-555.

[53] Batt AL, Kim S, Aga DS. Comparison of the occurrence of antibiotics in four full-scale wastewater treatment plants with varying designs and operations. Chemosphere. 2007; 68, 428-435.

[54] Ma Y, Li M, Wu M, Li Z, Liu X. Occurrences and regional distributions of 20 antibiotics in water bodies during groundwater recharge. Science of the Total Environment. 2015; 15, 518-519: 498-506.

[55] Krzeminski P, Tomei MC, Karaolia P, Langenhoff A, Almeida CMR, Felis E, et al. Performance of secondary wastewater treatment methods for the removal of contaminants of emerging concern implicated in crop uptake and antibiotic resistance spread: a review. Science of the Total Environment. 2019; 648, 1052-1081.

[56] Pereira, L.C. et al. A perspective on the potential risks of emerging contaminants to human and environmental health. Environmental Science and Pollution Research. 2015; 22:13800-13823.

[57] Loos R, Carvalho R, António DC, Comero S, Locoro G, Tavazzi S, et al. EU-wide monitoring survey on emerging polar organic contaminants in wastewater treatment plant effluents. Water Research. 2013; 47, 6475-6487.

[58] Martinez JL. Environmental pollution by antibiotics and by antibiotic resistance determinants. Environmental Pollution. 2009;157:2893-2902.

[59] Martin-Laurent F, Topp E, Billet L, Batisson I, Malandain C, Besse-Hoggan P, et al. Environmental risk assessment of antibiotics in agroecosystems: ecotoxicological effects on aquatic microbial communities and dissemination of antimicrobial resistances and antibiotic biodegradation potential along the soil-water continuum. Environmental Science and Pollution Research. 2019; 26(18):18930-18937.

[60] Rede D, Santos LHMLM, Ramos S, Oliva-Teles F, Antão C, Sousa SR, DelerueMatos, C. Individual and mixture toxicity evaluation of three pharmaceuticals to the germination and growth of *Lactuca sativa* seeds. Science of the Total Environment. 2019; 673, 102-109.

[61] Brandt KK, Amézquita A, Backhaus T, Boxall A, Coors A, Heberer T, et al. Ecotoxicological assessment of antibiotics: a call for improved consideration of microorganisms. Environment International. 2015; 85, 189-205.

[62] Luo Y, Xu L, Rysz M, et al. Occurrence and transport of tetracycline, sulfonamide, quinolone,

and macrolide antibiotics in the Haihe River Basin, China. Environmental Science & Technology. 2011; 45:1827-1833.

[63] Su H, Hu X, Wang L, Xu W, Xu Y, Wen G, et al. Contamination of antibiotic resistance genes (ARGs) in a typical marine aquaculture farm: source tracking of ARGs in reared aquatic organisms. Journal of Environmental Science and Health, Part B. 2020; 55(3):220-229.

[64] Petrovich ML, Zilberman A, Kaplan A, Eliraz GR, Wang Y, Langenfeld K, et al. Microbial and Viral Communities and Their Antibiotic Resistance Genes Throughout a Hospital Wastewater Treatment System Frontiers in Microbiology. 2020; 11: 153.

[65] Bielen A, Šimatović A, Kosić-Vukšić J, Senta I, Ahel M, Babić S, et al. Negative environmental impacts of antibiotic-contaminated effluents from pharmaceutical industries. Water Research. 2017;126:79-87.

[66] Bila DM, Dezotti M. Fármacos no meio ambiente. Química Nova. 2003; 26 (44):523-530.

[67] Zarei-Baygi A, Harb M, Wang P, Stadler LB, Smith AL. Evaluating Antibiotic Resistance Gene Correlations with Antibiotic Exposure Conditions in Anaerobic Membrane Bioreactors. Environmental Science & Technology. 2019; 53(7):3599-3609.

[68] Woodford N, Wareham DW, Guerra B, Teale C. Carbapenemase-producing Enterobacteriaceae and non-Enterobacteriaceae from animals and the environment: an emerging public health risk of our own making? Journal of Antimicrobial Chemotherapy. 2014; 69:287-291.

[69] Hladicz A, Kittinger C, Zarfel G. Tigecycline resistant *Klebsiella pneumoniae* isolated from Austrian river water. International Journal of Environmental Research and Public Health. 2017; 14:11-13.

[70] Marathe NP, Pal C, Gaikwad SS, Jonsson V, Kristiansson E, Larsson DGJ. Untreated urban waste contaminates Indian river sediments with resistance genes to last resort antibiotics. Water Research. 2017; 124:388-397.

[71] Nishiyama M, Ogura Y, Hayashi T, Suzuki Y. Antibiotic resistance profiling and genotyping of vancomycin-resistant enterococci collected from an urban river basin in the Provincial City of Miyazaki, Japan. Water. 2017; 9:79.

[72] Karkman A, Pärnänen K, Larsson DGJ. Fecal pollution can explain antibiotic resistance gene abundances in anthropogenically impacted environments. Nature Communications. 2019; 10:80.

[73] Niestępski S, Harnisz M, Korzeniewska E, Aguilera-Arreola MG, ContrerasRodríguez A, Filipkowska Z, Osińska A. The emergence of antimicrobial resistance in environmental strains of the *Bacteroides fragilis* group. Environment International. 2019; 124, 408-419.

[74] Osińska A, Korzeniewska E, Harnisz M, Niestępski S. Quantitative occurrence of antibiotic resistance genes among bacterial populations from wastewater treatment plants using activated sludge. Applied Sciences. 2019; 9:387.

[75] Alonso A, Sánchez P, Martínez JL. Environmental selection of antibiotic resistance genes. Environmental Microbiology. 2001; 3:1-9.

[76] Zhang T, Shao M-F, Ye L. Pyrosequencing reveals bacterial diversity of activated sludge from 14 sewage treatment plants. The ISME Journal. 2012; 6:1137-1147.

[77] Michael I, Rizzo L, McArdell CS, Manaia CM, Merlin C, Schwartz T, Dagot C, Fatta Kassinos D. Urban wastewater treatment plants as hotspots for the release of antibiotics in the environment: a review. Water Research. 2013; 47:957-95.

[78] Rizzo L, Manaia C, Merlin C, Schwartz T, Dagot C, Ploy MC, Michael I, Fatta-Kassinos D. Urban wastewater treatment plants as hotspots for antibiotic resistant bacteria and genes spread into the environment: a review. Science of the Total Environment. 2013b; 447:345-360.

[79] Davison J. Genetic exchange between bacteria in the environment. Plasmid. 1999; 42:73-91.

[80] Ali Abadi FS, Lee SP. Antibiotic treatment for animals: effect on bacterial population and dosage regimen optimization. International Journal of Antimicrobial Agents. 2000; 14:307-313.

[81] Wegener HC, Moller NF. Reducing the use of antimicrobial agents in animals and man. Journal of Medical Microbiology. 2000; 49:111-113.

[82] Shao S, Hu Y, Cheng J, Chen Y. Research progress on distribution, migration, transformation of antibiotics and antibiotic resistance genes (ARGs) in aquatic environment, Critical Reviews in Biotechnology. 2018; 38(8):1195-1208.

[83] Dickinson AW, Power A, Hansen MG, Brandt KK, Piliposian G, Appleby P, et al. Heavy metal pollution and co-selection for antibiotic resistance: A microbial palaeontology approach. Environment International. 2019;132:105117.

[84] Dahanayake PS, Hossain S, Wickramanayake MVKS, Heo G-J. Antibiotic and heavy metal resistance genes in Aeromonas spp. isolated from marketed Manila Clam (Ruditapes philippinarum) in Korea. Journal of Applied Microbiology. 2019;127(3):941-952.

[85] Gallo IFL, Furlan JPR, Sanchez DG, Stehling EG. Heavy metal resistance genes and plasmid-mediated quinolone resistance genes in Arthrobacter sp. isolated from Brazilian soils. Antonie Van Leeuwenhock. 2019;112(10):1553-1558.

[86] Zhang M, Wan K, Zeng J, Lin W, Ye C, Yu X. Co-selection and stability of bacterial antibiotic resistance by arsenic pollution accidents in source water. Environment International. 2020; 135:105351.

[87] Ohore OE, Addo FG, HAn N, Li X, Zhang S. Profiles of ARGs and their relationships with antibiotics, metals and environmental parameters in vertical sediment layers of three lakes in China. Journal of Environmental Management. 2020; 255:109583.

[88] Martins VV, Zanetti MOB, Pitondo-Silva A, Stehling EG. Aquatic environments polluted with antibiotics and heavy metals: a human health hazard. Environmental Science and Pollution Research. 2014; 21:5873-5878.

[89] Rasmussen LD, Sorensen SJ. The effect of long-term exposure to mercury on the bacterial community in marine sediment. Current Microbiology. 1998; 36: 291-297.

[90] Araújo LCAA, Purificação-Junior AF, Silva SM, Lopes ACS, Veras DL, et al. In vitro evaluation of Mercury (HG+2) effects on biofilm formation by clinical and environmental isolates de Klebsiella pneumoniae. Ecotoxicology and Environmental Safety. 2019; 169:669-677.

[91] Baker-Austin C, Wright MS, Stepanauskas R, mcarthur JV.

Co-selection of antibiotic and metal resistance. Trends in Microbiology. 2006; 14(4):176-182.

[92] Touchon M, Sousa JAM, Rocha EPC. Embracing the enemy: the diversification of microbial gene repertoires by phage-mediated horizontal gene transfer. Current Opinion in Microbiology. 2017, 38:66-73.

Section 4

Emerging Contaminants and Environment

Chapter 12

Polybrominated Diphenyl Ethers (PBDEs) as Emerging Environmental Pollutants: Advances in Sample Preparation and Detection Techniques

Japheth M. Nzangya, Elizabeth N. Ndunda, Geoffrey O. Bosire, Bice S. Martincigh and Vincent O. Nyamori

Abstract

Environmental pollution has been a challenging phenomenon in most developing countries, due to the weak enforcement of environmental regulations. As a result, humans and animals are exposed to different environmental pollutants, which threaten their very existence. Some of the emerging pollutants of great concern are polybrominated diphenyl ethers (PBDEs) since they are categorized as probable human carcinogens and are also known to bioaccumulate in fatty tissues of animals and humans, reaching toxic levels upon continued exposure. Monitoring of these pollutants is therefore paramount as it contributes to addressing the problem of human exposure and environmental pollution. Their monitoring involves sample preparation methods followed by quantification with various detection techniques. Sample preparation methods that aim at reducing matrix interferences, enriching analytes and transfer of analytes to a desirable solvent, have evolved from conventional methods to advanced methods that facilitate the detection of these chemicals at very low concentrations. Likewise, detection techniques have advanced from chromatographic detection techniques to miniaturized systems that involve sensors. This chapter discusses PBDEs as emerging pollutants, their sources, and toxicological implications on humans, as well as advances in sample preparation methods and detection techniques in the determination of PBDEs.

Keywords: polybrominated diphenyl ethers, persistent organic pollutants, pollution, emerging pollutants, detection techniques

1. Introduction

The preservation and conservation of the environment are of great significance for healthy living. However, efforts to conserve the environment have been futile due to escalated pollution from biogenic and anthropogenic sources, which constantly release pollutants to the environment [1]. In the recent past, increased industrial and agricultural activities have immensely contributed to the pollution of aquatic environments such as rivers and streams, which pose major detrimental environmental problems to humans [2]. It is evident that industrial development has generated a myriad of new chemicals produced and applied in daily activity, which is becoming a major concern for citizens, the research community, and authorities [3]. Among the pollutant chemicals that have been introduced into the environment are polybrominated diphenyl ethers (PBDEs). PBDEs are toxic, lipophilic, hydrophobic, and persistent artificial chemicals characterized by high physical and chemical stability [4]. They are commonly applied as flame retardants in polymer products such as electronics, plastics, textiles, and building materials [5, 6]. PBDEs have become a growing concern over the last two decades due to their ubiquity, persistence and accumulation capacity in the environment, as well as their potential risks to human health and wildlife [7, 8]. PBDEs are normally additive compounds, meaning they are not covalently bound to the polymeric products [9]. Therefore, they may leach out into the surrounding environment during their production, usage, disposal, or recycling process [10]. PBDEs can be transported away from their sources for long-ranges through aqueous and/or terrestrial environmental compartments [11, 12]. In this context, monitoring and assessment of environmental pollution by these compounds are very important.

Their determination involves a series of steps from sample pre-treatment to quantification of analytes using various detection systems. Different sample preparation strategies that range from conventional to advanced strategies have been applied for the determination of PBDEs in environmental samples. Some of the conventional sample enrichment methods include Soxhlet extraction [13, 14] and liquid-liquid extraction (LLE) [15]. More recently, ultrasound-assisted extraction (UAE) [16, 17], pressurized liquid extraction (PLE) [18, 19], microwave-assisted extraction (MAE), solid-phase extraction (SPE), and solid-phase microextraction (SPME) have exhibited successful extraction of PBDEs from environmental samples [20, 21]. The application of SPE and SPME has advanced from conventional adsorbent formats to the most improved formats which allow easy transfer of analytes from their complex matrices. This has been achieved by using novel adsorbent materials to replace conventional silica-based adsorbents which exhibit low selectivity towards targeted analytes [22]. Similarly, analytical techniques for the qualitative and quantitative determination of PBDEs have advanced from well-known gas chromatography-electron capture detection (GC-ECD) to sensor-based techniques that are more advantageous in terms of excellent selectivity, with opportunities for *in-situ* application. The following sections provide detailed information on PBDEs, advances in sample pre-treatment methods and detection techniques with a view of providing the current state-of-the-art as far as their monitoring is concerned.

2. Polybrominated diphenyl ethers

2.1 The chemistry of PBDEs

PBDEs comprise of two halogenated aromatic rings bonded by an ester bond and are classified in relation to the number and position of bromine atoms in a particular molecule [23]. They have a general molecular formula of $C_{12}H_{(10-x)}Br_xO$, where x is the number of bromine atoms in a molecule with numerical values [x = 1, 2, 3, ..., 10 = m + n] (**Figure 1**). Substitution of bromine atoms can take place at 10 possible positions on the two benzene rings resulting in 209 possible congeners [24].

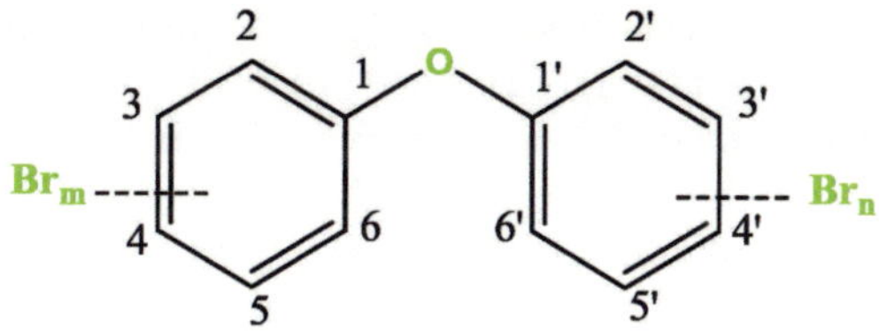

Figure 1.
General structural formula of PBDEs.

Different congeners are easily identified by their corresponding IUPAC numbers ranging from 1 to 209. In this case, 2,2′,4,4′-tetrabromodiphenyl ether is BDE-47, with bromine atoms in ortho and para positions on the first and second benzene rings, respectively (**Figure 2**).

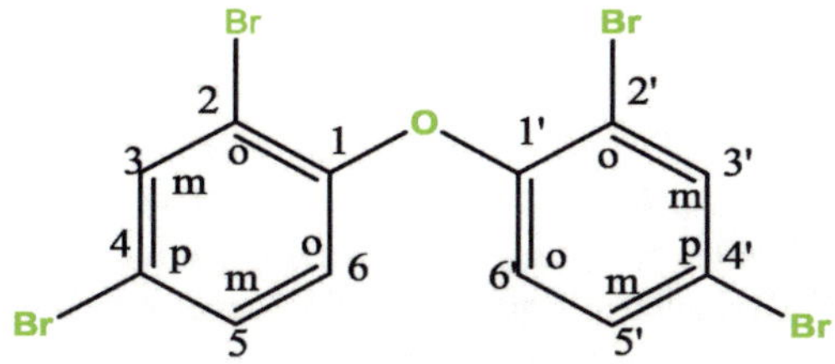

Figure 2.
Chemical structure of 2,2′,4,4′-tetrabromodiphenyl ether (BDE 47).

Molecules with one to four bromine atoms are classified as low molecular mass PBDEs, whereas the ones with five to ten bromine atoms are categorized as high molecular mass PBDEs. Less brominated PBDEs are more persistent and toxic than highly brominated diphenyl ethers [25]. The substitution pattern also affects the physicochemical properties of PBDEs, whereby the solubility of PBDEs decreases significantly with an increase in bromine substitution. The aqueous solubility (S_W) of low molecular mass PBDEs at room temperature ranges from 6.57×10^{-7} to 7.82×10^{-11} mol L^{-1} while those of high molecular mass have aqueous solubility values lower than 7.82×10^{-11} mol L^{-1} [26]. A wide range of PBDE congeners exhibit high lipophilic capacity and high resistance to degradation; a property that makes them bioaccumulate and magnify in biota [7]. PBDEs are also associated with high octanol-air partition coefficients (K_{OA}) with values ranging between 9.3 and 12.0 from BDE-17 to -126, which is approximately 1 to 2 orders of magnitude greater than PCBs [27]. Therefore, PBDEs are easily transported through air from one point to another, increasing their chances of exposure to humans. Dissolved organic matter has shown a high tendency to interact with hydrophobic compounds such as PBDEs, which hinders their mobility and degradation in the environment [28]. Reported binding coefficients of PBDEs (log K_{DOC}) towards organic matter range from 5.1 to 7.14, which implicates the high capability of PBDEs to adsorb and partition on organic matter [29].

2.2 Global production and regulation of PBDEs

PBDEs were commercially produced in three technical mixtures, typically known as pentaBDE, octaBDE, and decaBDE, basing on the number of bromine atoms [10]. By early 2000, the global production of commercial PBDE formulations was approximately 67,000 tons in the ratio 1:1.98:14.8 for octa-BDE, penta-BDE, and deca-BDE respectively, of which the United States production was approximately 50% of the global production [30]. Several governmental regulations and international environmental agencies have restricted and completely banned the use and production of some PBDE congeners [31]. In 2004, the European Union phased out the use and production of penta-BDE and octa-BDE. Consequently, in December 2004, Great Lakes Chemical Corporation, a sole manufacturer of penta-BDE and octa-BDE in North America, voluntarily phased out the production of these BDE formulations [32]. These efforts were boosted by the Stockholm Convention in 2012 when it listed commercial octa-BDE and penta-BDE among persistent organic pollutants that need to be eliminated. Despite the ban in the production of most PBDEs, they are still reported in air, soil and aquatic environments, which is attributed to their stability and subsequent release from techno-ecosystems, and production of deca-BDE, which still continues to be produced in some countries [33, 34].

2.3 PBDEs in the environment and their toxicological implications

There are diverse pathways by which PBDEs enter the environment. Major environmental sources of PBDE pollution comprise of leakage from consumer products and industrial facilities that synthesize PBDEs or PBDE-containing products [5]. Besides, PBDEs may enter the aquatic environment from illegal disposal of obsolete electrical appliances and electronic devices flame-retarded with PBDEs or other PBDEs-containing products [7]. They can also enter the aquatic environment through raw sewage and into the surrounding air through volatilization from products containing PBDEs and toxic fumes from e-waste recycling plants [35]. Since the first discovery of PBDEs in the aquatic environment on the West coast of Sweden in 1981, several studies have reported the presence of PBDEs in the environment [36]. This is despite the strict regulatory measures imposed by some governments and international environmental agencies to phase out some PBDE congeners and subsequent reduction in the production of particular PBDEs. BDE-47, 99, 100, and 153 are the ones that are frequently investigated because they are primary components of commercial mixtures, therefore, their ratios in the environment are expected to be significantly high. Moreover, less substituted BDE congeners such as BDE-28 and 47 are more toxic and non-biodegradable, hence their investigation in the environment and biota is of great significance in the monitoring of these pollutants [37]. Soil and sediment harbour higher concentrations of PBDEs, which is attributed to the organic carbon content, which makes them a sink for most organic pollutants [38]. Elevated levels of PBDEs have since been reported in agricultural soils after the application of sewage sludge at a concentration of 21 to 690 ng g^{-1} dry weight (dw) [39]. From statistics, human beings spend more than 70% of their lifetime indoors, in occupational offices, homes, learning institutions, and transport vehicles, and are therefore exposed to an array of contaminants from indoor dust [40]. The highest levels of PBDEs in dust samples have been reported in major industrialized cities in China and Europe at a concentration of 397–40,236 ng g^{-1} and 950–54,000 ng g^{-1}, respectively [41, 42], with comparably lower levels of 1710 ng g^{-1} in African regions [43]. **Table 1** presents a summary of reported PBDE levels in selected environmental matrices.

The principal route for PBDE exposure to humans was thought to be through food consumption [58]. However, inhalation of contaminated indoor and outdoor dust is also a significant pathway via which human beings may be exposed to PBDEs [46, 59]. Dermal absorption is another potential route for PBDE exposure [60]. Numerous studies have reported levels up to 160.3 ng g^{-1} of PBDEs in human samples, such as serum and milk. Increased application of PBDEs in electronics has significantly aroused more research work on the concentration of these pollutants

Country	Sample matrix	Concentration	Reference
South Africa	River water	2.60–4.83 ng L^{-1}	[44]
North America	River water	0.00013–0.01 ng L^{-1}	[45]
Great Britain	Indoor dust	950–54,000 ng g^{-1}	[46]
South Africa	Home dust	1710 ng g^{-1}	[43]
	Office dust	1520 ng g^{-1}	
Nigeria	Indoor dust	3700–19,000 ng g^{-1}	[47]
China	Indoor dust	397–40,236 ng g^{-1}	[41]
Uganda	Air	0.00340–0.00984 ng m^{-3}	[48]
Kenya	Soil	0.19–35.64 ng g^{-1}	[49]
China	Soil	4.8–533 ng g^{-1}	[50]
China	Sediment	0.03- 5.22 ng g^{-1}	[51]
China	Sediment	0.13–1.98 ng g^{-1}	[52]
Sweden	Sewage sludge	nd-450 ng g^{-1}	[53]
Spain	Sewage sludge	197–1185 ng g^{-1}	[39]
Kuwait	Sewage sludge	52.5–377* ng g^{-1}	[54]
USA	Serum	5.0–27.9	[55]
South Africa	Tigerfish	5.8	[56]
Uganda	Breast milk	0.59–8.11	[57]

nd, not detected.
**Mean concentration.*

Table 1.
Levels of PBDEs reported in the environment and biota from different locales worldwide.

in the blood of workers in e-waste processing plants and other exposed populations [61]. BDE 47, 153, and 209 are the most predominant congeners reported in human serum and milk [55, 62]. The toxicity of PBDEs is backed up by numerous epidemiological studies. Scientific research has linked PBDE exposure to an array of adverse health effects [63]. To mention a few, penta- and octa-BDEs at a concentration of 10,000 ng g^{-1} have been associated with disruption of thyroid hormone homeostasis [7]. Moreover, penta- and tetra-BDEs, within the range of 8000–18,000 ng g^{-1}, have been reported to affect the neurodevelopment of mice [64]. Exposure to high levels of deca-BDEs is likely to cause breast cancer [7]. PBDEs have been linked to developmental neurotoxicity and hence leading to severe effects on cognitive ability, behaviour, and health of both animals and humans [65, 66]. Several studies have also linked PBDEs with adverse effects on the human reproductive system. In particular, BDE-47, BDE-153, and BDE-154 in the range of 0.2–1.6 ng g^{-1} have been confirmed to have negative impacts on testosterone, luteinizing hormone, and estradiol [67]. Therefore, there is a need to have robust, accurate and reproducible methods to quantify PBDEs in different environmental matrices. The sections that follow will discuss these aspects with a particular focus on aquatic media.

3. Sample pre-treatment methods

Sample pre-treatment steps such as pre-concentration and clean-up are paramount before instrumental analysis [2, 68]. These steps ensure that analytes are enriched and converted into the right form/state to achieve their detection and any matrix that may interfere with the determination of the analytes is removed [69].

The choice of sample pre-treatment step is dependent on the physicochemical properties of the targeted analytes, their concentration in the environment, and the complexity of matrix interference [70, 71]. Soxhlet extraction, a traditional liquid-solid extraction method, has been used for decades in the extraction of analytes from their complex solid matrices. With the combination of polar and non-polar solvents, the Soxhlet extraction strategy has been proved to be efficacious, achieving extraction efficiencies greater than 70% [72, 73]. However, this method is hindered by several factors such as long extraction duration, excessive solvent consumption, and the need for subsequent clean-up steps [74]. With increasing demand for economical and fast sample extraction strategies with high enrichment factors, coupled with SPE clean-up procedures, techniques such as UAE, PLE, MAE, and supercritical fluid extraction (SFE) have been adopted in enrichment of analytes from solid matrices.

UAE encompasses the introduction of a finely divided sample contained in a sample holder in an ultrasonic bath with solvent and subjected to ultrasonic radiation. UAE is a vital technique in achieving sustainable green chemistry and is primarily employed in the extraction of analytes from solid sample matrices [75, 76]. This technique can achieve complete extraction with high reproducibility within a short duration. Moreover, small quantities of extraction solvents are used as compared to conventional Soxhlet extraction [77]. Methanol, acetonitrile, ethanol, and acetone are typical extractants used in this method in minimal volume. UAE based on ultrasound assisted-dispersive solid phase extraction (UAE-DSPE) coupled to GC-MS has been reported to achieve exemplary limits of detection and extraction efficiencies for 7 BDE congeners from dust samples collected from air conditioning filters in the range of 1.4–8.4 ng g^{-1} and 90–102%, respectively [78]. Some of the benefits of UAE include faster kinetics and an increase in extraction yield. Ultrasound can also reduce the operating temperature allowing the extraction of thermally labile compounds [79].

Unlike traditional Soxhlet extraction that consumes a large volume of solvent, PLE, also referred to as pressurized solvent extraction, has been of great interest due to its extraction effectiveness. Extraction of analytes from their environmental matrices is achieved via a synergistic mechanism that proceeds through liquid solvents at elevated temperature and pressure, which altogether enhance extraction throughput as compared with other techniques performed at ordinary atmospheric conditions [80]. PLE is viewed as another 'green' option for traditional sample extraction methods. High temperature accomplishes a higher dispersion rate, while high pressure keeps the extraction solvent below its boiling point. During the determination of brominated flame retardants in e-waste samples, PLE and UAE were evaluated in regard to extraction efficiencies. PLE demonstrated high extraction efficiencies of 95–100% as compared to 10–50% for UAE [81]. When contrasted with the conventional methods, PLE shows a decrease in extraction time and a significant decrease in the overall consumption of organic solvents [82].

Another type of extraction technique that enables a three-fold reduction in extraction time and solvent is MAE. This is a sample extraction method that employs microwave energy to extract analytes from solid sample matrices in contact with extraction solvents. Microwave energy directly generates heat which initiates molecular motion of the analytes in the solid-solvent complex mixture, hence facilitating the mass transfer of the target analyte from the solid matrix to the extracting solvent [83, 84]. MAE has been reported to achieve good recoveries of 80–106%, 72.4–108.4%, and 80–110% in the extraction/pre-concentration of PBDEs from airborne particulate matter [85], e-waste materials [86], and sewage sludge samples [87], respectively. Compared with Soxhlet extraction, MAE achieves better recoveries and uses small amounts of solvents (30 mL versus 200 mL for

Soxhlet extraction), at the same time allowing control of extraction parameters, such as extraction time and temperature [88]. However, MAE has some shortcomings, whereby the extracted sample usually contains some matrix interferences, such as lipids and lipophilic compounds, therefore, filtration and clean-up steps are required, which subsequently consume extra organic solvents.

Supercritical fluid extraction (SFE) is another method employed to extract PBDEs from solid matrices. Supercritical CO_2 is often used as an extracting solvent, which has the capability of attaining recoveries above 97%. Moreover, the extraction efficiency of SFE can be further improved by the use of modifiers such as acetonitrile, toluene, and tetrahydrofuran [89]. A successful application of SFE in the extraction of PBDEs from polymeric materials was reported by Peng et al. [90]. The authors used supercritical CO_2 as a solvent and SFE operating parameters such as temperature and pressure were optimized at 65°C and 20 MPa, respectively, achieving 97.6% extraction efficiency. This technique is a greener alternative to other techniques that use a large volume of solvents.

Numerous methodologies have been adopted in the determination of PBDE pollutants in liquid matrices. SPE and conventional LLE have been embraced as routine extraction techniques for PBDEs in liquid samples. The extractive capability of LLE is based on the transfer of analytes from an aqueous polar phase to a non-polar organic phase [91]. LLE coupled with GC-MS has been applied in the determination of 13 PBDEs and their metabolites in water, with recoveries of 77%-102% [92]. LLE has also been a desirable extraction method in the preparation of biota samples for the determination of PBDEs. Recently, a study aimed at assessing *in utero* exposure of 24 tri- to deca- BDE congeners on primiparous mothers in Kampala, Uganda reported a successful application of LLE, with appreciable recoveries of 81–91% [93]. However, LLE has some shortcomings; it suffers from low recovery, poor selectivity, high matrix interference in chromatographic analysis and increased sample loads [94]. In addition, the extraction of PBDEs from water samples requires extremely large volumes of solvents due to their hydrophobic character and low concentration in water, thus limiting its applications [95]. To overcome these challenges, different configurations of SPE have been adopted in sample enrichment strategies. SPE is a modern sample pre-treatment technique employed to concentrate analytes from liquid samples and to remove matrix interferents during the clean-up step, achieving exemplary recoveries and reproducible results over LLE [96, 97]. SPE protocols are usually performed by the use of a small column or separation cartridge packed with an appropriate sorbent material [98, 99]. Target analytes are adsorbed by the sorbent materials and later eluted with a solvent that has a greater affinity for the analytes. The chemistry behind this separation is based on intermolecular forces between the analytes, active sites of the adsorbent, and the liquid phase of the matrix [100]. SPE can be performed through an on-line or off-line approach. The on-line SPE configuration, which may enable automation, is directly coupled with specific analytical systems such as gas chromatography (GC) or high-performance liquid chromatography (HPLC). Whereas in the off-line protocol, a pre-concentration step is done separately using cartridges and further eluting the adsorbed analyte with an appropriate solvent for eventual chromatographic analysis [101]. Because of its robustness and flexibility, SPE has been widely employed in different analytical procedures in pre-concentration and clean-up steps in the determination of PBDEs [96, 102].

While SPE continues to be used because of its affordability and ease of use, other formats that offer high enrichment factors and shorter extraction times, such as SPME, stir-bar sorptive extraction (SBSE) and dispersive solid-phase extraction (DSPE), have been introduced [103]. SPME is an innovation and improvement of conventional SPE. Its stationary phase comprises of fused-silica fibers coated with a polydimethylsiloxane (PDMS) layer which are reusable. With this new formulation, the application of SPE has become versatile such that it can accommodate small volumes of samples. Furthermore, SPME has been considered an almost solvent-free extraction technique and can be easily automated as compared to conventional SPE [104, 105]. A miniaturized SPME has been applied in the extraction of PBDEs in environmental water samples followed by GC-MS quantitation, with low limits of detection and appreciable recoveries of 76.5–125.4% [106]. SBSE is a similar technique to SPME that has been adopted in the enrichment of PBDEs in liquid samples due to its improved extraction efficiency. The stir bars are coated with a thinner PDMS layer, as opposed to a thicker layer in SPME, a factor that allows improved enrichment efficiency [107, 108]. DSPE is another format of SPE based on the dispersion of solid sorbent materials in liquid samples to facilitate the isolation and extraction of target analytes from the complex sample matrix. In this process, matrix interferences remain embedded in the supernatant, which is later discarded while the target analyte is bound to the sorbent material and which is eventually eluted with a viable solvent [109]. DSPE has been employed in the enrichment and determination of PBDEs with recoveries within the range of 60–140% [110].

3.1 Advances in SPE sorbents

Complexity and matrix interferences encountered during sample preparation steps have attracted the invention of more selective sorbents to replace conventional silica sorbents that are associated with a number of drawbacks, such as instability at extreme pHs and low extraction efficiencies [111]. The new sorbents that include, nanocomposite materials, metal-organic frameworks, and molecularly imprinted polymers, among others, are characterized with high sensitivity and selectivity towards various environmental organic pollutants. They achieve fast dispersion and efficient recycling when applied in complex sample matrices [112, 113]. Reported nanocomposite sorbents in SPE for PBDE-containing samples include carbon nanotubes, graphene oxide (GO) [114, 115], and magnetic nanocomposite materials [113]. However, nanocomposite sorbents in classical SPE schemes have been associated with various drawbacks. A few of these challenges have been described in flow as well as batch systems, which originate from a slow flow rate of the sample through the packed SPE column and difficulty in separating the sorbent from the large volume of aqueous sample [113].

Other sorbent materials with fascinating properties are metal-organic frameworks (MOFs). These are hybrids of organic and inorganic materials characterized by a porous structure, large surface area, uniform nanoscale cavities, high adsorption capacity, and high thermal and chemical stability. Due to these advantageous properties, this class of materials has recently attracted enormous attention in the field of sample preparation [116]. The development of MOF adsorbents is still at its infancy stages, therefore, a limited number of studies have reported their application particularly in enrichment and determination of environmental PBDEs. A zirconium-based metal-organic framework material (UiO-66-OH) is a good example of a MOF. It has been synthesized and successfully applied as an adsorbent in SPME for enrichment and detection of 5 BDE congeners in milk samples using GC-MS, with low limits of detection in the range of 0.15–0.35 ng L^{-1} and excellent recoveries of 74.7%–118.0% [117]. A contrast study using silica-based sorbents in SPE for determination of 12 PBDEs in human serum, achieved mean recoveries of 64–95% and limits of detection in the range of 0.1–4.0 ng g^{-1} by using GC-MS [102], an evidence that MOF sorbents offer promising analytical results as compared with conventional sorbents.

With growing interest in sorbents that offer extraordinary extractive capability in SPE, molecularly imprinted polymers (MIPs) have been extensively explored as attractive options due to their robustness and selectivity towards particular target analytes providing exemplary substitute sorbents in sample clean-up and pre-concentration steps, especially in SPE and SPME [118]. MIPs are synthesized through molecular imprinting technology that involves polymerizing functional and cross-linking monomers in the presence of a target analyte, followed by the removal of the analyte to leave behind analyte-specific cavities. Their selectivity enables substantive removal of matrix interferents during the sample pre-treatment step [119]. MIP-based sorbents are readily available substitutes to silica-based adsorbents, which are reported to suffer from matrix interference, low selectivity, and sensitivity towards organic pollutants and may involve multiple steps that are labour-intensive for complete removal of interferences [120]. For example, commercial molecularly imprinted solid-phase extraction (MISPE) cartridges alongside alkaline extraction have been applied in aqueous enrichment and quantitation of PBDEs using GC-MS [121]. The extraction of PBDEs using MISPE gave recoveries above 60% compared to alkaline extraction which was below 60%. This confirms the selectivity capability of MIPs towards PBDEs from a complex environmental matrix. A more recent study has also reported recoveries of 60–87% in clean-up of soil and sediment samples using dummy molecularly imprinted polymers as SPE sorbent materials during determination of BDE-47 and BDE-99 [122].

Sample preparation technique	PBDE congeners	Sample analyzed	Analytical technique	% Recoveries	Reference
SPE	BDE-28, 47, 49, 66, 85, 99, 100, 138, 153, 154, 183 & 209	Human serum	GC-ECD	64–95	[102]
	BDE-47 and 99	Soil and bottom sediment	GC-MS	60–87	[125]
PLE	BDE-28, 47, 99, 100, 153, 154 & 183	Soil	GC-MS	95 ± 9	[68]
	BDE-28, 47, 99, 100, 154, 155 & 183	Soil and sediment	GC-MS	84–103	[92]
LLE	BDE-17, 28, 47, 66, 71, 85, 99, 100, 138, 153, 154, 183 & 190	Soil and sediment	GC-MS	85–103	[92]
Soxhlet extraction	42 mono- to deca-BDEs	Indoor dust sample	GC-MS	≥ 70	[72]
UAE	BDE-1, 3, 7, 8, 28 & 47	Industrial effluent	HPLC	98.7	[126]
SPME	BDE-49, 99, 100, 153 & 154	Milk and water	GC-ECD	90–119	[127]
MAE	BDE-47, 99, 100, 138, 153, 154, 184 & 209	Sewage sludge	GC-MS	80–110	[87]

Table 2.
Examples of sample preparation strategies.

However, a wide range of limitations still exist in MIPs, especially their poor water compatibility. Consequently, since MIPs and target analytes mainly interact through hydrogen-bonding, their recognition capability would be easily disturbed by polar solvents such as water. Therefore, the adsorption process is normally performed in non-polar or low-polar solvents such as dichloromethane and *n*-hexane rather than polar solvents. Additionally, polar solvents have a tendency to occupy binding sites, which affects the recognition capacity for the target analytes. In this context, it is necessary to continually invent new synthesis strategies for water-compatible MIPs [123, 124]. A summary of some of the sample pre-concentation strategies and their extraction efficiencies is presented in **Table 2**.

4. Analytical techniques for the detection of PBDEs

Sample pre-treatment steps are followed by quantification of the analytes using various detection systems. The choice of detection system depends on the physicochemical properties of the target analyte and the required detection levels. Detection techniques for quantification of PBDEs have evolved from liquid chromatography to gas chromatography and recently, miniaturized systems that involve the use of sensors. For chromatographic techniques, it's important to optimize the operational parameters to actualize reliable instrumental results. It is highly recommended to use a sample injector with programmed temperature vaporization (PTV) to avoid degradation of thermally labile BDE congeners. Additionally, the temperature of injection should be accurately defined, especially when using a split/splitless injector, which minimizes chances of thermal degradation of higher BDEs congers as well as discrimination of lower brominated congeners [95, 128]. The choice of a column is another important aspect in the analysis of PBDEs where lower brominated congeners are well separated on longer columns, whereas higher brominated congeners are well separated on shorter columns. In the case of a mixture comprising of a wide range of BDE congeners, a short column is highly recommended, which well separates nona- and deca-BDEs [129]. HPLC coupled with mass spectrometry (MS), is one of the chromatographic techniques which has rarely been applied in the quantification of some PBDE congeners. The HPLC separation is hindered by several factors such as poor solubility of highly brominated diphenyl ethers in the polar solvents of the mobile phase, especially in reversed-phase, and, thus, requiring the sample to be enriched with an organic modifier. Normal phase HPLC has offered better separation of some PBDEs though it still results in incomplete separation, especially when an electrospray ionization detector is incorporated [130]. One group used an automated on-line sample pre-concentration device coupled with HPLC-MS to determine decabrodiphenyl ether in human serum samples. This method achieved detection limits of 26.0 ng L^{-1} [130]. Otherwise, better detection limits of 0.2-25 ng L^{-1} were tenable when similar samples were analyzed for 12 PBDEs including decabromodiphenyl ether using gas chromatography-electron capture detection (GC-ECD) [102]. However, GC-ECD exhibits low selectivity and suffers from matrix interferences originating especially from halogenated species, as compared to GC-MS, which overcomes these challenges [131]. Fontana et al. [16] employed a coupled system, ultrasound-assisted emulsification microextraction-GC-MS (UAEMA-GC-MS) to determine PBDEs in water samples, with appreciably low detection limits of 1–2 ng L^{-1}. Moreover, lower limits of detection are achievable when tandem-mass spectrometry (MS^2) is utilized. For example, GC-MS^2 has been reported to achieve detection limits within the range of 0.002–0.0136 ng g^{-1} lipid weight (lw) in the determination of PBDEs in breast milk and serum samples [132].

With the recent technological revolution, a more sensitive mass spectrometer, a high-resolution mass spectrometer (HRMS), has been found to be a promising alternative to a conventional mass spectrometer as it identifies the analyte without mass fragmentation and at the lowest mass unit [133]. With this new format of detection, very low detection limits of 0.000262–0.046 ng g^{-1} for 23 PBDEs in dust samples were achieved [134]. However, GC-HRMS is more expensive than conventional GC-MS, compelling researchers to often rely on GC-MS since it is less expensive and readily available. Besides, the demand for techniques that provide rapid results at minimal cost has resulted in the introduction of sensor technology in the determination of PBDEs. In this context, various detection systems have been fabricated and shown a discerning capability in the detection of PBDEs. For instance, an immunoassay detection system based on graphene oxide-polydimethylsiloxane has demonstrated desirable limits of detection of 0.018 ng g^{-1} for PBDEs in a standard solution and environmental water samples [135]. Similarly, a novel electrochemical immunoassay sensor used for the detection of BDE-28, 47, 99, 100, 153, and 154 in food samples, achieved a detection limit of 0.00018 ng L^{-1} [136]. These limits are comparable with those obtained by HPLC, GC-MS, or GC-HRMS. A surface-enhanced Raman scattering-based sensor is another detection system that has been successfully applied for rapid detection of BDE 47 in aqueous media, with detection limits of 0.0364 ng L^{-1} [137]. The use of sensory techniques is cheaper and a low concentration of contaminants can be detected. Moreover, the analysis duration is reduced from 10 minutes to 3 minutes. Thus, these sensor methods offer scope for further evaluation.

5. Conclusion

This chapter has discussed PBDEs as emerging environmental pollutants, their sources, and toxicological implications on humans and their determination in the environment. Sample pre-concentration methods for PBDE-containing samples that include UAE, PLE, UAME, PLE, SFE, SPE, SPME, SBSE, and DSPE have been critically reviewed as preferred alternatives to LLE and Soxhlet extraction due to their enhanced extraction efficiency. Novel SPE and SPME sorbents that provide the desired selectivity in the determination of PBDEs have also been discussed. Though these sorbents are promising, their application in MISPE in the determination of PBDEs has been scantly employed and its dynamics are still at its infancy stages. Therefore, there is room for continuous introduction of highly selective materials for the quantification of PBDEs in the environment. Alongside the evolution of sample pre-treatment techniques for the detection of PBDEs, rapid sensor-based techniques that achieve the desired figures of merit similar to traditional instrumentation techniques have demonstrated great potential.

Acknowledgements

ENN is grateful to the FLAIR fellowship programme, which is a partnership between the African Academy of Sciences and the Royal Society, funded by the UK Government's Global Challenges Research Fund (GCRF), for financial support. BSM and VON thank the National Research Foundation (NRF) of South Africa, the University of KwaZulu-Natal (UKZN) and the UKZN Nanotechnology Platform for research support. BSM is also grateful for support from the European Union's Horizon 2020 research and innovation programme under the Marie Skłodowska-Curie grant agreement No 734522, and for funding from NAS, USAID and DST

(South Africa), under the PEER program cooperative agreement number: No. AID-OAA-A-11-00012.

Conflict of interest

The authors declare no conflict of interest.

Author details

Japheth M. Nzangya[1], Elizabeth N. Ndunda[1*], Geoffrey O. Bosire[2], Bice S. Martincigh[3] and Vincent O. Nyamori[3]

1 Department of Physical Sciences, School of Pure and Applied Sciences, Machakos University, Machakos, Kenya

2 Department of Chemistry, School of Physical Sciences, University of Nairobi, Nairobi, Kenya

3 School of Chemistry and Physics, University of KwaZulu-Natal, Westville Campus, Private Bag X54001, Durban, South Africa

*Address all correspondence to: elizabeth.ndunda@mksu.ac.ke

References

[1] Priti and K. Mandal, "Review on evolution of municipal solid waste management in India: practices, challenges and policy implications," *J. Mater. Cycles Waste Manag.*, vol. 21, pp. 1263-1279, 2019.

[2] Martín-Pozo L, de Alarcón-Gómez B, Rodríguez-Gómez R, García-Córcoles MT, Çipa M, Zafra-Gómez A. Analytical methods for the determination of emerging contaminants in sewage sludge samples. A review. *Talanta*. 2019;**192**:508-533

[3] Adeyemo OK. Consequences of pollution and degradation of Nigerian aquatic environment on fisheries resources. *Environmentalist*. 2004;**23**:297-306

[4] C. Bogdal, A. C. Gerecke, P. Schmid, N. V Heeb, and M. Sturm, "Temporal trends, congener patterns, and sources of octa-, nona-, and decabromodiphenyl ethers (PBDE) and hexabromocyclododecanes (HBCD) in Swiss lake sediments," *Environ. Sci. Technol.*, vol. 42, pp. 6378-6384, 2008.

[5] Vonderheide AP, Mueller KE, Meija J, Welsh GL. Polybrominated diphenyl ethers: Causes for concern and knowledge gaps regarding environmental distribution, fate and toxicity. *Sci. Total Environ.* 2008;**400**:425-436

[6] A. Besis and C. Samara, "Polybrominated diphenyl ethers (PBDEs) in the indoor and outdoor environments - A review on occurrence and human exposure," *Environ. Pollut.*, vol. 169, pp. 217-229, 2012.

[7] Darnerud PO, Eriksen GS, Jóhannesson T, Larsen PB, Viluksela M. Polybrominated diphenyl ethers: Occurrence, dietary exposure, and toxicology chemical and physical properties of PBDEs. *Environ. Health Perspect.* 2001;**109**:49-68

[8] Mariussen E et al. Elevated levels of polybrominated diphenyl ethers (PBDEs) in fish from Lake Mjøsa, Norway. *Sci. Total Environ.* 2007;**390**:132-141

[9] B. Li, M. N. Danon-schaffer, and J. R. Grace, "Occurrence of PFCs and PBDEs in landfill leachates from across Canada," *Water, Air Soil Pollut.*, vol. 223, pp. 3365-3372, 2012.

[10] Petreas M, Oros D. Polybrominated diphenyl ethers in California wastestreams. *Chemosphere*. 2009;**74**:996-1001

[11] Gouin T, Thomas GO, Chaemfa C, Harner T. Concentrations of decabromodiphenyl ether in air from Southern Ontario: Implications for particle-bound transport. *Chemosphere*. 2006;**64**:256-261

[12] Mandalakis M, Atsarou V, Stephanou EG. Airborne PBDEs in specialized occupational settings, houses and outdoor urban areas in Greece. *Environ. Pollut.* 2008;**155**:375-382

[13] Ocio AB, Lobet JML, Omingo JLD, Orbella JC, Eixido AT. Polybrominated diphenyl ethers (PBDEs) in foodstuffs: Human exposure through the diet. *J. Agric. Food Chem.* 2003;**51**:3191-3195

[14] Wang P et al. Evaluation of Soxhlet extraction, accelerated solvent extraction and microwave-assisted extraction for the determination of polychlorinated biphenyls and polybrominated diphenyl ethers in soil and fish samples. *Anal. Chim. Acta*. 2010;**663**:43-48

[15] Quiroz R, Arellano L, Grimalt JO, Fern P. Analysis of polybrominated diphenyl ethers in atmospheric deposition and snow samples by solid-phase disk extraction. *J. Chromatogr. A*. 2008;**1192**:147-151

[16] Fontana AR, Wuilloud RG, Martínez LD, Altamirano JC. Simple approach based on ultrasound-assisted emulsification-microextraction for determination of polibrominated flame retardants in water samples by gas chromatography-mass spectrometry. *J. Chromatogr. A*. 2009;**1216**:147-153

[17] E. Bizkarguenaga et al., "Focused ultrasound assisted extraction for the determination of PBDEs in vegetables and amended soil," *Talanta*, vol. 119, pp. 53-59, 2014.

[18] Ghosh R, Hageman KJ, Björklund E. Selective pressurized liquid extraction of three classes of halogenated contaminants in fish. *J. Chromatogr. A*. 2011;**1218**:7242-7247

[19] Vazquez-roig P, Picó Y. Pressurized liquid extraction of organic contaminants in environmental and food samples. *Trends Anal. Chem.* 2015;**71**:55-64

[20] M. Karlsson, A. Julander, B. Van Bavel, and G. Lindstro, "Solid-phase extraction of polybrominated diphenyl ethers in human plasma-comparison with an open column extraction method," *Chromatographia*, vol. 61, pp. 67-73, 2005.

[21] Liu X et al. Solid-phase extraction combined with dispersive liquid – liquid microextraction for the determination for polybrominated diphenyl ethers in different environmental matrices. *J. Chromatogr. A*. 2009;**1216**:2220-2226

[22] Zhang M, Zeng J, Wang Y, Chen X. Developments and trends of molecularly imprinted solid-phase microextraction. *J. Chromatogr. Sci.* 2013;**51**:577-586

[23] Stapleton HM. Instrumental methods and challenges in quantifying polybrominated diphenyl ethers in environmental extracts. *Anal Bioanal Chem*. 2006;**386**:807-817

[24] Rigét F, Vorkamp K, Dietz R, Rastogi SC. Temporal trend studies on polybrominated diphenyl ethers (PBDEs) and polychlorinated biphenyls (PCBs) in ringed seals from East Greenland. *J. Environ. Monit.* 2006;**8**:1000-1005

[25] A. Schecter, M. Pavuk, O. Päpke, J. J. Ryan, L. Birnbaum, and R. Rosen, "Polybrominated diphenyl ethers (PBDEs) in U.S. mothers' milk," *Environ. Health Perspect.*, vol. 111, pp. 1723-1729, 2003.

[26] Kuramochi H, Maeda K, Kawamoto K. Physicochemical properties of selected polybrominated diphenyl ethers and extension of the UNIFAC model to brominated aromatic compounds. *Chemosphere*. 2007;**67**:1858-1865

[27] Harner T, Shoeib M. Measurements of octanol-air partition coefficients (KOA) for polybrominated diphenyl ethers (PBDEs): Predicting partitioning in the environment. *J. Chem. Eng.* 2002;**47**:228-232

[28] Wei-haas ML, Hageman KJ, Chin Y. Partitioning of polybrominated diphenyl ethers to dissolved organic matter isolated from Arctic surface waters. *Environ. Sci. Technol.* 2014;**48**:4852-4859

[29] Li Y et al. Influences of binding to dissolved organic matter on hydrophobic organic compounds in a multi-contaminant system: Coefficients, mechanisms and ecological risks. *Environ. Pollut.* 2015;**206**:461-468

[30] Winid B. Environmental threats of natural water contamination with polybrominated diphenyl ethers (PBDEs). *Polish J. Environ. Stud.* 2015;**24**:47-55

[31] Siddiqi MA, Kurt RD. Polybrominate diphenyl ethers

(PBDEs): New pollutants-old diseases. *Clin. Med. Reaserach*. 2003;**1**:281-290

[32] Rodigari F, Crane D, Sericano J. Levels and Distribution of Polybrominated Diphenyl Ethers in Water, Surface Sediments, and Bivalves from the San Francisco Estuary. *Environ. Sci. Technol*. 2005;**39**:33-41

[33] Kodavanti PRS, Valdez MC, Yamashita N. *Brominated flame retardants and perfluorinated chemicals*. Third Edit: Elsevier Inc.; 2018

[34] Taylor P, Akutsu K. Food additives & contaminants: Part B: Surveillance dietary intake estimations of polybrominated diphenyl ethers (PBDEs) based on a total diet study in Osaka, Japan View Dataset. *Food Addit. Contam.* 2008;**1**:58-68

[35] Darnerud PO, Hallgren S. Polybrominated diphenyl ethers (PBDEs), polychlorinated biphenyls (PCBs) and chlorinated paraffins (CPs) in rats - Testing interactions and mechanisms for thyroid hormone effects. *Toxicology*. 2002;**177**:227-243

[36] Sjödin A, Hagmar L, Klasson-wehler E, Kronholm-diab K, Jakobsson E, Bergman A. Flame retardant exposure: Polybrominated diphenyl ethers in blood from Swedish workers. *Environ. Health Perspect.* 1999;**107**:643-648

[37] P. R. S. Kodavanti, J. E. Royland, C. Osorio, W. M. Winnik, and P. Ortiz, "Developmental exposure to a commercial PBDE mixture: Effects on protein networks in the cerebellum and hippocampus of rats," *Environ. Health Perspect.*, vol. 123, pp. 428-436, 2015.

[38] Gevao B, Ghadban AN, Uddin S, Jaward FM, Bahloul M, Zafar J. Polybrominated diphenyl ethers (PBDEs) in soils along a rural-urban-rural transect: Sources, concentration gradients, and profiles. *Environ. Pollut.* 2011;**159**:3666-3672

[39] Eljarrat E, Labandeira A, Marsh G. Effect of sewage sludges contaminated with polybrominated diphenylethers on agricultural soils. *Chemosphere*. 2008;**71**:1079-1086

[40] Huang Y, Chen L, Peng X, Xu Z, Ye Z. PBDEs in indoor dust in South-Central China : Characteristics and implications. *Chemosphere*. 2010;**78**:169-174

[41] Kang Y, Wang HS, Cheung KC, Wong MH. Polybrominated diphenyl ethers (PBDEs) in indoor dust and human hair. *Atmos. Environ.* 2011;**45**:2386-2393

[42] Harrad S, Ibarra C, Diamond M, Melymuk L, Robson M, Douwes J. Polybrominated diphenyl ethers in domestic indoor dust from Canada, New Zealand, United Kingdom and United States. *Environ. Int.* 2008;**34**:232-238

[43] Abafe OA, Martincigh BS. Polybrominated diphenyl ethers and polychlorinated biphenyls in indoor dust in Durban, South Africa. *Environ. Sci. Pollut. Res.* 2015;**25**:547-556

[44] Daso AP, Fatoki OS, Odendaal JP. Occurrence of polybrominated diphenyl ethers (PBDEs) samples from the Diep River, Cape Town, South Africa. *Environ. Sci. Pollut.* 2013;**20**:5168-5176

[45] Streets SS, Henderson SA, Stoner AD, Carlson DL, Simcik MF, Swackhamer DL. Partitioning and bioaccumulation of PBDEs and PCBs in Lake Michigan. *Environ. Sci. Technol.* 2006;**40**:7263-7269

[46] Banasik M, Hardy M, Stedeford T. An assessment of the human risks from exposure to polybrominated diphenyl ethers (PBDEs) in house dust. *Chemosphere*. 2009;**77**:704-705

[47] Harrad S, Abdallah MA, Oluseyi T. Polybrominated diphenyl ethers and polychlorinated biphenyls in dust from cars, homes, and offices in Lagos, Nigeria. *Chemosphere*. 2016;**146**:346-353

[48] Arinaitwe K, Muir DCG, Kiremire BT, Fellin P, Li H, Teixeira C. Polybrominated diphenyl ethers and alternative flame retardants in air and precipitation samples from the Northern Lake Victoria region, East Africa. *Environ. Sci. Technol.* 2014;**66**:1453-1461

[49] H. Sun, Y. Qi, D. Zhang, Q. X. Li, and J. Wang, "Concentrations, distribution, sources and risk assessment of organohalogenated contaminants in soils from Kenya, Eastern Africa," *Environ. Pollut.*, vol. 209, pp. 177-185, 2016.

[50] Wang Y, Luo C, Li J, Yin H, Li X, Zhang G. Characterization of PBDEs in soils and vegetations near an e-waste recycling site in South China. *Environ. Pollut.* 2011;**159**:2443-2448

[51] Chen L et al. PBDEs in sediments of the Beijiang River, China: Levels, distribution, and influence of total organic carbon. *Chemosphere*. 2009;**76**:226-231

[52] Zhao X, Zhang H, Ni Y, Lu X, Zhang X, Su F. Polybrominated diphenyl ethers in sediments of the Daliao River Estuary, China: Levels, distribution and their influencing factors. *Chemosphere*. 2011;**82**:1262-1267

[53] Karin O, Warman K, Tomas O. Distribution and levels of brominated flame retardants in sewage sludge. *Chemosphere*. 2002;**48**:805-809

[54] Gevao B, Muzaini S, Helaleh M. Occurrence and concentrations of polybrominated diphenyl ethers in sewage sludge from three wastewater treatment plants in Kuwait. *Chemosphere*. 2008;**71**:242-247

[55] Sjödin A, Wong L, Jones RS, Park A, Zhang Y, Hodge C. Serum concentrations of polybrominated diphenyl ethers (PBDEs) and polybrominated biphenyl (PBB) in the United States population: 2003-2004. *Environ. Sci. Technol*. 2008;**42**:1377-1384

[56] Wepener V, Smit N, Covaci A, Dyke S, Bervoets L. Seasonal bioaccumulation of organohalogens in Tigerfish, Hydrocynus vittatus castelnau, from Lake Pongolapoort, South Africa. *Bull Env. Contam Toxicol*. 2012;**88**:277-282

[57] Matovu H, Sillanpää M, Ssebugere P. Polybrominated diphenyl ethers in mothers' breast milk and associated health risk to nursing infants in Uganda. *Sci. Total Environ.* 2019;**692**:1106-1115

[58] Ni H, Ding C, Lu S, Yin X, Olatunbosun S. Food as a main route of adult exposure to PBDEs in Shenzhen, China. *Sci. Total Environ.* 2012;**437**:10-14

[59] N. Wu, T. Herrmann, O. Paepke, J. Tickner, R. Hale, and E. Harvey, "Human exposure to PBDEs: Associations of PBDE body burdens with food consumption and house dust concentrations," vol. 41, pp. 1584-1589, 2007.

[60] Abdallah MA, Pawar G, Harrad S. Effect of bromine substitution on human dermal absorption of polybrominated diphenyl ethers effect of bromine substitution on human dermal absorption of polybrominated diphenyl ethers. *Environ. Sci. Technol.* 2015;**49**:10976-10983

[61] Kuo L, Cade SE, Cullinan V, Schultz IR. Polybrominated diphenyl ethers (PBDEs) in plasma from e-waste recyclers, outdoor and indoor workers in the Puget Sound, WA region. *Chemosphere*. 2019;**219**:209-216

[62] Wang C, Lin Z, Dong Q, Lin Z, Lin K, Wang J. Ecotoxicology and environmental safety polybrominated diphenyl ethers (PBDEs) in human serum from Southeast China. *Ecotoxicol. Environ. Saf.* 2012;**78**:206-211

[63] He Y, Murphy MB, Yu RM, Lam MH, Hecker M, Giesy JP. Effects of 20 PBDE metabolites on steroidogenesis in the H295R cell line. *Toxicol. Appl. Pharmacol.* 2008;**176**:230-238

[64] Ji K, Choi K, Giesy JP, Musarrat J, Takeda S. Genotoxicity of several polybrominated diphenyl ethers (PBDEs) and hydroxylated PBDEs, and their mechanisms of toxicity. *Environ. Sci. Technol.* 2011;**45**:5003-5008

[65] Eriksson P, Jakobsson E, Fredriksson A. Brominated flame retardants: A novel class of developmental neurotoxicants in our environment? *Environ. Health Perspect.* 2001;**109**:903-908

[66] Chen L, Huang C, Yu L, Zhu B, Lam J, Zhou B. Prenatal transfer of polybrominated diphenyl ethers (PBDEs) results in developmental neurotoxicity in zebra fish larvae. *Environ. Sci. Technol.* 2012;**46**:9727-9734

[67] Roze E, Meijer L, Bakker A, Van Braeckel KNJA, Sauer PJJ, Bos AF. Prenatal exposure to organohalogens, including brominated flame retardants, influences motor, cognitive, and behavioral performance at school age. *Environ. Health Perspect.* 2009;**117**:1953-1958

[68] Zhang Z, Shanmugam M, Rhind SM. PLE and GC-MS determination of polybrominated diphenyl ethers in soils. *Chromatographia*. 2010;**72**:535-543

[69] Kole PL, Venkatesh G, Kotecha J, Sheshala R. Recent advances in sample preparation techniques for effective bioanalytical methods. *Biomed. Chromatogr.* 2011;**25**:199-217

[70] Covaci A, Voorspoels S, Ramos L, Neels H, Blust R. Recent developments in the analysis of brominated flame retardants and brominated natural compounds. *J. Chromatogr. A*. 2007;**1153**:145-171

[71] Berton P, Lana NB, Ríos JM, García-reyes JF, Altamirano JC. State of the art of environmentally friendly sample preparation approaches for determination of PBDEs and metabolites in environmental and biological samples : A critical review. *Anal. Chim. Acta*. 2016;**905**:24-41

[72] Gevao B et al. House dust as a source of human exposure to polybrominated diphenyl ethers in Kuwait. *Chemosphere*. 2006;**64**:603-608

[73] Castro L, Capote P. Soxhlet extraction : Past and present panacea. *J. Chromatogr. A*. 2010;**1217**:2383-2389

[74] Da C, Wang R, Ye J, Yang S. Sediment records of polybrominated diphenyl ethers (PBDEs) in Huaihe River, China: Implications for historical production and household usage of PBDE-containing products. *Environ. Pollut.* 2019;**254**:112955-112963

[75] Hemwimol S, Pavasant P, Shotipruk A. Ultrasound-assisted extraction of anthraquinones from roots of Morinda citrifolia. *Ultrason. Sonochem.* 2006;**13**:543-548

[76] Tadeo L, Rosa AP, Albero B. Ultrasound-assisted extraction of organic contaminants. *Trends Anal. Chem.* 2019;**118**:739-750

[77] Chemat F, Rombaut N, Sicaire A, Meullemiestre A, Abert-vian M. Ultrasound assisted extraction of food and natural products. Mechanisms, techniques, combinations, protocols and applications. *Ultrason. Sonochem.* 2017;**34**:540-560

[78] Śmiełowska M, Zabiegała B. Determination of polybrominated diphenyl ethers (PBDEs) in dust samples collected in air conditioning filters of different usage – method development. *J. Chromatogr. A*. 2018;**1565**:57-67

[79] A. Jorge Moreda-Pineiro and Moreda-pineiro, "Combined assisted extraction techniques as green sample pre-treatments in food analysis," *Trends Anal. Chem.*, vol. 118, pp. 1-18, 2019.

[80] Mustafa A, Turner C. Pressurized liquid extraction as a green approach in food and herbal plants extraction : A review. *Anal. Chim. Acta*. 2011;**703**:8-18

[81] Vilaplana F, Karlsson P, Ribes-Greus A, Ivarsson P, Karlsson S. Analysis of brominated flame retardants in styrenic polymers. Comparison of the extraction efficiency of ultrasonication, microwave-assisted extraction and pressurised liquid extraction. *J. Chromatogr. A*. 2008;**1196-1197**:139-146

[82] Andreu V, Pic Y. Pressurized liquid extraction of organic contaminants in environmental and food samples. *Trends Anal. Chem.* 2019;**118**:709-721

[83] Flórez N, Conde E, Domínguez H. Microwave assisted water extraction of plant compounds. *J. Chem. Technol. Biotechnol.* 2015;**90**:590-607

[84] Llompart M, Celeiro M, Dagnac T. Microwave-assisted extraction of pharmaceuticals, personal care products and industrial contaminants in the environment. *Trends Anal. Chem.* 2019;**116**:136-150

[85] Beser MI, Beltrán J, Yusà V. Design of experiment approach for the optimization of polybrominated diphenyl ethers determination in fine airborne particulate matter by microwave-assisted extraction and gas chromatography coupled to tandem mass spectrometry. *J. Chromatogr. A*. 2014;**1323**:1-10

[86] Li Y, Wang T, Hashi Y, Li H, Lin J. Determination of brominated flame retardants in electrical and electronic equipments with microwave-assisted extraction and gas chromatography-mass spectrometry. *Talanta*. 2009;**78**:1429-1435

[87] Shin M, Svoboda ML, Falletta P. Microwave-assisted extraction (MAE) for the determination of polybrominated diphenylethers (PBDEs) in sewage sludge. *Anal Bioanal Chem*. 2007;**387**:2923-2929

[88] Pérez-Lemus N, López-Serna R, Pérez-Elvira SI, Barrado E. Analytical methodologies for the determination of pharmaceuticals and personal care products (PPCPs) in sewage sludge: A critical review. *Anal. Chim. Acta*. 2019;**1083**:19-40

[89] Shao M, Jiang J, Li M, Wu L, Hu M. Recent developments in the analysis of polybrominated diphenyl ethers and polybrominated biphenyls in plastic. *Rev. Anal. Chem.* 2016;**35**:133-143

[90] Peng S, Liang S, Yu M, Li X. Extraction of polybrominated diphenyl ethers contained in waste high impact polystyrene by supercritical carbon dioxide. *J. Mater. Cycles Waste Manag.* 2014;**16**:178-185

[91] Pedersen-Bjergaard S, Rasmussen KE, Grønhaug Halvorsen T. Liquid-liquid extraction procedures for sample enrichment in capillary zone electrophoresis. *J. Chromatogr. A*. 2000;**902**(1):91-105

[92] Hu X, Hu D, Chen W, Wu B, Lin C. Simultaneous determination of methoxylated polybrominated diphenyl ethers and polybrominated diphenyl ethers in water, soil and sediment from China by GC-MS. *J. Chromatogr. Sci.* 2015;**53**:1239-1249

[93] Matovu H, Ssebugere P, Ssebugere P. Prenatal exposure levels of polybrominated diphenyl ethers in mother-infant pairs and their transplacental transfer characteristics in Uganda (East Africa). *Environ. Pollut.* 2020;**258**:113723-113767

[94] Zhou S, Song Q, Tang Y, Naidong W. Critical review of development, validation, and transfer for high throughput bioanalytical LC-MS/MS methods. *Curr. Pharm. Anal.* 2006;**1**:3-14

[95] Fulara I, Czaplicka M. Methods for determination of polybrominated diphenyl ethers in environmental samples – review. *J. Sep. Sci.* 2012;**35**:2075-2087

[96] Chen Y, Xia L, Liang R, Lu Z, Li L, Huo B. Advanced materials for sample preparation in recent decade. *Trends Anal. Chem.* 2019;**120**:115652-115666

[97] Han F et al. Solid-phase extraction of seventeen alternative flame retardants in water as determined by ultra-high-performance liquid chromatography-tandem mass spectrometry. *J. Chromatogr. A*. 2019;**1602**:64-73

[98] Wille SMR, Lambert WEE. Recent developments in extraction procedures relevant to analytical toxicology. *Anal Bioanal Chem*. 2007;**388**:1381-1391

[99] Huck CW, Bonn GK. Recent developments in polymer-based sorbents for solid-phase extraction Recent developments in polymer-based sorbents for solid-phase extraction. *J. Chromatogr. A*. 2000;**9673**:51-72

[100] García-Córcoles MT et al. Chromatographic methods for the determination of emerging contaminants in natural water and wastewater samples: A Review. *Crit. Rev. Anal. Chem.* 2019;**49**:160-186

[101] Vasapollo G et al. Molecularly imprinted polymers: Present and future prospective. *Int. J. Mol. Sci.* 2011;**12**:5908-5945

[102] Covaci A, Voorspoels S. Optimization of the determination of polybrominated diphenyl ethers in human serum using solid-phase extraction and gas chromatography-electron capture negative ionization mass spectrometry. *J. Chromatogr. B*. 2005;**827**:216-223

[103] Vallecillos L, Pocurull E, Borrull F. A simple and automated method to determine macrocyclic musk fragrances in sewage sludge samples by headspace solid-phase microextraction and gas chromatography – mass spectrometry. *J. Chromatogr. A*. 2013;**1314**:38-43

[104] Seidi S, Yamini Y, Rezazadeh M. Electrochemically assisted solid based extraction techniques: A review. *Talanta*. 2015;**132**:339-353

[105] Madikizela LM, Ncube S, Chimuka L. Recent Developments in Selective Materials for Solid Phase Extraction. *Chromatographia*. 2019;**82**:1171-1189

[106] Liu L, Meng WK, Zhou YS, Wang X, Xu GJ, Wang ML. B-Ketoenamine-linked covalent organic framework coating for ultra-high-performance solid-phase microextraction of polybrominated diphenyl ethers from environmental samples. *Chem. Eng. J.* 2019;**356**:926-933

[107] Kelemen H, Hancu G, Papp LA. Analytical methodologies for the determination of endocrine disrupting compounds in biological and environmental samples. *Biomed. Chromatogr.* 2019;**2013**:1-23

[108] Rocha-Gutiérrez BA, Lee WY, Shane Walker W. Mass balance and mass loading of polybrominated diphenyl ethers (PBDEs) in a tertiary wastewater

treatment plant using SBSE-TD-GC/MS. *Water Sci. Technol.* 2016;**73**:302-308

[109] Rejczak T, Tuzimski T. A review of recent developments and trends in the QuEChERS sample preparation approach. *Open Chem.* 2015;**13**:980-1010

[110] Lu D et al. Determination of polybrominated diphenyl ethers and polychlorinated biphenyls in fishery and aquaculture products using sequential solid phase extraction and large volume injection gas chromatography/tandem mass spectrometry. *J. Chromatogr. B.* 2014;**945-946**:75-83

[111] Fontanals N, Marc RM. Materials for solid-phase extraction of organic compounds. *Separations*. 2019;**6**:56-82

[112] Ayazi Z. Application of nanocomposite-based sorbents in microextraction techniques. *Analyst*. 2017;**142**:721-739

[113] Yu M, Wang L, Hu L, Li Y, Luo D, Mei S. Recent applications of magnetic composites as extraction adsorbents for determination of environmental pollutants. *Trends Anal. Chem.* 2019;**119**:115611

[114] Zhang W, Sun Y, Wu C, Xing J, Li J. Polymer-functionalized single-walled carbon nanotubes as a novel sol-gel solid-phase micro-extraction coated fiber for determination of poly-brominated diphenyl ethers in water samples with gas chromatography-electron capture detection. *Anal. Chem.* 2009;**81**:2912-2920

[115] Xiang L, Sheng H, Bian Y, Kang J, Yang X, Herzberger A. Optimization of sample pretreatment based on graphene oxide dispersed acid silica gel for determination of polybrominated diphenyl ethers in vegetables near an e-waste recycling plant. *Bull. Environ. Contam. Toxicol.* 2019;**103**:23-27

[116] Li X, Ma W, Li H, Bai Y, Liu H. Metal-organic frameworks as advanced sorbents in sample preparation for small organic analytes. *Coord. Chem. Rev.* 2019;**397**:1-13

[117] H. L. Jiang, N. Li, X. Wang, X. Y. Wei, R. S. Zhao, and J. M. Lin, "A zirconium-based metal-organic framework material for solid-phase microextraction of trace polybrominated diphenyl ethers from milk," *Food Chem.*, vol. 317, no. 126436, 2020.

[118] Chidambara A, Duong D, Version D, Chidambara A. Molecularly imprinted polymers for sample preparation and biosensing in food analysis: progress and perspectives. *Biosens. Bioelectron.* 2017;**91**:606-615

[119] Li G, Row KH, Li G, Row KH. Recent applications of molecularly imprinted polymers (MIPs) on micro-extraction techniques. *Sep. Purif. Rev.* 2018;**47**:1-18

[120] He Y, Concheiro-Guisan M. Microextraction sample preparation techniques in forensic analytical toxicology. *Biomed. Chromatogr.* 2019;**33**:634-645

[121] Roszko M, Szymczyk K, Renata J. Simultaneous separation of chlorinated/brominated dioxins, poly- chlorinated biphenyls, polybrominated diphenyl ethers and their methoxylated derivatives from hydroxylated analogs on molecularly imprinted polymers prior to gas/liquid chromatography. *Talanta*. 2015;**144**:171-183

[122] Marć M, Wieczorek PP. Application potential of dummy molecularly imprinted polymers as solid-phase extraction sorbents for determination of low-mass polybrominated diphenyl ethers in soil and sediment samples. *Microchem. J.* 2019;**144**:461-468

[123] Hu Y, Pan J, Zhang K, Lian H, Li G. Novel applications of molecularly-imprinted polymers in sample preparation. *Trends Anal. Chem.* 2013;**43**:37-52

[124] Ansari S, Karimi M. Novel developments and trends of analytical methods for drug analysis in biological and environmental samples by molecularly imprinted polymers. *Trends Anal. Chem.* 2017;**89**:146-162

[125] Marć M, Panuszko A, Namieśnik J, Wieczorek PP. Preparation and characterization of dummy-template molecularly imprinted polymers as potential sorbents for the recognition of selected polybrominated diphenyl ethers. *Anal. Chim. Acta*. 2018;**1030**:77-95

[126] He K, Lv Y, Chen Y. Optimized determination of polybrominated diphenyl ethers by ultrasound-assisted liquid-liquid extraction and high-performance liquid chromatography. *J. Sep. Sci*. 2014;**37**:2874-2881

[127] Wang JX, Jiang DQ, Gu ZY, Yan XP. Multiwalled carbon nanotubes coated fibers for solid-phase microextraction of polybrominated diphenyl ethers in water and milk samples before gas chromatography with electron-capture detection. *J. Chromatogr. A*. 2006;**1137**:8-14

[128] Blanco SL, Vieites JM. Single-run determination of polybrominated diphenyl ethers (PBDEs) di- to deca-brominated in fish meal, fish and fish feed by isotope dilution: Application of automated sample purification and gas chromatography/ion trap tandem mass spectrometry (GC/ITMS). *Anal. Chim. Acta*. 2010;**672**:137-146

[129] Vonderheide AP. A review of the challenges in the chemical analysis of the polybrominated diphenyl ethers. *Microchem. J.* 2009;**92**:49-57

[130] Lin X, Li H, He X, Hashi Y, Lin J, Wang Z. Automated online pretreatment and cleanup recycle coupled with high-performance liquid chromatography-mass spectrometry for determination of deca-bromodiphenyl ether in human serum. *J. Sep. Sci*. 2012;**35**:2553-2558

[131] Stapleton HM et al. Determination of polybrominated diphenyl ethers in indoor dust standard reference materials. *Anal Bioanal Chem*; 2005

[132] Butryn DM, Gross MS, Chi LH, Schecter A, Olson JR, Aga DS. 'One-shot' analysis of polybrominated diphenyl ethers and their hydroxylated and methoxylated analogs in human breast milk and serum using gas chromatography-tandem mass spectrometry. *Anal. Chim. Acta*. 2015;**892**:140-147

[133] Schecter A, Päpke O, Kuang CT, Joseph J, Harris TR, Dahlgren J. Polybrominated diphenyl ether flame retardants in the U.S. population: Current levels, temporal trends, and comparison with dioxins, dibenzofurans, and polychlorinated biphenyls. *J. Occup. Environ. Med.* 2005;**47**:199-211

[134] Gou YY, Hsu YC, Chao HR, Que DE, Tayo LL, Lin CH. Pollution characteristics and diurnal variations in polybrominated diphenyl ethers in indoor and outdoor air from vehicle dismantler factories in Southern Taiwan. *Aerosol Air Qual. Res.* 2016;**16**:1931-1941

[135] Chałupniak A, Merkoçi A. Toward integrated detection and graphene-based removal of contaminants in a lab-on-a-chip platform. *Nano Res.* 2017;**10**:2296-2310

[136] Bettazzi F, Martellini T, Shelver WL, Cincinelli A, Lanciotti E, Palchetti I. Development of an

electrochemical immunoassay for the detection of polybrominated diphenyl ethers (PBDEs). *Electroanalysis*. 2016;**28**:1817-1823

[137] Sun Z, Du J, Yan L, Jing C. Rapid detection of 2,2′,4,4′-tetrabromodiphenyl ether (BDE-47) using a portable Au-colloid SERS sensor. *J. Raman Spectrosc.* 2014;**45**:745-749

Chapter 13

Sources, Fate, and Impact of Microplastics in Aquatic Environment

Sukanya Mehra, Khushboo Sharma, Geetika Sharma, Mandeep Singh and Pooja Chadha

Abstract

Over the past decade, enhanced scientific interest has produced an expanding knowledge base for microplastics. The highest abundance of microplastics is typically associated with coastlines and oceans but the fate of these microplastics is elusive. Microplastics sink following fragmentation which is further ingested by marine biota thus imposes threat to them. Thus, the present review focuses on properties and sources of microplastics, its impact on environment, the bioaccumulation and trophic transfer of microplastics and its impact on living biota. This study would be helpful for the development and implementation of risk management strategies for managing the disposal of microplastics.

Keywords: microplastics, sources, accumulation, toxicity, aquatic organisms

1. Introduction

Due to the permanence and robustness, plastic has infiltrated every aspect of life like in clothing, electronics, cleaning products as well as in building materials [1]. World production of synthetic organic polymer plastic has skyrocketed from 1950 to 2013, showing an escalation from 1.5 to 299 million tons. Around 8-16 million tons plastic waste invades sea and oceans annually, substantial section of which comes from land borne sources [1, 2]. In the very beginning more attentiveness was towards large plastic debris; however prevalence of smaller plastic particles in the marine environment elucidated in early 1970's [3, 4]. Due to minuscule proportion of microplastics they are ingested by protozoans to marine mammals and by many filter feeders [5]. Amphipods, polychaete worms, barnacles and sea cucumber ingest microplastic which gets accumulated in food web [6]. According to Setälä et al. [7] and Green et al. [8] microplastics are omnipresent in nature and possess high potential to interrelate with environment (biotic and abiotic) thus menacing with biogenic domain of flora and fauna. Presences of microplastics were perceived more in aquatic ecosystems, surface waters, sediments and water column. Deep seas and mountain lakes were also sullied by the presence of microplastics and thus scrutinized as global pollutant [9, 10]. Worldwide pollution provoked by plastic is dispersed maximum across seas and oceans. Longevity and buoyancy are some properties that have led these pollutants fall under the category of hazardous waste [11–13]. In the environment microplastics are present in heterogeneous group, according to varying size and shape, specific density and composition. Prodigious plastic wastes are easily perceptible [14, 15]. Although microplastics are inconspicuous, their dissemination into the oceans has profound repercussions leading to cumulative effect in the food chain [16].

Microplastics may pose a risk to aquatic environments due to their documented ubiquity in marine ecosystems, long residence time, and propensity to be ingested by biota. As Microplastics from different sources ultimately reaches water bodies and from here microplastics disperse into surface water, underground water, and benthic sediment, etc. and their bioavailability gets affected [17]. After consumption or ingestion, microplastics can remain in the digestive tracts of aquatic organisms for periods of days to weeks before excretion. The more time of excretion likely allows the transfer of microplastics both up the food web and to new geographic locations. Exposure of individual aquatic organisms to microplastics may negatively impact feeding, growth, reproductive capabilities, or survival [18]. While studies and reviews on plastic pollution in the marine environment are increasingly common, to date, few studies have assessed the sources, fate and impact of microplastics in freshwater as well as marine environment. Thus, the present article has been made in order to fill the lacuna in this regard.

2. Types of microplastics

Two primary or foremost types of microplastics are: primary and secondary microplastics. According to Sundt et al. [19] plastics that are instantaneously propagated into the environment compose primary microplastics. They are produced in relatively micro size. Secondary microplastics pioneered from the deterioration or fragmentation of larger plastics.

2.1 Primary microplastics

Plastic ranging between size of 1 μm and 5 mm are considered primary microplastics. Microplastics are affixed to certain products due to their discrete functions. They are also operated as mordant in cosmetics and soap products and also act as conveyor of pigments. Plastic powder, granulates, pellets are some examples of primary microplastics [1, 20]. Primary microplastics are also used as exfoliates. They are the main protagonist of several day to day products like hand cleaners, toothpastes, face washes [20]. Primary microplastics are also used in dental polishes. If they are not discarded in the efficient way possible, they end up blemishing the environment. Primary microplastics also possess diverse industrial implementations like for gas and oil analysis they are required as drilling fluids, for cleaning metal surfaces to eliminate the paint, rinsing of engines etc. [21–23]. Microplastics like polystyrene, acrylic, polyester are used in industries [24].

2.2 Secondary microplastics

Repudiated plastic bags or fishing nets, household items and discarded plastics undergo weathering and photo degradation process and get transfigured into smaller plastic particles, thus constituting secondary microplastics. Abrasion of plastics is manifested by UV light at soil surface and by ocean waves. Secondary microplastics are also fabricated by washing machines [24, 25]. It also includes fragments of textile fiber originated from synthetic fibers, and released during the laundering process [26]. The root sources of secondary microplastics are discarded plastic debris from household items and industrial products. Secondary microplastics are considered to be the most dominating microplastics [27, 28]. Their copiousness in water bodies increases with elevating discharge of plastic debris and its continuous transformation into secondary microplastics. Textile industry products, tires, decorative paints all contribute to the genesis of secondary microplastics [20].

3. Sources of microplastics

3.1 Personal care products

According to Leslie [29] plastic microbeads are utilized as additives in countless cosmetics and personal care products. They act as sorbent as well as exfoliators. Plastic microbeads are incorporated in several cosmetic products such as 350,000 plastic particles were observed in a facial scrub tube in USA [20]. A study from USA also evaluated the presence of 1700-6400 particles of plastic per g in toothpaste [1]. According to Strand [30] toothpaste contains 0.1-0.4% microbeads in accordance with weight, facial scrubs contain around 1.6-3.0%. Facial scrub possesses 0.9-4.2%, exfoliating scrub contains 10.6% and shower gel around 0.5-3.0%. One of the plastic types that are frequently perceived in microplastics is polyethylene (PE). In European country PE plastic microbeads were considered most dominating, with around 4073 tons usage.

3.2 Blasting abrasives and cleaning products

For cleansing surfaces primary microplastics are used as abrasive. Plastic such as polyester, polycarbonate, polyamide is used in blast media. Main purpose of blasting abrasives is in cleansing of rims, removal of paints and cleansing of ships. These abrasives are also used in marine industries for cleaning the tanks. Blasting is done in different cabins, closed or semi-closed. Area must be encrusted properly. Emission rate are quite high if done in open premises. Turbines blades are cleansed by this process; this leads to release of primary microplastics in aquatic environment. Microplastics are even used in maintenance and cleaning products as abrasive material. To remove grease, paints, oil from hands primary microplastics are used [20].

3.3 Synthetic textile and tyres

Browne et al. [25] assorted that process of laundries in household and industries leads to mass production of primary microplastics via scraping and dispersion of fibers, which was then emitted out in sewage water and culminates in the ocean [31]. Tyres contain profuse mixture of several synthetic polymers in addition with natural rubber. Tyres get deteriorated when used and tyre dust that contains synthetic rubber circulated by wind or swiped away by rain. Large segments of such particles were reported to congregate in the sea [19, 23].

3.4 Paint and wood preservatives

Primary microplastics are appended to paint and preservative. This gives matting effect and acts as color amplifier. This improves longevity of wood, provides hardened and abrasion resistance. It is also used to diminish the density of paints [20]. A study done by Poulsen et al. [32] stated that 8-30% of waste generation is triggered by paint spillage and other paint jobs. Approximately 65-97% of waste culminates as solid waste and around 35% ended up in sewer system. 1.0-5.3 t/y is the estimated amount of microplastics release from paints, out of which 0.1-0.5 t/y is the guesstimate amount of primary microplastics in paints that ends up in aquatic environment.

3.5 Synthetic waxes and oil-gas industry

According to Essel et al. [23] synthetic waxes are scrutinized as primary microplastics. They are used in dyeing, food coating, as lubricants and also in processing of plastics. Synthetic waxes are also used to coat papers they are extensively operative in textile processing by providing smoother surface. Polytetrafluroethylene (PTFR) is used for drilling purpose. Drilling fluids composed of microbeads, used in oil-gas industry [19, 32]. This chemical gets directly discharge into the oceans, contributing to microplastics accumulation.

3.6 Plastic pellets manufactures

Pellets are the primary form of many plastics around 2.55 mm in diameter. These pellets are used to generate plastic products. These pellets, spherules contribute to 79.4% of total plastic debris in the water of river Danube [33, 34]. According to Dhodapkar et al. [35], in addition to pellets, plastic dust also gets accumulated during manufacturing process or generated due to relocation and transportation [20]. Pellets contain certain perilous additives like plasticizers and flame retardants that promote the eco toxicity. These additives mixed prior to the production or added during conversion. These pellets are often termed as nibs and nurdles [19].

3.7 Weathering and abrasion

Metropolitan environment often encounters with city dust. Synthetic cooking utensils abrasion, footwear soles abrasion, infrastructure abrasion, blasting abrasives all culminates into city dust. Independently importance of these factors is insignificant but together they are accountable for sizeable losses in the country [20, 31]. For the advancements of roads, road markings are administered. Thermoplastic, paint, polymer tapes are preferred in this process. Weathering by vehicles induces microplastics loss, which is washed off by rain or wind and ultimately outstretched to oceans [20]. Coatings of boats are done by various anticorrosive paints, mostly polyurethane, lacquers and vinyl [36].

3.8 Packaging material and litter

Higher preference is given to packaging materials as a source of microplastics. It contributes to about 62% of all plastic collected. It usually involves secondary microplastics. Plastic bags, soft drink bottles all culminates into it [1]. Packaging materials constitutes major portion of litter. Toys, cutlery, shoes, clothing are other forms of litter [15]. Litter from agricultural plastics is non-biodegradable, although biodegradable plastics are also prevalent nowadays. Addition of preservatives in such plastics make them less biodegradable, and these plastics get perished into smaller fragments, eventually via nearby streams microplastics enters surface water [1].

3.9 Domestic items, food stuffs and toys

According to United Nations Environment Programme (UNEP), 2014 [15], domestic items are considered as the mighty source of pollution in the sea. Items like cups, plastic cutlery, bottles and straws are present in abundance in oceans.

Food stuffs and snacks also contribute to microplastics. Chewing gum contains microplastics fillers. A study done at Dutch coast revealed the presence of 105 particles of microplastics per gram in mussels and for oyster it reaches up to 87 microplastics particles per gram [37]. Party items like balloons, confetti firework wastes, fragments of toys all culminate to waterborne litter. Loom bracelets contain microplastics which can adulterate the environment via surface runoff [38].

3.10 Medical resources

In medical sector, microplastics are profoundly used [23]. Microplastics treat the reverse flow of gastric juices. Aluminum coated compounds tend to possess deleterious effect on human health. They are replaced by microplastics. Nappies, sanitary towels, plasters constitute litter. Capsules used in the edicine field contain plastic. Spectacles, contact lenses are one of the define sources of microplastics [1].

4. Microplastics in marine ecosystems

The pollution of microplastics varies geographically with location because as the consumption of plastic increases, there is increase in production of MP. Marine life is more disturbed by this plastic waste because ocean become a dump yard for running water system either directly via riverine system as river ultimately end up meet up with the ocean or indirectly as waste water treatment plant dispose of their waste directly in the ocean or in river which end up by meeting the marine water body. However, the size of sediment and distribution of MP is influenced by oxidative degradation (either photo- or thermal initiated), friction and bio-degradation [39, 40]. The typical shape of microplastics consists of pellets, fibers and fragments but according to literature, majority of microplastics in Oceans are microfibers [41, 42]. Distribution and abundance of microplastics is chiefly determined by environmental [42–44] and anthropogenic factors [45]. Environmental factors include runoff, infiltration, river discharge, wind action, ocean currents, cyclones, river hydrodynamics, wave current, tides and movement/dispersion of animals. On the other hand, anthropogenic activities either they are for industrial or tourism or transport purpose which further led to accumulation of plastic debris in environment. According to literature, these environmental factors play vital role in determining the distribution of microplastics more intensely than anthropogenic activities, however anthropogenic activities are the core source of production of these plastic wastes.

Abundance of microplastics in oceans distribute across various strata of Ocean. In the sediments- water systems, microplastics only sink and accumulate in the sediment when their density exceed seawater (>1.02 g/cm^3); otherwise it tends to float on the sea surface or in the water column [46], hence low density microplastics float on surface layer of ocean water whereas high density microplastics sinks down to benthos layer [5]. Buoyancy of microplastics can depend on befouling in which former biomass accumulation led to increase in microplastics density and later can decrease microplastics density which is responsible for sinking, neutral or floating action of microplastics. But in case of High density microplastics, there distribution depends on other factors also like change in tidal fronts, high flow rates or larger surface area of High density microplastics.

Beaches are a reservoir of highly fragmented plastic debris that transport microplastics back to costal water and finally to open ocean [47]. It is based on observation of Wang et al. [48] that concentration of microplastics is usually higher in upper layer i.e. epipelagic layer than the immediate lower mesopelagic layer this may be due to preferential flow or animal movement. Even, mesoscale ocean dynamics have impact on distribution of plastic debris at sea surface within subtropical gyres [49]. Usually, sea platforms and marine trafficking are responsible for microplastics in far off Ocean, whereas microplastics in near shore originate mostly from waste water, runoffs, rivers etc. [50, 51]. However terrestrial environment also determines the concentration of microplastics as harbor and industries add huge amount of plastic debris either directly or indirectly which add up to the acumen concentration of microplastics in the ocean. The dire situation of disturbance in aquatic ecosystem is becoming huge day by day as these not only affect flora but fauna as well; even coral beds are not far away from disturbance. This plastic debris also includes mesoplastics accumulation. Use and through plastic items are becoming huge threat to the aquatic organisms hence number of reports are increasing in this area of research which indicate the negative impact of microplastics and is alarming accumulation rate in ecosystem which is also eye catching for researchers and environmentalists.

5. Microplastics in freshwater ecosystem

To understand the impact of microplastics pollution in freshwater environment, various aspects are to be analyzed i.e. source, distribution, type and effect of microplastics. Source of freshwater pollution is usually synthetic textile, personal care products, industrial raw material, whereas the main source of microplastics in riverine system are wastewater discharge which may be industrial or household untreated waste water disposal. Microplastics are of serious concern because their accumulation potential increases with decreasing size [52]. It is also noted that, there may be change of composition of MP as they accumulate with waterborne contaminates which includes metals and persistent, bioaccumulative and toxic compounds this is possible due to larger surface-to-volume ratio. Studies by Engler [53] showed relationship between plastic debris and PBTs (e.g. PCBs and DDT) similarly a number of studies exist for polycyclic aromatic hydrocarbons [54–56]. Since the spectrum of contaminants is different between freshwater and marine system. In stagnant riverine system like ponds and lakes water pollution is more severe problems because of the irresponsible behavior of the inhabitants or by various tourism related activities which disturb the ecosystem due to accumulation of degradable or non- degradable waste. They float in the surface water and stay in the water sink into sediments of lake. For stagnant system rate of accumulation of microplastics is higher, since there is no efflux. Therefore, it can be concluded that there is a direct correlation between distance of contamination source and microplastics pollution levels in sediments [57]. Various studies confirmed the presence of microplastics in drinking water system which makes it a serious issue [58]. Research about river system and watersheds can provide the knowledge to the people to understand the alarming situation of microplastics accumulation in freshwater system [59]. Further the flow of river plays significant role in removal of plastic fragments. It is also observed that after precipitation high amount of MP is observed in sediments and running water [60]. Eventually, freshwater system also contributes to the pollutant content of marine ecosystems because ultimately riverine merge with the ocean resulting merger of mineral, sediments, soil content as well as pollutants. Hence the fact that freshwater system act as strong source of microplastics to marine ecosystem cannot be neglected. Although distribution of microplastics in freshwater system is not uniform, it depends on nearby source of waste water disposal.

It is observed that the condemnation in water is observed higher in riverine near industrial area as compared to the residential area. However waste water treatment plants are established by organizations but they remove large plastic waste more efficiently than meso and micro plastic waste, as various technologies are installed to remove large size particle but these are not specified to retain microplastics [61, 62]. Discharge from waste water treatment plant contain may hazardous compounds along with micro –and Nano- plastic particles which enter the food web and cause diverse effect in biotic ecosystem.

6. Fate of microplastics

The threat of microplastics is rapidly increasing, and as the global plastic production projected to reach an accumulative 25 billion tons by 2050, things are going to be worse [63]. Although these plastic materials are key factors for innovation and development of various fields such as healthcare, energy generation, aerospace, automotive industries, construction, electronics, packaging, textile and many others [64]. However, instead of recycling or incineration or utilizing any other way of removing unused or discarded plastic from environment, these plastic wastes enters the environment from year to year and it is accumulated in Marine, freshwater and terrestrial ecosystem worldwide, even from densely populated countries like India and China to cold desert like Antarctica. And this became a matter of concern for scientists across the world. The reliance on plastic for huge number of consumer products, many of them being single-use, results in continuous entry into environment.

No doubt, with due course of time via biotic and abiotic degradation pathway, plastic loose its mechanical integrity but it may take several years to degrade completely. With gradual degradation this immortal plastic emits smaller size particle in environment i.e. macroplastics, mesoplastics and microplastics. Plastic particles <5 mm size are considered Microplastic, although there are efforts to redefine them as <1 mm in size, as recommended by [65]. However minimum size of Microplastics has not yet been specified and it depends on the sampling and processing as well as on the applied method for Microplastics identification.

7. Effect of microplastics on aquatic biota

Microplastics are of special concern because they can be ingested throughout the food web more readily than larger particles. It is to be noted that the impact of microplastics on public health and aquatic ecosystems is not yet fully understood, but there is increasing number of reports which indicate negative impact of microplastics on marine and freshwater biota.

With increased focus on microplastics debris, several groups have studied the influence of microplastics uptake by different organisms. As microplastics invasion appear to occur across all ecosystem from terrestrial to marine environment in different trophic level not only invertebrates but vertebrates also seem to be affected by their presence [66, 67]. Organism ingests these microplastics debris while swallowing their food. And with due course of time bioaccumulation of microplastics results in diverse negative impact on various organism like disruption of organ system, rupturing of digestive system, weakening of immune system, impotency, various respiratory and circulatory problems, even failure of organ and in extreme cases led to death of organism [51, 68, 69]. However, continuous accumulation of these deadly microplastics in various systems of the body is possible through food chain via ingestion as well as by accumulation around gill aperture (or around respiratory apertures) and appendages of body by diverse aquatic organisms [70, 71]. However, situation become direr for the predators and humans which directly or indirectly consume microplastics affected aquatic organisms [72]. As reported by Wright et al. [5], there are various consequences from ingestion of plastics and MPs by various species such as planktons, copepods, zooplanktons, crabs, small fishes, turtles, fish larvae, sea birds and mammals.

7.1 Effect of microplastics on marine ecosystem

Marine microplastics debris is a global threat because of its abundance, persistence and mobility across scale, with subsequent widespread distribution potential, geophysical and biological impact [73]. Across the globe, research on the ingestion of microplastics by biota has predominantly focused on wide range of marine species with different feeding strategies [74–76]. As microplastics have been shown to obstruct feeding appendages and limit food intake, physical injury and oxidative stress, reduced energy allocation in various aquatic organisms and in some cases damages in the alimentary canal were also observed [77]. Alteration in the feeding behavior of some group of crustaceans was also studied such as in copepods which feed on algae, but when these copepods feed on natural assemblage of algae with the addition of polystyrene microbeads they showed a significant decrease in herbivory which further results in decrease in growth rate of organism [5, 78]. However, it is not just growths which microplastics injection can disrupt, but also observable change in physical development of organism. An alternation was observed in life cycle of sea urchin *Paracentrotus lividus* which depicted alterations in shape of pelagic planktotrophic pluteus larva when Microplastics were ingested [79]. Another study of Kaposi et al. [80] by examining short term exposure of Polyethylene on the sea urchin *Tripneustes gratilla*, which was done by using fluorescent labels green PE Microspheres having diameter 10-45 μm, with exposure of time ranged between 15 min to 5 d. There was decrease in injection rate even when phytoplankton food was provided.

While some of the chemicals associates with microplastics, which possess endocrine disruptive activity and are responsible for the hormonal imbalance in organisms [81]. In a study by Sussarellu et al. [82] on Oysters which is a keystone species with high ecological and economic value. When adult oysters were exposed to microplastics polystyrene of diameter about 2 μm during a critical point of their reproductive stage adults were preparing for production of gametes. And after the exposure, there was an alteration in the feeding as well as absorption efficiency of food. Reproductive changes were also observed that there is reduction in the quality of oocytes and sperm swimming speed as well as fecundity. Moreover, these impacts had clear carryover effect on offspring quality and further reduced growth in their larval progeny. Similar effects were observed in planktonic copepods when exposed to micropolystyrene for prolonged period followed by reduced food consumption and resulting in reduced reproductive outputs [83].

However, Corals which occur in both deep sea and Antarctic system are not untouched by the effect of microplastics as some of the corals known to ingest microplastics and demonstrably negatively impact occurs both in terms of energy level, growth and pathogen frequency of reefs [84, 85].

7.2 Effect of microplastics on fresh water ecosystem

It's is not only marine wildlife that can take up microplastics, bioavailability of microplastics for freshwater fauna (for both invertebrates and fishes) has also been observed. Although there are few freshwater studies so far, A study by Rehse et al. [86] shows that immobilization has occurred in freshwater zooplankton (*Daphnia magna)* after ingestion of polyethylene microplastics of about 1 mm, however due to the smaller size of this freshwater zooplankton, it was not able to engulf microplastics of more than 1 mm size. It is also observed that small size microplastics usually possess large surface area to volume ratio which differentiates the property of microplastics from meso and macroplastics.

In freshwater habitat, the different POPs (persistent organic pollutants) that is PCBs (polychlorinated biphenyls), HCBs (hexachlorobenzenes), PBDEs (polybrominated diphenyl ethers) and metals are present in significantly higher concentrations. And the adsorption ratio of POPs to microplastics is different in freshwater as compared to marine ecosystems due to the proximity to the sources and use of these chemicals. As organism in freshwater ecosystem are more exposed to POPs and microplastics due to occurrence of industrial and populous area nearby. Study by Rochman et al. [87] revealed that freshwater fishes experience hepatic stress due to ingestion of polyethylene which ultimately led to bioaccumulation and toxicity in fishes. A significant amount of POPs to microplastics could accumulate in adult zebrafish gills and zebra fish embryos [88]. Another observation in the study of European perch *Perca fluviatilis* by Lönnstedt and Eklöv [89], suggested the effect of microplastics when larva of European perch were exposed to different concentrations of 10,000 or 80,000 particles/meter which resulted in inhibition of hatching and reduced growth rates. There were alterations in the feeding and innate behavior as compared to normal individuals which were not exposed with microplastics.

Microplastics can also act as an artificial substrate for microorganisms. This has raised concern about the potential ecological effect on freshwater habitat, which is utilized for anthropogenic activities as well as by the wildlife organisms. Microplastics form biofilms by providing surface for microorganisms and rafting the colonized organisms over long distances. It has also been shown that biofilms containing potentially pathogenic microorganisms which can develop on plastic in the marine system. Some microorganisms in the biofilm are assumed to be potentially opportunistic (human) pathogens, for example, members of the genus *Vibrio* have been found on the particles and making microplastics vector for pathogens, toxic algae, bacteria and invasive species. Various studies are performed on different rivers for the estimation of assimilation of microplastics by aquatic organisms. A study conducted by McGoran et al. [90], in the river Thames, revealed that up to 75% of sampled European flounder (*Platichthys flesus)* has plastic fibers in their gut compared to 20% of European smelt (*Osmeruseperlanus)* however it is estimated that this huge difference in the concentration is due to the feeding habitats of both the fishes as European flounder are benthic feeders while European smelt are pelagic feeder and these observations also suggested the relative distribution of microplastics in different strata of riverine system [91].

In study of microplastics in freshwater, Au et al. [92], investigated the ingestion and effects of PE (fluorescent blue PE microplastics particles, 10–27 μm) and PP (black polypropylene microplastics fibers from marine rope, 20–75 mm in length) on the growth and mortality of the freshwater amphipod *Hyalella azteca*. The LC50 of PE and PP in *H. azteca* after a 10-d exposure were 4.6 × 104 and 71 microplastics/mL respectively. The effects of chronic exposure to PE and its influence on the reproduction of amphipods were analyzed. Chronic exposure of *H. azteca* to PP fibers, even at a low concentration, significantly decreased growth and reproduction.

7.3 Toxic effects of microplastics shown by trophic transfer

To evaluate the process of trophic transfer and toxicological effects of microplastics at different trophic levels, a number of factors need to be considered that are involved in ingestion, bioaccumulation and biomagnification of microplastics and their associated chemicals.

Resemblance in shape and size of microplastics with many species of planktons and other type of food particles is usually observed. Hence sometimes microplastics are normally ingested by aquatic filter feeders along with some associated contaminants led to bioaccumulation and trophic transfer to higher organisms [93]. The size and shape of plastic particles are the two most important parameters which determine the extent of microplastics retention. This is because smaller particles are more likely to be ingested and particles with angular shapes may be harder to egest. The available body of evidence indicated that trophic transfer of microplastics may occur [94, 95]. Hence, pollutants may be transferred along with microplastics by means of oral ingestion as well as other pathways, which include ventilation or simple microplastics attachment and resuspension into the water column [88, 96].

Setälä et al. [7], observed the trophic transfer of polystyrene microplastics to macrozooplankton occurred after only 3 hrs of exposure to mesozooplaktons that had previously infested PS microplastics. Studies also revealed that uptake of microplastics can be influenced by the surface characteristics of plastic particles. As MP that was neutrally or positively charged had a higher binding affinity for algal cell wall than negatively charged microplastics. And hence they adhere to surface of seaweeds like *Fucus vesiculosus,* resulting in their consumption by grazing gastropods which further eventually led to trophic transfer of microplastics [97]. Microplastics are ingested by organism at lower trophic level and are further transferred to higher trophic level and ultimately results in bioaccumulation in higher organism and causes ill effects which may be life threating for them.

The impact of microplastics on humans is not yet fully understood, however many studies depict that there are many chemicals that are used in plastic production show toxic effect on living organism some of these chemicals are bisphenol A (BPA), polybrominated diphenyl ethers (PBDE), and tetrabromobisphenol (TBBPA). Studies have already detected these chemicals in human tissues and biological fluids [27]. It has also been reported that additives, for example, di(2-ethylhexyl)phthalate (DEHP), can leach from medical supplies made of PVC and accumulate in the blood of hemodialysis patients [98]. Moreover, the presence of microplastics in seafood, for example, bivalves cultured for human consumption has already been shown [28, 99]. It should be further investigated whether beverage or food products act as possible microplastics sources which is can further enter food web and results in bioaccumulation in living organism.

8. Management of microplastics

To get the problem under control, the society has to take initiatives which includes significantly curtailing unnecessary single-use plastic items such as water bottles, plastic shopping bags, straws and utensils, stringent policies should be implemented by the governments ensuring the need to strengthen garbage collection and recycling systems to prevent waste from leaking into the environment to improve recycling rates. New ways to break plastic down into its most basic units, which can be rebuilt into new plastics or other materials should be considered.

9. Conclusion

Production and applications of microplastics resulted to an enhanced incidence of plastics debris and microplastics, in the aquatic environment. Not only one mechanism such as the weathering-related fracturing and surface embrittlement of plastics in beach environments is the root cause of generation of microplastics but industrial waste also constitute the major sources of them. As microplastics are recalcitrant in nature, only small fraction of the microplastics present in aquatic body imposes a serious threat to aquatic life. As microplastics are potentially ingestible by aquatic organisms including micro and nano plankton species, the delivery of toxins across trophic levels via this mechanism is very common. The efficiency of such transfer will depend on the bioavailability of microplastics and the residence time of meso or microplastics in the organisms. Endocytosis of plastic nanoparticles by micro- or nanofauna can also result in adverse toxic endpoints. As aquatic species constitute the very foundation of the aquatic food web, any threat to these can have serious and far-reaching effects in the world oceans. There is an urgent need to quantify the magnitude of these potential outcomes and assess the future impact of increasing microplastics levels on the world's aquatic bodies.

Author details

Sukanya Mehra, Khushboo Sharma, Geetika Sharma, Mandeep Singh and Pooja Chadha*
Cytogenetics Laboratory, Department of Zoology, Guru Nanak Dev University, Amritsar, India

*Address all correspondence to: poojachadha77@yahoo.co.in

References

[1] Verschoor A, de Poorter L, Roex E, Bellert B. Quick scan and prioritization of microplastic sources and emissions. RIVM Letter report. 2014; 5:156

[2] Duis K and Coors A. Microplastics in the aquatic and terrestrial environment: sources (with a specific focus on personal care products), fate and effects. Environmental Sciences Europe. 2016;28(1):1-25.

[3] Carpenter EJ and Smith KL Jr. Plastics on the Sargasso Sea surface. Science. 1972;175:1240-1241

[4] Carpenter EJ, Anderson SJ, Harvey GR, Miklas HP, Beck BB. Polystyrene spherules in coastal waters. Science.1972;178:749-750.

[5] Wright SL, Thompson RC, Galloway TS. The physical impacts of microplastics on marine organisms: a review. Environmental pollution. 2013;178:483-92.

[6] Thompson RC, Olsen Y, Mitchell RP, Davis A, Rowland SJ, John AW, McGonigle D, Russell AE. Lost at sea: where is all the plastic? Science (Washington). 2004;304:838.

[7] Setälä O, Fleming-Lehtinen V, Lehtiniemi M. Ingestion and transfer of microplastics in the planktonic food web. Environmental pollution. 2014 Feb 1;185:77-83.

[8] Green DS, Boots B, O'Connor NE, Thompson R. Microplastics affect the ecological functioning of an important biogenic habitat. Environmental Science & Technology. 2017;51(1):68-77.

[9] Free CM, Jensen OP, Mason SA, Eriksen M, Williamson NJ, Boldgiv B. High-levels of microplastic pollution in a large, remote, mountain lake. Marine pollution bulletin. 2014;85(1):156-63.

[10] Woodall LC, Sanchez-Vidal A, Canals M, Paterson GL, Coppock R, Sleight V, Calafat A, Rogers AD, Narayanaswamy BE, Thompson RC. The deep sea is a major sink for microplastic debris. Royal Society open science. 2014;1(4):140317.

[11] Teuten E, Rowland S, Galloway T, Thompson R. Potential for plastics to transport hydrophobic contaminants. Environ Sci Technol.2007; 41: 7759-7764.

[12] Mato Y, Isobe T, Takada H, Kanehiro H, Ohtake C, et al. Plastic resin pellets as a transport medium for toxic chemicals in the marine environment. Environ Sci Technol.2001; 35: 318-324.

[13] Rochman C, Browne M, Halpern B, Hentschel B, Hoh E, et al. Classify plastic waste as hazardous. Nature.2013; 494: 169-171.

[14] GESAMP. Sources, fate and effects of microplastics in the marine environment: a global assessment. 2015

[15] Raynaud J. Valuing plastics: The business case for measuring, managing and disclosing plastic use in the consumer goods industry. UNEP. 2014.

[16] Eriksen M, Lebreton LC, Carson HS, Thiel M, Moore CJ, Borerro JC, Galgani F, Ryan PG, Reisser J. Plastic pollution in the world's oceans: more than 5 trillion plastic pieces weighing over 250,000 tons afloat at sea. PloS one. 2014;9(12):1-15.

[17] Prakash V, Dwivedi S, Gautam K, Seth M, Anbumani S. Occurrence and Ecotoxicological Effects of Microplastics on Aquatic and Terrestrial Ecosystems. In Springer, Berlin, Heidelberg; 2020. p. 1-21.

[18] Foley CJ, Feiner ZS, Malinich TD, Höök TO. A meta-analysis of the effects of exposure to microplastics on fish and aquatic invertebrates. Science of the Total Environment. 2018;631-632:550-9.

[19] Sundt P, Schulze PE, Syversen F. Sources of microplastic-pollution to the marine environment. Mepex for the Norwegian Environment Agency. 2014 ;86.

[20] Lassen C, Hansen SF, Magnusson K, Hartmann NB, Jensen PR, Nielsen TG, Brinch A. Microplastics: occurrence, effects and sources of releases to the environment in Denmark.2015;1-208.

[21] Gregory MR. Plastic 'scrubbers' in hand cleansers: a further (and minor) source for marine pollution identified. Marine pollution bulletin. 1996;32(12):867-71.

[22] Derraik JG. The pollution of the marine environment by plastic debris: a review. Marine pollution bulletin. 2002;44(9):842-52.

[23] Essel R, Engel L, Carus M, Ahrens RH. Sources of microplastics relevant to marine protection in Germany. Texte. 2015;64.

[24] Eriksen M, Mason S, Wilson S, Box C, Zellers A, Edwards W, Farley H, Amato S. Microplastic pollution in the surface waters of the Laurentian Great Lakes. Marine pollution bulletin. 2013;77(1-2):177-82.

[25] Browne MA, Crump P, Niven SJ, Teuten E, Tonkin A, Galloway T, Thompson R. Accumulation of microplastic on shorelines woldwide: sources and sinks. Environmental science & technology. 2011;45(21):9175-9.

[26] Fendall LS and Sewell MA. Contributing to marine pollution by washing your face: microplastics in facial cleansers. Marine pollution bulletin. 2009;58(8):1225-8.

[27] Cole M, Lindeque P, Halsband C, Galloway TS. Microplastics as contaminants in the marine environment: a review. Marine pollution bulletin. 2011;62(12):2588-97.

[28] Eerkes-Medrano D, Thompson RC, Aldridge DC. Microplastics in freshwater systems: a review of the emerging threats, identification of knowledge gaps and prioritisation of research needs. Water research. 2015;75:63-82.

[29] Leslie HA. Plastic in Cosmetics: Are we polluting the environment through our personal care?: Plastic ingredients that contribute to marine microplastic litter. 2015.

[30] Strand J. Contents of polyethylene microplastic in some selected personal care products in Denmark. In International Conference on Plastics in Marine Environments. 2014.

[31] Magnusson K, Eliasson K, Fråne A, Haikonen K, Hultén J, Olshammar M, Stadmark J, Voisin A. Swedish sources and pathways for microplastics to the marine environment. A review of existing data. IVL, C. 2016;183:1-87.

[32] Poulsen PB, Stranddorf HK, Hjuler K, Rasmussen JO. Vurderingafmalingsmiljø belastningianvendelsesfasen [Assessment of the environmental impact of paint in the use phase]. Miljøprojekt nr. Danish Environmental Protection Agency.2002;662.

[33] Fabbri D, Tartari D, Trombini C. Analysis of poly (vinyl chloride) and other polymers in sediments and suspended matter of a coastal lagoon by pyrolysis-gas chromatography-mass spectrometry. Analytica Chimica Acta. 2000;413(1-2):3-11.

[34] Norén F and Naustvoll LJ. Survey of microscopic anthropogenic particles in Skagerrak. Report commissioned by Klima-ogForurensningsdirektoratet. 2010.

[35] Dhodapkar S, Trottier R, Smith B. Measuring Dust and Fines In Polymer Pellets. Chemical Engineering. 2009 :1;116(9):24.

[36] OECD Series on emissions documents. Emission Scenario documents on coating industry (Paints, Lacquers and Varnishes). 2009.

[37] Leslie HA, Van Velzen MJ, Vethaak AD. Microplastic survey of the Dutch environment. Novel data set of microplastics in North Sea sediments, treated wastewater effluents and marine biota, The Netherlands. 2013 ;1-30.

[38] Siegle, L. Are loom bands the next environmental disaster?, in: The Observer. 2014.

[39] Mason SA, Garneau D, Sutton R, Chu Y, Ehmann K, Barnes J, Fink P, Papazissimos D, Rogers DL. Microplastic pollution is widely detected in US municipal wastewater treatment plant effluent. Environmental Pollution. 2016 Nov 1;218:1045-54.

[40] Hidalgo-Ruz V, Gutow L, Thompson RC, Thiel M. Microplastics in the marine environment: a review of the methods used for identification and quantification. Environmental science & technology. 2012 Mar 20;46(6):3060-75.

[41] Mathalon A, Hill P. Microplastic fibers in the intertidal ecosystem surrounding Halifax Harbor, Nova Scotia. Marine pollution bulletin. 2014 Apr 15;81(1):69-79.

[42] Veerasingam S, Saha M, Suneel V, Vethamony P, Rodrigues AC, Bhattacharyya S, Naik BG. Characteristics, seasonal distribution and surface degradation features of microplastic pellets along the Goa coast, India. Chemosphere. 2016 Sep 1;159:496-505.

[43] Dris R, Imhof HK, Löder MG, Gasperi J, Laforsch C, Tassin B. Microplastic contamination in freshwater systems: Methodological challenges, occurrence and sources. In Microplastic Contamination in Aquatic Environments 2018 Jan 1 (pp. 51-93). Elsevier.

[44] Kim IS, Chae DH, Kim SK, Choi S, Woo SB. Factors influencing the spatial variation of microplastics on high-tidal coastal beaches in Korea. Archives of environmental contamination and toxicology. 2015 Oct 1;69(3):299-309.

[45] Shahul Hamid F, Bhatti MS, Anuar N, Anuar N, Mohan P, Periathamby A. Worldwide distribution and abundance of microplastic: how dire is the situation?. Waste Management & Research. 2018 Oct;36(10):873-97.

[46] Van Cauwenberghe L, Devriese L, Galgani F, Robbens J, Janssen CR. Microplastics in sediments: a review of techniques, occurrence and effects. Marine environmental research. 2015 Oct 1;111:5-17.

[47] Fok L, Lam TW, Li HX, Xu XR. A meta-analysis of methodologies adopted by microplastic studies in China. Science of The Total Environment. 2020 May 20;718:135371.

[48] Wang J, Liu X, Li Y, Powell T, Wang X, Wang G, Zhang P. Microplastics as contaminants in the soil environment: A mini-review. Science of The Total Environment. 2019 Nov 15;691:848-57.

[49] Brach L, Deixonne P, Bernard MF, Durand E, Desjean MC, Perez E, van Sebille E, ter Halle A. Anticyclonic eddies increase accumulation of

microplastic in the North Atlantic subtropical gyre. Marine pollution bulletin. 2018 Jan 1;126:191-6.

[50] Lu K, Qiao R, An H, Zhang Y. I nfluence of microplastics on the accumulation and chronic toxic effects of cadmium in zebrafish (*Danio rerio*). Chemosphere. 2018 Jul 1;202:514-20.

[51] Zhang Y, Gao T, Kang S, Sillanpää M. Importance of atmospheric transport for microplastics deposited in remote areas. Environmental Pollution. 2019 Nov 1;254:112953.

[52] Frias JP, Nash R. Microplastics: finding a consensus on the definition. Marine pollution bulletin. 2019 Jan 1;138:145-7.

[53] Engler RE. The complex interaction between marine debris and toxic chemicals in the ocean. Environmental science & technology. 2012 Nov 20;46(22):12302-15.

[54] Antunes J, Frias J, Sobral P. Microplastics on the Portuguese coast. Marine pollution bulletin. 2018 Jun 1; 131:294-302.

[55] Fisner M, Taniguchi S, Moreira F, Bícego MC, Turra A. Polycyclic aromatic hydrocarbons (PAHs) in plastic pellets: Variability in the concentration and composition at different sediment depths in a sandy beach. Marine pollution bulletin. 2013 May 15;70(1-2):219-26.

[56] Fries E, Dekiff JH, Willmeyer J, Nuelle MT, Ebert M, Remy D. Identification of polymer types and additives in marine microplastic particles using pyrolysis-GC/MS and scanning electron microscopy. Environmental Science: Processes & Impacts. 2013;15(10):1949-56.

[57] Qu X, Su L, Li H, Liang M, Shi H. Assessing the relationship between the abundance and properties of microplastics in water and in mussels. Science of the total environment. 2018 Apr 15;621:679-86.

[58] Li J, Green C, Reynolds A, Shi H, Rotchell JM. Microplastics in mussels sampled from coastal waters and supermarkets in the United Kingdom. Environmental pollution. 2018 Oct 1;241:35-44.

[59] Miller ME, Kroon FJ, Motti CA. Recovering microplastics from marine samples: A review of current practices. Marine Pollution Bulletin. 2017 Oct 15;123(1-2):6-18.

[60] Lima AR, Costa MF, Barletta M. Distribution patterns of microplastics within the plankton of a tropical estuary. Environmental Research. 2014 Jul 1;132:146-55.

[61] Mani T, Hauk A, Walter U, Burkhardt-Holm P. Microplastics profile along the Rhine River. Scientific reports. 2015 Dec 8;5(1):1-7.

[62] Prata JC. Microplastics in wastewater: State of the knowledge on sources, fate and solutions. Marine pollution bulletin. 2018 Apr 1;129(1):262-5.

[63] Geyer R, Jambeck JR, Law KL. Production, use, and fate of all plastics ever made. Science advances. 2017 Jul 1;3(7):e1700782.

[64] Ivleva NP, Wiesheu AC, Niessner R. Microplastic in aquatic ecosystems. Angewandte Chemie International Edition. 2017 Feb 6;56(7):1720-39.

[65] Hartmann NB, Hüffer T, Thompson RC, Hassellöv M, Verschoor A, Daugaard AE, Rist S, Karlsson T, Brennholt N, Cole M, Herrling MP. Are we speaking the same language? Recommendations for a definition and categorization framework for plastic debris.2019

[66] Rezania S, Park J, Din MF, Taib SM, Talaiekhozani A, Yadav KK, Kamyab H. Microplastics pollution in different aquatic environments and biota: A review of recent studies. Marine pollution bulletin. 2018 Aug 1;133:191-208.

[67] Wright SL, Ulke J, Font A, Chan KL, Kelly FJ. Atmospheric Microplastic deposition in an urban environment and an evaluation of transport. Environment international. 2020 Mar 1;136:105411.

[68] Triebskorn R, Braunbeck T, Grummt T, Hanslik L, Huppertsberg S, Jekel M, Knepper TP, Krais S, Müller YK, Pittroff M, Ruhl AS. Relevance of nano-and microplastics for freshwater ecosystems: a critical review. TrAC Trends in Analytical Chemistry. 2019 Jan 1;110:375-92.

[69] James E, Turner A. Mobilization of antimony from microplastics added to coastal sediment. Environmental Pollution. 2020 May 1:114696.

[70] Cressey D. The plastic ocean. Nature. 2016 Aug 18;536(7616):263-5.

[71] Abidli S, Toumi H, Lahbib Y, El Menif NT. The first evaluation of microplastics in sediments from the complex lagoon-channel of Bizerte (Northern Tunisia). Water, Air, & Soil Pollution. 2017 Jul 1;228(7):262.

[72] Guilhermino L, Vieira LR, Ribeiro D, Tavares AS, Cardoso V, Alves A, Almeida JM. Uptake and effects of the antimicrobial florfenicol, microplastics and their mixtures on freshwater exotic invasive bivalve *Corbicula fluminea*. Science of the Total Environment. 2018 May 1;622:1131-42.

[73] Zalasiewicz J, Waters CN, do Sul JA, Corcoran PL, Barnosky AD, Cearreta A, Edgeworth M, Gałuszka A, Jeandel C, Leinfelder R, McNeill JR. The geological cycle of plastics and their use as a stratigraphic indicator of the Anthropocene. Anthropocene. 2016 Mar 1;13:4-17.

[74] Silva-Cavalcanti JS, Silva JD, de França EJ, de Araújo MC, Gusmao F. Microplastics ingestion by a common tropical freshwater fishing resource. Environmental Pollution. 2017 Feb 1;221:218-26.

[75] Gasperi J, Wright SL, Dris R, Collard F, Mandin C, Guerrouache M, Langlois V, Kelly FJ, Tassin B. Microplastics in air: are we breathing it in?. Current Opinion in Environmental Science & Health. 2018 Feb 1;1:1-5.

[76] Redondo-Hasselerharm PE, Gort G, Peeters ET, Koelmans AA. Nano-and microplastics affect the composition of freshwater benthic communities in the long term. Science advances. 2020 Jan 1;6(5):eaay4054.

[77] Cole M, Lindeque P, Fileman E, Halsband C, Galloway TS. The impact of polystyrene microplastics on feeding, function and fecundity in the marine copepod *Calanus helgolandicus*. Environmental science & technology. 2015 Jan 20;49(2):1130-7.

[78] Lindeque PK, Cole M, Coppock RL, Lewis CN, Miller RZ, Watts AJ, Wilson-McNeal A, Wright SL, Galloway TS. Are we underestimating microplastic abundance in the marine environment? A comparison of microplastic capture with nets of different mesh-size. Environmental Pollution. 2020 May 3:114721.

[79] Messinetti S, Mercurio S, Parolini M, Sugni M, Pennati R. Effects of polystyrene microplastics on early stages of two marine invertebrates with different feeding strategies. Environmental Pollution. 2018 Jun 1;237:1080-7.

[80] Kaposi KL, Mos B, Kelaher BP, Dworjanyn SA. Ingestion of microplastic has limited impact on a

marine larva. Environmental science & technology. 2014 Feb 4;48(3):1638-45.

[81] Sazakli E, Leotsinidis M. Possible effects of microplastics on human health. Microplastics in Water and Wastewater, Hrissi K. Karapanagioti, Ioannis K. Kalavrouziotis Download citation file: Ris (Zotero) Reference Manager EasyBib Bookends Mendeley Papers EndNote RefWorks BibTex Close Search. 2019.

[82] Sussarellu R, Suquet M, Thomas Y, Lambert C, Fabioux C, Pernet MEJ, et al. Oyster reproduction is affected by exposure to polystyrene microplastics. Proceedings of the National Academy of Sciences of the United States of America. 2016;113(9):2430-5.

[83] Clark JR, Cole M, Lindeque PK, Fileman E, Blackford J, Lewis C, Lenton TM, Galloway TS. Marine microplastic debris: a targeted plan for understanding and quantifying interactions with marine life. Frontiers in Ecology and the Environment. 2016 Aug;14(6):317-24.

[84] Lamb JB, Willis BL, Fiorenza EA, Couch CS, Howard R, Rader DN, True JD, Kelly LA, Ahmad A, Jompa J, Harvell CD. Plastic waste associated with disease on coral reefs. Science. 2018 Jan 26;359(6374):460-2.

[85] Reichert J, Arnold AL, Hoogenboom MO, Schubert P, Wilke T. Impacts of microplastics on growth and health of hermatypic corals are species-specific. Environmental Pollution. 2019 Nov 1;254:113074.

[86] Rehse S, Kloas W, Zarfl C. Short-term exposure with high concentrations of pristine microplastic particles leads to immobilisation of *Daphnia magna*. Chemosphere. 2016 Jun 1;153:91-9.

[87] Rochman CM, Hoh E, Kurobe T, Teh SJ. Ingested plastic transfers hazardous chemicals to fish and induces hepatic stress. Scientific reports. 2013 Nov 21;3:3263.

[88] Batel A, Borchert F, Reinwald H, Erdinger L, Braunbeck T. Microplastic accumulation patterns and transfer of benzo[a]pyrene to adult zebrafish (*Danio rerio*) gills and zebrafish embryos. Environmental Pollution. 2018 Apr 1;235:918-30.

[89] Lönnstedt OM, Eklöv P. Environmentally relevant concentrations of microplastic particles influence larval fish ecology. Science. 2016 Jun 3;352(6290):1213-6.

[90] McGoran AR, Clark PF, Morritt D. Presence of microplastic in the digestive tracts of European flounder, Platichthys flesus, and European smelt, Osmeruseperlanus, from the River Thames. Environmental Pollution. 2017 Jan 1;220:744-51.

[91] Meng Y, Kelly FJ, Wright SL. Advances and challenges of microplastic pollution in freshwater ecosystems: A UK perspective. Environmental Pollution. 2020 Jan 1;256:113445.

[92] Au SY, Lee CM, Weinstein JE, van den Hurk P, Klaine SJ. Trophic transfer of microplastics in aquatic ecosystems: identifying critical research needs. Integrated environmental assessment and management. 2017 May 1;13(3):5059.

[93] Desforges JP, Galbraith M, Ross PS. Ingestion of microplastics by zooplankton in the Northeast Pacific Ocean. Archives of environmental contamination and toxicology. 2015 Oct 1;69(3):320-30.

[94] Nelms SE, Galloway TS, Godley BJ, Jarvis DS, Lindeque PK. Investigating microplastic trophic transfer in marine top predators. Environmental Pollution. 2018 Jul 1;238:999-1007.

[95] Welden NA, Abylkhani B, Howarth LM. The effects of trophic

transfer and environmental factors on microplastic uptake by plaice, Pleuronectes plastessa, and spider crab, *Maja squinado*. Environmental Pollution. 2018 Aug 1;239:351-8.

[96] Gray AD, Weinstein JE. Size-and shape-dependent effects of microplastic particles on adult daggerblade grass shrimp (*Palaemonetes pugio*). Environmental toxicology and chemistry. 2017 Nov;36(11):3074-80.

[97] Nolte TM, Hartmann NB, Kleijn JM, Garnæs J, Van De Meent D, Hendriks AJ, Baun A. The toxicity of plastic nanoparticles to green algae as influenced by surface modification, medium hardness and cellular adsorption. Aquatic toxicology. 2017 Feb 1;183:11-20.

[98] Dong CD, Chen CW, Chen YC, Chen HH, Lee JS, Lin CH. Polystyrene microplastic particles: In vitro pulmonary toxicity assessment. Journal of hazardous materials. 2020 Mar 5;385:121575.

[99] Van Cauwenberghe L, Janssen CR. Microplastics in bivalves cultured for human consumption. Environmental pollution. 2014 Oct 1;193:65-70.

Chapter 14

Screening and Potential Uses of Contaminated Spent Mushroom (*Pleurotus* spp.)

Mathipriya Shanmugavelu and Ganesan Sevugaperumal

Abstract

The commercial production of mushrooms generates a co-product, a virtually inexhaustible supply of spent mushroom substrate (SMS). It represents an ideal growth medium for plants and plant disease suppressive quality. Here we discussed about the contaminated microbial flora of SMS, potential antifungal and plant growth promoting activities, the results of these findings were also discussed in relation to the usage of SMS as a potential product for organic farming. SMS contained moisture content 72%, EC 1.75 mmho.cm^{-1} and had pH of 6.1. The cellulose and hemicellulose content of paddy straw substrate were 30.25%, 23.18% and 15.31% dry weight respectively. Growth in terms of root and shoot weight of the seedlings of green gram, black gram, tomato and chili were significantly higher when grown in 60% SMS amended soil. Spent mushroom compost from *Pleurotus eous* used in this study harbored bacterial population including, *Bacillus* sp., *Clostridium* sp., *Pseudomonas* sp. and *E. coli*. Bacterial isolate B1 was identified as *Bacillus* sp., isolate B2 was identified as *Clostridium* sp., isolate B3 as *Pseudomonas* sp. and B4 as *Escherichia coli*. These bacterial strains showed significant antagonistic activity against soil borne pathogenic fungi viz., *Fusarium* sp., *Alternaria* sp., *Phytophthora* sp. and *Aspergillus* sp.

Keywords: SMS, *Pleurotus*, antifungal activity, contaminants

1. Introduction

Mushrooms have been recognized as the alternate source of good quality protein. They are capable of producing the highest quantity of protein per unit area and time from agro-wastes which are available to the tune of more than 300 million tons per annum in India. Many species of mushrooms are cultivated world-wide. Seventy percent of the global mushroom production is derived from three mushroom groups, *Agaricus bisporus, Pleurotus* spp., and *Lentinula edodes*. The remaining mushroom volume is generated by at least a dozen species [1].

Edible mushrooms commonly have insignificant lipid level with higher proportion of polyunsaturated fatty acids. All these result in low calorific yield from mushroom foods. Mushrooms do not have cholesterol. Instead, they have ergosterol that acts as a precursor for Vitamin D synthesis in human body. Similarly, ergosterol in button mushroom is converted in to vitamin D2 when exposed to UV radiation or sunlight. The protein content of edible mushrooms is usually high, but varies greatly. The crude protein content of mushrooms varied from 12 to 35% dry weight depending upon the species. The free amino acids composition differs widely but in general they are rich in theronine and valine but deficient in sulfur containing amino acids (Methionine and cysteine) [2].

Antibiotic resistance has become a global concern [3]. The clinical efficacy of many existing antibiotics is being threatened by the emergence of multidrug resistant pathogens [4–6]. Already, a number of antibacterial compounds have been isolated from Basidiomycetes fungi, including Collybial and Frustulosin [7–9].

Pleurotus spp. are mushrooms which belong to the class basidiomycetes. They are generally understood to be called white rot fungi, because of their ability to degrade lignocellulosic materials. The oyster mushroom consists of a number of several edible *Pleurotus* species. This species represented 14% of the world production in 1997 [1]. *Pleurotus* mushrooms are edible with excellent flavor and taste. They have nutritional as well as medicinal properties [10]. They are low in calories, fats, sodium, carbohydrates and cholesterol, while being rich in proteins, minerals, vitamins and fibers [11]. *Pleurotus* spp.is promising as medicinal mushrooms, exhibiting hematological, antiviral, antitumor, antibiotic, antibacterial, hypocholesterolic and immunomodulation activities [12].

1.1 Spent mushroom substrate

After the cultivated mushroom have exhausted the nutrients within the substrates, and there were no more fruitbodies harvest, the so called remains, regarded as "the useless material" is known as spent mushroom substrate (SMS) [13].

Several agro industrial wastes could be used to prepare mushroom composts. These growing substrates may be composed from different wastes materials such as sawdust, rice straw, bedded horse manure, cotton wastes, paper wastes, cocoa shells, wheat straw, maize husks and various other wastes [14]. Additives such as rice bran, calcium carbonate or wheat bran may be added to enhance mushroom fructification [15].

Compost is considered "spent substrate" when one full crop of mushroom has been taken and further extension becomes unremunerative [16–18]. Mushroom industry needs to dispose off more than 50 million tons of used mushroom compost each year called Spent Mushroom Substrate (SMS) [19]. Recently, the term spent compost or spent mushroom substrate has been replaced by a more appropriate term, "post mushroom substrate" because it is not "spent" and is ready to be further attacked by a new set of microorganisms. The large dumped piles of spent mushroom substrate become anaerobic and give off offensive odor. The run-off from such piles contaminates nearby water sources and pollutes them [20]. Under normal circumstances, the spent mushroom substrate is discarded as waste without considering environmental repercussion.

The disease suppressive properties of composting materials are known for many decades and much scientific evidence have revealed favorable properties of composts for the management of plant diseases [21, 22]. Due to the unique chemical constitution and the microflora present in SMS, its application can be more diversified than what is normally predicted.

1.2 The Management of Spent Mushroom Substrate (SMS)

Actinomycetes, bacteria and fungi inhabiting the compost, not only play role in its further decomposition but also exert antagonism to the normal pathogens surviving and multiplying in the soil ecosystem.

2. Review of literature

Mushroom is a macrofungus with a distinctive fruiting body, which can be either hypogeous or epigeous, large enough to be seen with the naked eye and to be picked by hand [23]. The number of mushroom species on the earth is estimated to be 1,40,000 suggesting that only 10% are known. Assuming that the proportion of useful mushrooms among the undiscovered and unexamined mushrooms will be 5%, which implies 7000 yet undiscovered species will be of possible benefit to mankind [24].

2.1 *Pleurotus* spp.

Mushrooms are considered as a functional food, which can provide health benefits beyond the traditional nutrients they contain [25, 26]. Nowadays, several species of Pleurotus are cultivated commercially because of their rich mineral contents and medicinal properties, short life cycle, reproducibility in the recycling of certain agricultural and industrial wastes and low demand on resources and technology.

2.2 Anti-microbial activity of *Pleurotus* spp.

Water and alcoholic extracts from *P. ostreatus* mycelium have been used in studies on antimicrobial activities against numerous types of microbes. The highest potency was shown by water extract, especially towards fungi, *Candida albicans, Cryptococcus humicola, Trichosporon cutaneum* and bacteria *Staphylococcus aureus* and *Escherichia coli* [27, 28].

The antimicrobial properties of mushroom extracts and highlighted some of the active compounds identified, including low- and high-molecular weight compounds which showed antagonistic activity against gram positive bacteria. LMW compounds are mainly secondary metabolites, such as sesquiterpenes and other terpenes, steroids, anthraquinones, benzoic acid derivatives, and quinolines, but also primary metabolites such as oxalic acid. HMW compounds are mainly peptides and proteins.

2.3 Spent mushroom substrate (SMS)

After mushroom cultivation, the partially degraded paddy or wheat straw and other agricultural waste, which form as valuable by-products of edible mushroom cultivation, have been termed as Spent Mushroom Substrate (SMS). Antibacterial activity of *H. erinaceus* SMS against phyto-pathogenic bacteria and evaluated the role of this extract in improving plant defense and growth [29, 30].

2.4 Composition of SMS

The macro and micronutrients of the raw material and the initial and spent substrates of *Pleurotus ostreatus* [31–33]. The mineral composition of the fruiting body varied with the substrates, which made possible the production of a fruiting body rich in K, P, Mg and Fe. Potassium was the mineral with the highest content in the fruiting body in all substrates tested. There was an increase in protein and mineral content in the spent substrate in relation to the initial one.

The pH of the compost was found to be 7.58 and the electrical conductivity of the compost was found to be 0.71 dms^{-1}. Chemical analysis of the compost showed varying organic matter and nutrient content. The carbon to nitrogen (C: N) ratio of a product was 13:1 and the bioavailability of total potassium (2.64%), magnesium

(2.26%) and calcium (5.16%) were comparatively higher to the availability of total phosphorous (0.48%) and sodium (0.29%).

2.5 Anti-microbial activity of SMS

The uncontrolled use of antibiotics has caused serious problems in human and animal health, causing that pathogens develop resistance to them, so World Health Organization considered the infections caused by pathogens resistant to drugs as a public health problem; therefore, it is necessary to find new pharmacological strategies, among which we can find natural products such as plants and fungi.

The results showed that in the case of *Escherichia coli*, the greatest inhibition zone was of 12.66 mm at a concentration of 6 mg ml^{-1}, with treatment of *Lentinula edodes*/cedar; *Salmonella typhimurium* showed a greatest inhibition zone of 31.10 mm to a concentration of 5.12 mg mL-1, with treatment of *Pleurotus ostreatus*/ Barley straw.

2.6 Microbial flora of SMS

The bacterial diversity in SMC by using molecular techniques in order to reveal the origin of SMC microflora and its potential effect on soil microbial communities after incorporation into agricultural soils [30, 34, 35]. Fifty bacterial isolates were classified into 14 operational taxonomic units (OTUs) following ARDRA-PCR of the 16S rDNA gene. Sequencing of the 16S rDNA amplicon assigned 12 of the 14 OTUs to Gram-positive bacteria, associated with the genera *Bacillus*, *Paenibacillus*, *Exiguobacterium*, *Staphylococcus*, *Desemzia*, *Carnobacterium*, *Brevibacterium*, *Arthrobacter* and *Microbacterium* of the bacterial divisions *Firmicutes* and *Actinobacteria*.

Objectives of the study.

With the background research and the literature review on spent mushroom substrate (SMS), an attempt was made to isolate, identify and characterize the microbial flora of spent mushroom substrate of *Pleurotus eous*. The research involved the followings objectives:

- Analysis of the composition and physicochemical properties of spent mushroom substrate (SMS) of *Pleurotus eous*.
- Isolation of bacteria from SMS.
- Antimicrobial activity of the bacterial isolates against selected soil pathogenic fungi.

3. Materials and methods

Fresh Spent Mushroom Substrate (SMS) of *Pleurotus eous* was used for the present study. SMS was obtained by growing *Pleurotus eous* mushroom using paddy straw in the mushroom house at Post Graduate and Research Department of Botany, Thiagarajar College, Madurai. The spent paddy straw substrate after three yields of the mushroom *P. eous* was used fresh for characterization studies, dried in the sun and powdered for use in further analyses.

Laboratory tests measured the following properties: P^H, EC, soluble salts, moisture, organic matter, carbon, nitrogen and phosphorus, potassium, Carbon:Nitrogen (C:N) ratio, cellulose and hemicelluloses.

3.1 Growth studies

For pot culture studies, the powdered SMS was mixed with garden soil in different ratio as 20%,40%,60%80%100% and control.

Twenty seeds of green gram and black gram, fifty seeds of tomato and chili were sown in each pot containing garden soil with SMS in various concentrations. Pot contained only the garden soil was maintained as control for each treatment. The treatments were laid out in a completely randomized pattern with three replicates per treatment. Germination was performed under ambient conditions in the net house, and pots were irrigated daily. Growth was monitored up to 30 days after seeding.

3.2 Isolation of microbial flora from SMS

Sample of SMS of *Pleurotus eous* was mechanically stirred with distilled water. The SMS suspension was clarified by centrifugation, serially diluted (10^{-2} to 10^{-7}) with sterile water and inoculated on potato dextrose agar medium (PDA) and nutrient agar (NA) medium. All the strains observed growing on plates were arbitrarily selected, transferred and maintained in new PDA or NA plates for further use.

A representative sample colony of each visually differentiable bacterium was selected using a sterile inoculating loop. Each colony was transferred by streaking an inoculating loop in parallel lines over four quadrants of NA plate. The plates were incubated at 37°c for twenty-four hour. The isolated colonies were used for initial observations about the shape, color, size and other visual properties of each isolate were recorded. The bacterial strains isolated from the SMS of *Pleurotus eous* were characterized and identified by using the standard procedures.

3.3 Antifungal activity of bacteria isolated from SMS in dual culture plate assay

The fungal pathogens viz., *Fusarium sp., Alternaria sp., Phytophthora sp.*, and *Aspergillus sp*., were obtained from Biotechnology lab at Post Graduate and Research Department of Botany, Thiagarajar College, Madurai.

Antifungal activity of bacteria isolated from SMS of *Pleurotus eous* mushroom was evaluated in a dual plate assay against four fungal pathogens of plants viz., *Fusarium sp.*, *Alternaria sp.*, *Phytophthora sp.* and *Aspergillus sp.* The bacteria and fungi were cultured in opposing fashion on PDA plates. Bacterial isolate broth (100 μl) of approximately 108 cells/ml was inoculated on PDA plates using spread plate technique. Mycelial agar plugs of one day old culture of the test fungi and the bacterial isolate were inoculated opposite to each other. All the plates were maintained at room temperature and observed for the appearance of zone of inhibition surrounding the bacterial colony where the fungal mycelium failed to grow.

4. Observations and results

4.1 Composition of spent mushroom substrate (SMS)

The empirical data depicted in **Tables 1** and **2** show the composition of Spent Mushroom Substrate (SMS) of *Pleurotus* spp., compared with the substrate, paddy straw. The cellulose and hemicellulose content of paddy straw substrate were 30.25% and 23.18% and dry weight respectively. Cultivation of *Pleurotus eous* on the substrate had lowered the cellulose content by 24.89%. The hemicellulose

Component	Untreated substrate (paddy straw)	Spent mushroom substrate (paddy straw)	Degradation %
Cellulose (% dry wt.)	30.25 ± 0.73	23.97 ±0.65	24.89
Hemicellulose (% dry wt.)	23.18 ±0.43	22.23 ±0.64	3.35

Table 1.
Biochemical properties of fresh SMS.

pH	6.1
EC	1.75 mmho.cm^{-1}
Moisture content	72%
N	0.87%
P	0.26%
K	0.19%
Organic carbon	15%
C:N	17.24:1

Table 2.
Physico chemical properties of fresh SMS.

utilization was less comparatively, showing a degradation rate of 3.35% from the initial content. The physico-chemical properties of SMS of Fresh SMS had an acidic pH 6.1 and its conductivity was 1.75 mmho.cm^{-1}. It had 72% moisture content. The fresh SMS had 0.87% nitrogen, 0.26% phosphorus, 0.19% potassium and 15% organic carbon. The C/N ratio of fresh SMS was 17.24:1.

4.2 Growth studies

4.2.1 Green gram

The effect of different concentrations of SMS extract on seedling fresh weight and dry weight of green gram revealed that it is also concentration dependent, as in case of seed germination. Green gram seedlings grown in 60% w/v aqueous SMS extract showed the highest shoot weight (2.112 g). The lowest fresh weight of shoot was observed in 100% of SMS extract (0.621 g). The highest root fresh weight (0.498 g) was found in the green gram seedlings treated with 60% of SMS extract and the lowest root fresh weight (0.157 g) in 100% SMS extract treatment. In case of whole plant, fresh weight of the whole plant was also the highest in 60% of SMS treated seedlings and the lowest in 100% concentration of SMS extract. Control seedlings had 2.074 g as whole plant fresh weight.

The observations regarding the dry weight of shoot, root and whole plant of green gram showed the same response shown for fresh weight. i.e. the seedlings treated with 60% SMS extract produced the highest dry weight values of 0.234 g, 0.056 g and 0.291 g for shoot, root and whole plant respectively. The lowest values of 0.069 g, 0.019 g and 0.069 g for shoot, root and whole plant respectively, were observed on 100% SMS extract treated seedlings. The control seedlings showed 0.187 g dry weight for shoot, 0.043 g for root and 0.230 g for whole plant (**Table 3**).

Aqueous SMS conc. (%) (w/v)	Fresh weight (g)			Dry weight (g)		
	Shoot	Root	Whole plant	Shoot	Root	Whole plant
20	1.693 ± 0.028	0.430 ± 0.014	2.124 ± 0.038	0.188 ± 0.043	0.046 ± 0.003	0.234 ± 0.005
40	1.620 ± 0.173	0.445 ± 0.012	2.066 ± 0.168	0.180 ± 0.019	0.049 ± 0.001	0.229 ± 0.019
60	2.112 ± 0.053	0.498 ± 0.012	2.610 ± 0.058	0.234 ± 0.005	0.056 ± 0.003	0.291 ± 0.008
80	1.891 ± 0.067	0.453 ± 0.013	2.344 ± 0.079	0.210 ± 0.017	0.049 ± 0.002	0.259 ± 0.010
100	0.621 ± 0.081	0.157 ± 0.033	0.778 ± 0.114	0.069 ± 0.009	0.019 ± 0.003	0.069 ± 0.009
Control	1.683 ± 0.020	0.390 ± 0.007	2.074 ± 0.016	0.187 ± 0.002	0.043 ± 0.002	0.230 ± 0.001

Table 3.
Effect of SMS on seedling growth of green gram.

4.2.2 Black gram

The results on the effect of SMS extracts on biomass of black gram are given and shown in **Table 4**. In shoot, the fresh weight, 1.824 g was the highest in 60% of SMS treated seedlings and the lowest fresh weight, 1.4 g was observed in 100% concentration of SMS extract. Similarly 60% SMS extract treated seedlings produced the highest fresh root weight value (0.536 g) and 100% of SMS extract treatment showed the least value (0.353 g). The highest whole plant fresh weight value (2.361 g) was found in the seedlings treated with 60% of SMS extract and the lowest value of whole plant fresh weight (1.753 g) was observed in 100% SMS extract treatment compared with the untreated control experiments (1.917 g).

In case of the dry weight of black gram seedlings, the seedlings treated with 40% SMS extract produced the highest dry weight values of 0.212 g and 0.265 g for shoot and whole plant respectively. In root, 0.059 g was found as the highest dry weight in seedlings treated with 60% of SMS extract, followed by 0.053 g in 40% SMS extract treatment. The lowest values of 0.155 g, 0.039 g and 0.194 g for shoot, root and whole plant respectively, were observed on 100% SMS extract treated seedlings. The control seedlings showed 0.171 g dry weight value for shoot, 0.041 g value for root and 0.213 g for whole plant (**Table 4**).

4.2.3 Tomato

Table 5 presents the results of the effect of aqueous extract of SMS in different concentrations on biomass of tomato. The highest shoot fresh weight value (4.446 g) was found in the seedlings treated with 60% of SMS extract, followed by 4.048 g in 80% SMS extract treated seedlings. Control seedlings had the lowest value of fresh weight (1.259 g). 1.108 g of root fresh weight was found as the highest in 60% of SMS treated seedlings and 0.301 g as the lowest root fresh weight in control. The 60% SMS treated seedlings produced the highest whole plant fresh weight value (5.554 g) followed by 80% treated seedlings (5.061 g) compared to the control seedlings (1.560 g).

Tomato seedlings, the seedlings treated with 60% of SMS extract produced the highest dry weight values of 0.264 g, 0.065 g and 0.330 g for shoot, root and whole plant respectively. The lowest dry weight values of 0.129 g in shoot of 100% SMS extract treated seedling, 0.039 g and 0.173 g for root and whole plant seedlings respectively, were observed in the seedlings treated as control. (**Table 5**).

4.2.4 Chili

Table 6 show the results on the effect of aqueous extract of SMS in different concentrations on biomass of chili. The highest shoot fresh weight value (8.887 g) was found in the seedlings treated with 60% of SMS extract, followed by 5.560 g in 40% treated seedlings and the lowest value of shoot fresh weight (0.779 g) in control. 1.488 g of root fresh weight was found as the highest in 60% of SMS treated seedlings and 0.253 g as the lowest fresh weight of root in control. The 60% SMS extract treated seedlings produced the highest whole plant fresh weight value (10.375 g) followed by 40% treated seedlings (6.848 g). The lowest fresh weight of shoot was observed in control seedlings (1.033 g).

The seedlings treated with 60% SMS extract produced the highest dry weight values of 1.903 g, 0.297 g and 2.200 g for shoot, root and whole plant respectively. The lowest values of 0.086 g, 0.028 g and 0.114 g for shoot, root and whole plant respectively, were observed for the seedlings treated as control (**Table 6**).

Aqueous SMS conc. (%) (w/v)	Fresh weight (g)			Dry weight (g)		
	Shoot	Root	Whole plant	Shoot	Root	Whole plant
20	1.513 ± 0.063	0.388 ± 0.015	1.902 ± 0.078	0.185 ± 0.004	0.048 ± 0.007	0.233 ± 0.004
40	1.612 ± 0.066	0.482 ± 0.007	2.094 ± 0.066	0.212 ± 0.002	0.053 ± 0.008	0.265 ± 0.003
60	1.824 ± 0.072	0.536 ± 0.008	2.361 ± 0.070	0.202 ± 0.008	0.059 ± 0.009	0.262 ± 0.007
80	1.691 ± 0.066	0.458 ± 0.013	2.149 ± 0.070	0.187 ± 0.007	0.050 ± 0.001	0.238 ± 0.007
100	1.400 ± 0.042	0.353 ± 0.008	1.753 ± 0.050	0.155 ± 0.004	0.039 ± 0.009	0.194 ± 0.005
Control	1.544 ± 0.042	0.373 ± 0.009	1.917 ± 0.048	0.171 ± 0.004	0.041 ± 0.001	0.213 ± 0.005

Table 4.
Effect of SMS on seedling growth of black gram.

Aqueous SMS conc. (%) (w/v)	Fresh weight (g)			Dry weight (g)		
	Shoot	Root	Whole plant	Shoot	Root	Whole plant
20	3.182 ± 0.038	0.828 ± 0.014	4.010 ± 0.048	0.189 ± 0.049	0.047 ± 0.018	0.237 ± 0.021
40	3.622 ± 0.031	0.898 ± 0.011	4.520 ± 0.043	0.215 ± 0.006	0.053 ± 0.024	0.269 ± 0.013
60	4.446 ± 0.094	1.108 ± 0.024	5.554 ± 0.119	0.264 ± 0.053	0.065 ± 0.027	0.330 ± 0.016
80	4.048 ± 0.095	1.012 ± 0.025	5.061 ± 0.120	0.240 ± 0.012	0.060 ± 0.038	0.300 ± 0.025
100	2.252 ± 0.258	0.658 ± 0.072	2.910 ± 0.330	0.129 ± 0.096	0.048 ± 0.039	0.177 ± 0.019
Control	1.259 ± 0.011	0.301 ± 0.009	1.560 ± 0.019	0.139 ± 0.001	0.033 ± 0.001	0.173 ± 0.002

Table 5.
Effect of SMS on seedling growth of tomato.

Aqueous SMS conc. (%) (w/v)	Fresh weight (g)			Dry weight (g)		
	Shoot	**Root**	**Whole plant**	**Shoot**	**Root**	**Whole plant**
20	5.178 ± 0.201	1.103 ± 0.037	6.282 ± 0.239	1.035 ± 0.040	0.220 ± 0.027	1.256 ± 0.047
40	5.560 ± 0.284	1.288 ± 0.037	6.848 ± 0.266	1.520 ± 0.125	0.257 ± 0.007	1.777 ± 0.129
60	8.887 ± 0.192	1.488 ± 0.028	10.375 ± 0.218	1.903 ± 0.134	0.297 ± 0.015	2.200 ± 0.135
80	4.517 ± 0.219	1.140 ± 0.051	5.658 ± 0.182	1.223 ± 0.098	0.228 ± 0.010	1.452 ± 0.104
100	3.921 ± 0.301	0.418 ± 0.048	4.339 ± 1.166	0.880 ± 0.022	0.083 ± 0.019	0.963 ± 0.254
Control	0.779 ± 0.022	0.253 ± 0.007	1.033 ± 0.029	0.086 ± 0.002	0.028 ± 0.001	0.114 ± 0.003

Table 6.
Effect of SMS on seedling growth of chili.

Bacterial isolate*	Inhibition zone (cm)			
	Fusarium sp.	*Phytophthora* sp.	*Alternaria* sp.	*Aspergillus* sp.
B1	0.45	1.17	1.12	1.66
B2	0.77	1.4	1.00	1.51
B3	1.05	1.25	0.75	0.82
B4	0.95	1.02	0.69	1.05

**Bacterial isolates - B1, B2, B3 and B4.*

Table 7.
Antifungal activity of bacteria isolated from SMS in dual culture plate assay.

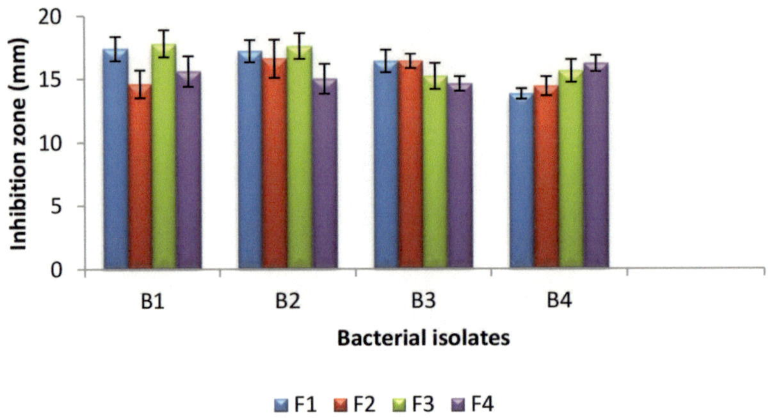

Figure 1.
Antifungal activity of bacteria isolated from SMS in dual culture plate assay. F1 - Fusarium *sp. F2 -* Alternaria *sp. F3 -* Phytophthora *sp. F4 -* Aspergillus *sp.*

4.3 Isolation of microbial flora from SMS

In this study, four different bacteria were isolated from the fresh spent mushroom substrate after *Pleurotus* spp., cultivation. The colony morphological variations among the four bacterial isolates of the bacterial strain 1 (B1) were irregular in shape with a colony size of 1.2 cm dia. The colonies had eros type margin rough or dry texture and were white in color. Cells of bacterial isolate B2 showed a diameter of 0.7 cm. The margin of B2 colonies were entire with smooth texture, mucoid consistency and cream in color. Colony morphology of B3 isolate showed irregular shaped colonies of 0.5 cm dia. B3 isolate colonies showed serrated margin with smooth or glistening texture and cream color. The B4 bacterial isolate had circular shaped colonies of size 0.2 cm with entire margin and yellow in color. The cellular morphology of B1, B2 and B3 isolates was observed as rods while B4 isolate had short rod shaped cells.

The biochemical properties of four bacterial strains isolated from the SMS obtained after *P. eous* cultivation showed significant variations in the biochemical characteristics studied. Based on the observations and results obtained by subjecting the bacterial colonies in different identification techniques, bacterial isolate B1 was identified as *Bacillus* sp., isolate B2 was identified as *Clostridium* sp., isolate B3 as *Pseudomonas* sp. and B4 as *Escherichia coli*. The findings on antifungal activity of four bacterial strains isolated from SMS are shown in **Table 7** and **Figure 1**.

5. Discussion

Mushroom growing is an ecofriendly activity as it utilizes the waste from agriculture, horticulture, poultry, brewery etc. for its cultivation. However, piling up of "spent mushroom substrate" released after mushroom crop harvesting may cause various environmental problems, including ground water contamination and nuisance [20, 36].

Production of 1 kg of mushrooms will generate 5 kg of spent residual material called spent mushroom substrate (SMS). An average farm discards about 24 t of SMS per month [37]. In Ireland, approximately 254,000 t of SMS is generated each year [38] and in The Netherlands, more than 800,000 t of SMS is produced per year [39].

In some countries, waste management of SMS is a major problem faced by farmers. Apparently, the obvious solution is to increase the demand for SMS through exploration of new applications for utilization. It would be more economical and favorable if SMS is to be recycled and reused. Considering the high organic matter of SMS, rapid advances have been made and the number of scientific research has increased in the past few years.

5.1 Composition of spent mushroom substrate (SMS)

Potting medium is an important factor for the production of crop in containers, and component and properties of the potting media are very crucial for higher and quality yields of potted plants. Chemical properties of growth media are very crucial from the point of view of nutrient availability to the plants. SMS used in our study in **Tables 1** and **2** showed that pH range 6.10 and EC range 1.75 dsm^{-1} are suitable for normal growth of plant. The average moisture content of SMS in our study has been measured as 72% which was also reported. [40].

Ability to provide essential nutrients to plants is one of the most fundamental criteria while judging the suitability of a growth medium [41]. Primary nutrients like nitrogen, phosphorus, and potassium are more available at pH 5.5–6.5 for substrates of organic and mineral origins [42]. Moreover, with the increasing pH, the solubility of many nutrients is reduced and some nutrients are precipitated as solid materials that plant cannot use [43]. In contrast many researchers reported high salinity of SMS, which is mostly responsible for the limited use of SMS as a potting media [9].

Results of our study showed that SMS had 0.87% nitrogen, 0.26% phosphorus content, 0.19% of potassium. SMS has been shown to increase the nutrient availability of growth media [44]. In general, most mushroom substrates have low N content, typically in the 1% to 3% range [45]. The overall nutrients of SMS were not enough to support normal plant growth without external fertilizer application. It is well known that physical properties of soil were directly related to crop yield [46]. SMS by maintaining high organic matter content in the soil and by providing the three primary nutrients e.g. nitrogen, phosphorus and potassium helps to provide soil fertility [41, 47].

5.2 Effect of SMS on growth and biomass of selected plants

Growth of root and shoot weight of the seedlings of green gram, black gram, tomato and chili were significantly higher when grown in 60% SMS mended soil than the control (**Tables 3–6**). Higher nutrient availability provided by SMS might have contributed to the better growth in 60% amended soil. Spent mushroom compost (SMC) of *Pleurotus ostreatus* improved the agronomic characters and yield (pod no, fresh weight and dry weight) when it was added as soil conditioner to soybean at different levels of its concentrations [14].

Reason for poor root and shoot growth of the seedlings of the selected plants in 100% SMS may be that the paddy straw based. SMS may only be used as an amendment and not as a basic growth medium. While growth on straw, *Pleurotus* releases humic acids like fractions which when added to soil would increase its fertility. In addition, humic substances may affect the plant biochemical process [48]. Present findings confirm the efficacy of SMS in growth promotion in terms of seedling shoot and root weight.

SMS from *Agaricus bisporus, Hericium erinaceus* and *Pleurotus ostreatus* are effective to the growth promotions of pea, pepper and tomato plants respectively [29, 36].

The results of the present study revealed that the compost has a good impact in promoting better growth and yield. Further in order to promote growth and yield, it becomes imperative to optimize the usage of organic manure according to the crop requirement.

5.3 Characterization of bacterial strains isolated from SMS

Four bacteria species were isolated from spent mushroom substrate used in this study. They were coded isolate B1, B2, B3 and B4. These microorganisms were identified and characterized as *Bacillus, Clostridium, Pseudomonas,* and *Escherichia coli*.

The isolation of these bacteria from composting agricultural substrates suggests that a form of fermentation had taken place during the composting process. Addition of straw in the soil caused an increase in the number of total bacteria, actinomycetes and fungi of the rhizosphere [49].

Different substrate harbor different kind and number of microorganisms and the variation in microbial population in different substrates is due to nutritional or chemical composition of the substrate.

Similar isolation of *Bacillus* and *Clostridium* species were reported from fermenting cocoa beans [50, 51]. The presence of *Pseudomonas* sp. (Isolate B3) in the fermenting SMS may be related to its ability to survive in vast number of habitats. From the results obtained, it can be concluded that various bacteria genera were involved in the decomposition of further microbial SMS. The pure cultures of these bacteria could be incorporated into agricultural wastes in a controlled fermentation unit.

Effect of rice straw compost on soil microbial population reported that, compost application resulted in marked increase of organic matter content in the soil in relation to initial value of plain sandy soil which affirmatively exaggerated the bacterial and fungal populations and that microbial population increased with the increase in dosage of compost [52].

Similar results were reported in maize by using *Agaricus bisporus* spent mushroom compost [53] using recomposted button mushroom spent substrate with wheat crop and with *Pleurotus florida* spent substrate on tomato crop.

5.4 Anti-fungal activity of bacterial strains isolated from SMS extracts

Spent mushroom compost from *Pleurotus* sp., used in this study harbored bacterial population including, *Bacillus, Clostridium, Pseudomonas* and *E. coli*. This is in support of the findings on the microbial composition of spent mushroom compost of *Pleurotus* sp. [53]. SMS used for soil amendment has been found to be more efficient than commercial fungicides and nematicide in controlling soil borne pathogens like *Meloidogyne* sp. in tomato, *Venturia inaequalis* in apple [53]. Some of these microorganisms have been reported to possess antagonistic property in several studies [54] and this was confirmed in this study from the result obtained from the dual culture assay involving the SMS microbial isolates and plant pathogenic fungi.

Fungal pathogens such as *Fusarium oxysporum* and *Phytophthora* cause severe plant diseases, limiting plant yields as well as the quality of the products. Moreover, they have wide host spectra, causing diseases in economically important agricultural crops worldwide [55]. These fungal phytopathogens are difficult to control not only because of their wide host spectra, but also because of their soil borne nature [56]. *Aspergillus flavus* is the most deleterious fungus in stored rice grains, and it receives particular attention because of its ability to produce potent carcinogenic aflatoxins [57].

Alternaria solani, *Alternaria alternata*, *Fusarium solani*, *Phytophthora megasperma* and *Verticillium dahlia* are ubiquitous and cosmopolitan phytopathogens causing severe diseases in wide range of crops [58, 59].

The findings on antifungal activity of four bacterial strains *Bacillus* sp., *Clostridium* sp., *Pseudomonas* sp. and *E. coli* isolated from SMS are shown in **Table 7** and **Figure 1**. Using antibiotic producing bacteria to control plant fungal diseases is a popular topic and has extensively been studied [60]. Compared with chemical biocides, many antibiotics produced by antagonistic strains have the advantage of being easily decomposed in nature, leaving no harmful residues behind. The results, of the present study on the **in vitro** sensitivity of phytopathogenic fungi to antagonistic bacteria revealed that the isolates of *B. subtilis* were suppressive, though with different degrees, to the tested isolates of phytopathogenic fungi, are consistent with those obtained by others [61, 62].

Bacillus and *Pseudomonas* sp. show antifungal effect against soil borne plant pathogenic fungi *c, Fusarium solani* and *Fusarium oxysporum* [63].

Bacillus subtilis showed strong ability against many common plant fungal pathogens **in vitro** [61]. The investigation of bio control activity of the strains in this study revealed that they could produce extracellular secondary metabolites with antifungal activity against the tested fungi. These strains inhibited the mycelial growth of the fungus in dual-culture assays. The Pseudomonas isolates obtained from soil were shown to reduce growth of *Aspergillus niger, Fusarium sp., Alternaria solani, Drechslera oryzae* etc., [64].

Our results on the antifungal activity of the microbes isolated from SMS (**Table 7** and **Figure 1**) which showed that the culture filtrates of bacterial strains, *Bacillus megaterium* KU143, *Microbacterium testaceum* KU313, and *Pseudomonas protegens* significantly inhibited the growth of *A. flavus*.

The SMS isolate of *Pseudomonas* sp. was tested positive for antifungal activity against *Phytophthora* sp. (**Table 7** and **Figure 1**). Similarly this bacterium was found efficient in inhibiting the mycelial growth and the antifungal compounds extracted were found inhibitory to the growth of *Rhizoctonia* sp., *Phytophthora parasitica, P. palmivora* and *Fusarium solani*.

The inhibitory properties of spent mushroom substrate remained unaffected even after autoclaving and filter sterilization of extract [54]. Unsterilized spent mushroom compost had a better inhibition potential than sterilized compost, this suggested that the pathogen inhibitory properties of spent mushroom compost could be more due to the biotic components than the abiotic components, i.e., more due to the activities of the inherent microorganisms rather than the chemical properties or the organic matter content. Similar views were reported [65].

The results of the present study confirm that spent mushroom substrate contains a large number of indigenous beneficial microbes capable of suppressing soil-borne pathogens. This character of the SMS can be utilized as an alternate substrate for the in mass production of biocontrol agents for field application which may lead to suppression of diseases leading to increased crop productivity.

Author details

Mathipriya Shanmugavelu and Ganesan Sevugaperumal*
Centre for Research and PG Department of Botany, Thiagarajar College of Arts and Science (Autonomous), Madurai, Tamil Nadu, India

*Address all correspondence to: sganesan76@mail.com

References

[1] Chang, S. T, 1999. *Ganoderma lucidum* (Curt.: Fr.). P. Karst. (Aphyllophoromycetideae) – A mushrooming medicinal mushroom. *Int. J. Med. Mushrooms,* 1: 139-146.

[2] Jin-Ming, G, 2006. New biologically active metabolites from Chinese higher fungi. *Curr. Org. Chem.*, 10: 849-871.

[3] Westh, H., Zinn, C.S. and Rosahi, V. T, 2004. An international multicenter study of antimicrobial consumption and resistance in *Staphylococcus aureus* isolates from 15 hospitals in 14 countries. *Microb. Drug Resist.*, 10: 169-179.

[4] Alves, M.J., Ferreira, I.C., Dias, J., Teixeira, V., Martins, A., Pintado, M, 2012. A review on antimicrobial activity of mushroom (*Basidiomycetes*) extracts and isolated compounds. *Planta Med.*, 78: 1707-1718.

[5] Bandow, J.E., Brotz, H., Leichert, L.I. , 2003. Proteomic approach to understanding antibiotic action. *Chemother.*, 47: 948-955.

[6] Hoitink, H. A., Stone, A. G. and Han, D. Y, 1997. Suppression of plant diseases by composts. *Hortsci., 32*, 184-187.

[7] Armstrong, F. A. J., Strens, C. R. and Strickland, J. K. R., 1967. Mineral component analysis, *Deep Sea Res.*, 14: 381.

[8] Jonathan, S.G., Oyetunji, O.J., Olawuyi, O. J and Uwukhor, P. O, 2013. Application of *Pleurotus ostreatus* SMC as soil conditioner for the growth of soybean (*Glycine max). Academia Arena.*, 5: 54-61.

[9] Jordan, S.N., Mullen, G. J and Murphy, M. C, 2008. Composition variability of spent mushroom compost in Ireland. *Biores. Technol.*, 99: 411-418.

[10] Mahmood Khan, S., Nawaz, A., Malik, W., Javed, N., Yasmin, T., Rehman, M., Qayyum, A., Iqbal, Q., Ahmad, T. and Ali Khan, A., 2011. Morphological and molecular characterization of Oyster mushroom (*Pleurotus* spp.), *African J. Biotechnol.*,10: 2638-2643.

[11] Gupta, B.P., Reddy, N. and Kotasthane, A. S. 2011. Molecular characterization and mating type analysis of oyster mushroom (*Pleurotus* spp.) using single basidiospores for strain improvement, *World J. Microbiol. Biotechnol.*, 27: 1-9.

[12] Cohen, R., Persky, L. and Hadar, Y, 2004. Biotechnological applications and potential of wood-degrading mushrooms of the genus *Pleurotus*. *Appl. Microb. Biotech.*, 58: 582-594.

[13] Fasidi, I. O., Kadiri, M., Jonathan, S. G., Adenipekun, C. O. And Kuforiji, O. O. , 2008. *Cultivation of Tropical Mushrooms*, Ibadan University Press, Ibadan, Nigeria.

[14] Jonathan, S. G, 2002. Vegetative growth requirements and antimicrobial activities of some higher fungi in Nigeria [dissertation] Ibadan: University of Ibadan.

[15] Jayasinghe, C., Imtiaj, A., Hur, H., Lee, G.W., Lee, T.S. and Lee, U. Y, 2008. Favorable culture conditions for mycelial growth of Korean wild strains of *Ganoderma lucidum*. *Mycobiol.*, 36: 28-33.

[16] Wuest, P.J., H.K. Fahy, J. Ferdinand, K. Long and C.C. Bahler, 1991. Development of procedures for using and storing spent mushroom compost to reduce the risk of lowering water quality. Final Research Project Report. The Pennsylvania State University. University Park, PA 16802.

[17] Yohalem, D.S., Nordheim, E.V. and Andrews, J. H, 1996. The effect of water extracts of spent mushroom compost on apple scab in the field. *Phytopathol.*, 86, 914-922.

[18] Simon, B., Anke, T., Anders, U., Neuhaus, M. and Hansske, F, 1995. Collybial, a new antibiotic sesqui terpenoid from *Collybia confluens* (Basidiomycetes). *Z. Natur. forsch. C.*, 50: 173-180.

[19] Fox, R. and Chorover, J, 1999. Seeking a cash crop in spent substrate. *http: /aginfo. psu.edu/psa/sgg/mushrooms. html.*

[20] Beyer M, 1996. The impact of the mushroom Industry on the environment. *Mushroom News*, 44: 6-13.

[21] Hoitink, H. A. J. and Fahy, P. C, 1986. Basis for the control of soil-borne plant pathogens with composts. *Ann. Rev. Phytopathol., 24*, 93-114.

[22] Kolawole, D.O. and Okonkwo, B. A, 1985. Microbiolgy of Traditional Fermented Oil Bean Seeds for Upaka, a Nigerian Condiment *NIFO J.* , 3 (1, 2 and 3): 149-152.

[23] Chang, S.T. and Miles, P.G., 1992. Mushroom biology- a new discipline. *Mycologist*, 6: 64-65.

[24] Hawksworth, D.L., 2001. Mushrooms: the extent of the unexplored potential. *Int. J. Med. Mush.*, 3: 333-337.

[25] *Cheung, P. C,* 2008. *Mushrooms as functional foods.* Wiley Press, New Jersey. pp: 1: 268.

[26] Koch, B.L. 1980. Influence of a compost-peat mixture on the establishment of apple seedlings. *Goodfruit Grower*, *31*: 11, 14-15.

[27] Piska, K., Sułkowska-Ziaja, K. and Muszyńska, B, 2017. Edible mushroom *Pleurotus ostreatus* (Oyster mushroom) – its dietary significance and biological activity. *Acta Sci. Pol., Hortorum Cultus.*, 16: 151-161.

[28] Verma, R.R. 1986. Efficacy of organic amendments against *Meloidogyne incognita* infesting tomato. *Ind. J. Nematol.,16*: 105-106.

[29] Kwak, A.M., Kang, D.S., Lee, S.Y. and Kang, H. W, 2015. Effect of spent mushroom substrates on *Phytophthora* blight disease and growth promotion of pepper. *J. Mushrooms,* 13: 16-20.

[30] Popoola TOS and Ekueshi, C. O, 1985. Microorganisms Associated with Fermentation of Soybean for Production of Daddawa (a condiment) *NIFOJ.*, 3: 194-196.

[31] Maynard, A. A, 1994. Sustained vegetable production for three years using composted animal manures. *Compost Sci. Util.*, 2: 88-96.

[32] Okafor, N, 1977. Microorganisms Associated with Cassava Fermentation for Gari Production. *J. Appl. Bac.*, 42: 279-284.

[33] SALES-CAMPOS, C, 2008. Aproveitamento de resíduos madeireiros e da agroindústria regional para o cultivo de fungos comestíveis de ocorrência na região amazônica. 182. f. Tese (Doutorado em Biotecnologia), Universidade Federal do Amazonas, Manaus.

[34] Ntougias, S., Zervakis, G. I., Kavroulakis, N., Ehaliotis, C., and Papadopoulou, K. K, 2004. Bacterial diversity in spent mushroom compost assessed by amplified rDNA restriction analysis and sequencing of cultivated isolates. *Syst. Appl. Microbiol.*, 27: 746-754.

[35] Tariq, U., Rehman, S.U., Khan, M.A. and Yunus, A., 2012. "Agricultural and municipal waste as potting media components for the growth and

flowering of *Dahlia hortensis Figaro*, "*Turkey J. Bot.*, 36: 378-385.

[36] Ahlawat, O.P., Manikandan K., Sagar M.P., Dev, R. and Vijai, B, 2011. Effect of composted button mushroom spent substrate on yield, quality and disease incidence of Pea (*Pisum sativum*). *Mushroom Res.*, 20: 87-94.

[37] Singh, A.D., Vikineswary, S., Abdullah, N., Sekaran, M, 2011. Enzymes from spent mushroom substrate of *Pleurotus sajor-caju* for the decolourisation and detoxification of textile dyes. *World J. Microbiol. Biotechnol.*, 27: 535-545

[38] Barry, J., Doyle, O., Grant, J., Grogan, H, 2012. Supplementing of spent mushroom substrate (SMS) as a casing material to improve the structure and productivity. In: 18th Congress of the International Society for Mushroom Science. Beijing, China. pp. 735-742.

[39] Oei, P. and Albert, G, 2012. *Recycling casing soil*. In: 18th Congress of the International Society for Mushroom Science. Beijing, China. pp. 757-765.

[40] Maher, M., Lenehan, J. and Staunton, W, 1993. Spent Mushroom Compost-Options for Use. Teagasc, Dublin. pp. 1-44.

[41] Idowu, O.O. and Kadiri, M. 2013. "Growth and yield response of okra to spent mushroom compost from the cultivation of *Pleurotus ostreatus* an edible mushroom," *Academic J. Agri. Res.*,1: 39-44.

[42] Munita, J., 2001. "Characteristics and classification of soils," In Chemical and Mining Company of Chile, Santiago, Chile,11th edition. pp. 1515.

[43] Altland, J. E. , 2006. "Substrate pH, a tricky topic," *Digger*, 50: 42-47.

[44] Medina, E., Paredes, C., Bustamante, M.A., Moral, R. and Moreno-Caselles, J., 2012. Relationships between soil physico-chemical, chemical and biological properties in a soil amended with spent mushroom substrate. *Geoderma.* , 173-174, 152-161.

[45] Polat, E., Onus, A.N. and Demir, H, 2004. The Effects of Spent Mushroom Compost on Yield and Quality in Lettuce Growing. *J. Fac. Agric. Akdeniz Univ.*, 17: 149-154.

[46] Stewart, D., Cameron, K. and Cornforth, I, 1998. Effects of spent mushroom substrate on soil chemical conditions and plant growth in an intensive horticultural system: a comparison with inorganic fertilizer. *Aus. J. Soil. Res.*, 36: 185-198.

[47] Rinker, D.L., ZERI and Kang, S.W., 2004. Recycling of oyster mushroom substrate. In: Mushroom Growers' Handbook-1, 9: 187-191.

[48] Vaughan, D., Malcolm, R.E., Ord, B.G, 1985. Influence of humic substances on biochemical processes in plants. In: Vaughan D, Malcolm RE. (Eds). *Soil Organic Matter and Biological Activity*. Kluwer Academic Publishers, pp. 77-100.

[49] Shukry, W.M., El-Fallal, A.A. and H.M.S. El-Bassiouny, 1999. Effect of spent wheat straw growth, growth hormones, metabolism and rhizosphere of *Cucumis sativa*. *Egypt. J. Physiol. Sci.*, 23: 39-69.

[50] Ojey, O. A, 1981. *Nig. J. of Nutritional Sciences,* 2: 1-5.

[51] Rombouts, J. E. 1952. Critical review of the yeast species previously described from cocoa. *Tropical Agriculture, Trinidad*, 30: 34-39.

[52] Rashad, F.M., Saleh, W.D. and Moselhy, M. A, 2010. Bioconversion of rice straw and certain agro-industrial wastes to amendments for organic farming systems: 1. Composting,

quality, stability and maturity indices. *Bioresour. Technol.*, 101: 5952-5960.

[53] Ahlawat, O.P., Sagar, M.P., Raj, D., Indu Rani, C., Gupta, P. and Vijay, B., 2007. Effect of spent mushroom substrate on yield and quality of capsicum (*Capsicum annuum). Indian J Hort.*, 64: 430-434.

[54] Yohalem, D.S., Harris, R.F. and Andrews, J. H, 1994. Aqueous extracts of spent mushroom substrate for foliar disease control. *Compost Sci. Util.*, 2: 67-74.

[55] Agrios, G.N., 2005. Plant Pathology. 5th Edn., Elsevier Academic Press, New York, USA. pp. 922.

[56] Summeral, B.A., Salleh, B. and Leslie, J. F, 2003. A utilitarian approach to *Fusarium* identification. *Plant Dis.*, 87: 117-128.

[57] Scheidegger, K.A. and Payne, G. A, 2003. Unlocking the secrets behind secondary metabolism: a review of *Aspergillus flavus* from pathogenicity to functional genomics. *J. Toxicol.*, 22: 423-459

[58] Kucuk, C. and Kivanc, M., 2003. Isolation of *Trichoderma* spp. and determination of their antifungal, biochemical and physiological features. *Turk. J. Biol.*, 27: 247-253.

[59] Okami, B. and Hotta, A.K., 1988. Search and Discovery of New Antibiotics. In: *Actinomycetes* in Biotechnology, Goodfellow, M., S.T. Williams and M. Mordarski (Eds.). Pergamon Press, Oxford. pp: 33-67.

[60] Raaijmakers, J.M., M. Vlami and J.T. de Souza, 2002. Antibiotic production by bacterial biocontrol agents. *Anton. Leeuwen.,* 81: 537-547.

[61] Gong, M., Wang, J.D., Zhang, J., Yang, H., Lu, X.F., Pei, Y, and Cheng J. Q, 2006. Study of the antifungal ability of *B.subtilis* strain PY-1 *in vitro* and identification of its antifungal substance (iturin A). *Acta. Biochim. Biophys. Sin. (Shanghai)*., 38: 233-240.

[62] Wagih, E.E., Shehata, M.R.A., Farag, S.A. and Dawood, M.K., 1989. *Dracaena* leaf proliferosis, a newly recorded disease affecting *Dracaena sanderiana* in *Egypt. J. Phytopathol.*, 126: 7-16.

[63] Hameeda, B., Harini, G., Rupela, O.P., Wani, S.P., and Reddy, G. 2006. Growth promotion of maize by phosphate-solubilizing bacteria isolated from composts and macrofauna. *Microbiol. Res.,* 163: 234-242.

[64] Ray, S., Ghosh, M. and Mukherjee, N, 1990. Fluorescent *Pseudomonas* for plant disease control. *J. Mycopathol. Res.*, 28: 135-140.

[65] Suárez, E.F., Bustamante, M.A. and Moral, R, 2012. In vitro control of Fusarium wilt using agroindustrial Subproduct–based composts. J. Plant Pathol., 94: 59-70.

Chapter 15

Determination of Concentration for Some Priority Substances in Paddy Fields of Ergene River, Meriç River, and Yenikarpuzlu Dam, Turkey

Barış Can Körükçü and Cemile Ozcan

Abstract

This study was intended in paddy (rice husk and rice), sediment, and irrigation water samples taken from the paddy fields of Ergene River, Meriç River, and Yenikarpuzlu Dam reservoir which are frequently grown in the river basin in Thrace region and endosulfan (EN) and PAHs were investigated. For analysis, EN and PAHs were studied by GC-MS. The data obtained as a result of the analyses were compared with the results of the standard reference items, and the accuracy of the results was determined. The standard addition method was used to prove the accuracy of EN and PAHs. The recovery parameters on the extraction efficiency of EN and PAHs were optimized, and the recoveries ranged from 82 to 105%. The methods showed good linearity for EN and PAHs, and the LOD and LOQ for methods were found 0.03–63.1 and 0.1–210 $\mu g\ kg^{-1}$, respectively.

Keywords: Thrace region, paddy, sediment, water, endosulfan, PAHs

1. Introduction

Industrial and agricultural activities along with increasing industrialization are polluted very quickly of water and soil resources. When these sources of pollution are taken into consideration, it is necessary to follow the industrial and agricultural residues. These pollutants, which are used in agriculture and industry, interfere with the natural environment and threaten the ecological environment. Some pollutants can be found in the environment even after years of prohibition, and others can be transported over long distances. Pollutants are spread to the environment as industrial, agricultural and domestic sources [1, 2].

Direct transport of pesticides on the soil surface or on the plant play a role factors such as evaporation, surface flow, soil penetration and adsorption. Evaporation is on the soil, water and plant surface and the most important factor affecting the evaporation of pesticide is its evaporation pressure. In addition, high temperature, low relative humidity and air movement are environmental factors that accelerate evaporation. Pesticides strongly absorbed by soil particles are much less likely to evaporate [3–5].

Endosulfan (EN) is an organochlorine and acaricide group. Acute toxicity is a colorless, solid agricultural chemical prohibited due to its bioaccumulative potential and endocrine disrupting effects [4–6]. EN residues in nature are also known to remain in the soil for at least 6 years [7, 8]. Therefore, it requires examination of the soil contaminated with EN, the product grown in soil, sediment and the water used. Polycyclic systems occur when one ring is sharing two carbons with another ring, or the rings are connected to each other by a C–C bond [9]. PAHs are from the group of compounds which show unsaturation in molecular formulas and do not give addition reactions which are characteristic for them. In the cyclic structure, PAHs from the class of planar molecules are resistant to oxidation. In addition, PAHs can be found in petrochemical, rubber, plastic, mineral oil, rust oil, paint, leather and other products. Rubber and plastic materials are high-risk materials containing PAH. In the Ergene Basin, where the industry is intense, these compounds are likely to be found. As the molecular weights of PAHs increase, their solubility in water decreases. However, their toxic and carcinogenic properties increase [9–12]. Contaminated soil, air and aquatic products may also contain PAH. The cooking meat or other food on the grill or at high temperatures increases the amount of PAH in food [11]. In this study, analysis of EN and PAHs compounds shown in **Figure 1** was performed.

The organic components (PAHs and EN) we analyzed are considered among the primary pollutants [9, 13–17]. When the pesticides in our ecological environment are taken in high concentration, they can cause deformations on the biological structure of the organism. In the event of prolonged exposure to certain pesticides, cancer can be seen or short-term exposure may result in direct death [13, 15–19].

The extensively used high sensitivity analytical techniques for the determination of EN and PAHs at low concentrations in environmental samples are

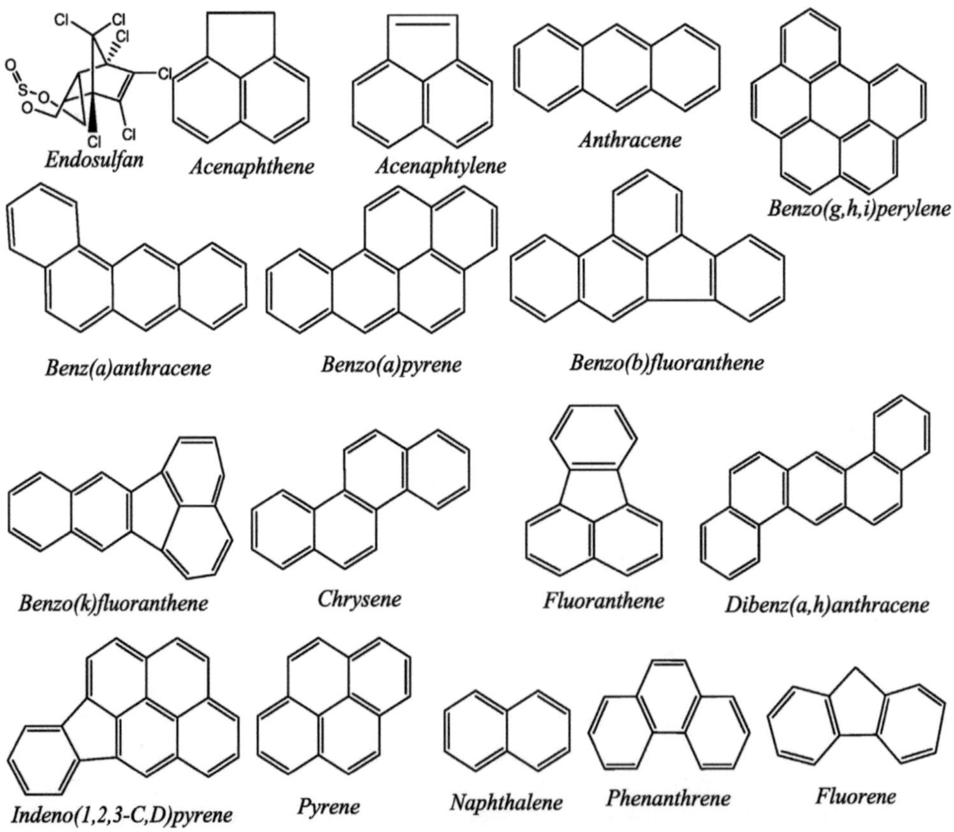

Figure 1.
Structural formulas of analyzed compounds.

GC-MS [6, 8, 20, 21], HPLC-MS [10, 22]. The separation and preconcentration techniques for pesticides are solid phase extraction (SPE) [21, 23, 24], solid phase micro extraction (SPME) [21, 22, 25] and sonication [22, 26], which are used to solve these problems in analysis of EN and PAHs.

The aim of this study was intended in paddy (rice husk and rice), sediment, and irrigation water samples taken from the paddy fields of Ergene River, Meriç River, and Yenikarpuzlu Dam reservoir which are frequently grown in the river basin in Thrace region and EN and PAHs were investigated. Thrace region is a place where industry and agriculture are intense, so the analysis with real examples will be performed of great importance here. For this reason, EN and PAHs, which developed method validation, were studied by GC-MS.

2. Materials and methods

Within the scope of this study, the paddy production areas in the agricultural areas of Thrace Region, direct irrigation from Ergene-Meriç rivers and dam ponds, Ergene and Meriç Basin were evaluated and three regions were determined as the study area. The sampling points are shown in **Figure 2**. Endosulfan and PAH analyses were made in paddy plant, sediment and irrigation water samples taken from paddy fields, which irrigated from the Ergene River, the Meriç River, and the Yenikarpuzlu Dam Reservoir, considering the distinction of irrigation resources in the rice plants often grown in river basins in the Thrace region. In the Yenikarpuzlu village of Edirne province, around Sığırcı Dam, and from the irrigated area from Edirne-İpsala-Yenikarpuzlu, Edirne-Merkez-Üyüklütatar and Edirne-Uzunköprü-Muhacirkadı Village were collected the paddy, sediment and paddy irrigation water (**Figure 2**). Endosulfan and PAH analysis were performed method validation by GC-MS.

Agilent GC-MS was used in the determination phase for all studies. The instrument used is the HP-5 MS UI capillary column (30 m × 250μm × 0.25 μm) and the 5990C (Agilent) inert MSD mass detector with 7890A (Agilent) model GC-MS. The electron ionization (EI) system with 70 eV ionization energy in GC-MS, and the He gas was used as carrier gas.

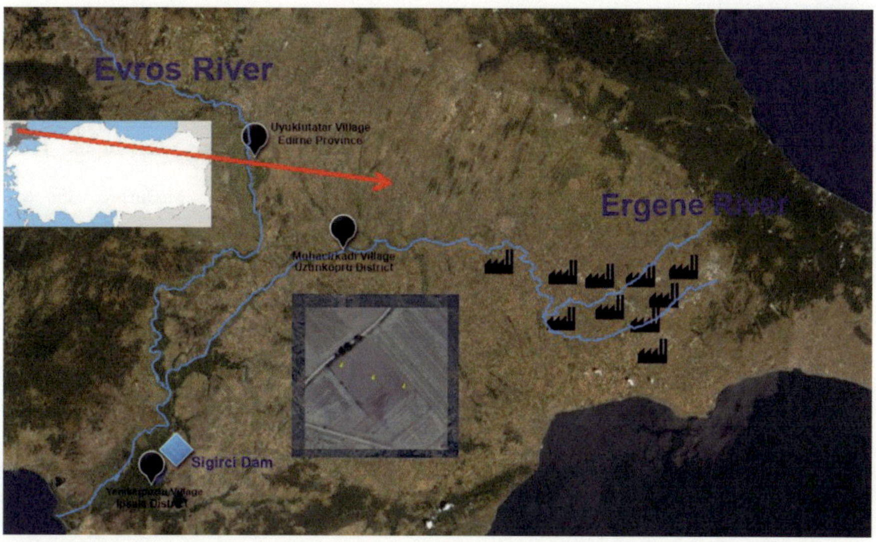

Figure 2.
Representation of the sampling points on the map.

2.1 Preparation of standard solutions

Dilution for Endosulfan (EN) was carried out on pure standard (SIGMA-ALDRICH, Product: 45852, EN solution 100 ng μL^{-1} in n-hexane, PESTANAL®). As a standard for PAHs, Dr. Ehrenstorfer 2095009 product PAH-mix 9 was used. In the standard, there are mixtures of 16 polycyclic aromatic hydrocarbons (PAHs); naphthalene (NAP), acenaphthalene (ACE), acenaphtylene (ACY), fluorene (FLU), phenantherene (PHN), anthracene (ANT), fluoranthene (FLR), pyrene (PYR), benz(a)anthracene (BaA), chrysene (CRY), benzo(b)fluoranthene (BbF), benzo(k)fluoranthene (BkF), benzo(a)pyrene (BaP), dibenz(a,h)anthracene (DahA), benzo(g,h,i)perylene (BghiP) and indeno(1,2,3,c,d)pyrene (IcdP). Optimum conditions of EN and PAHs were shown in **Tables 1** and **2**. According to the mass spectra of the chromatograms of the EN and PAH pesticides, column retention times and ion inputs were studied in SIM mode. Pesticide standards were prepared (0.1, 0.25, 0.5, 1, 2, 2.5, 5, 7.5, 10, 25, 50, 100, 250, 500, 750 and 1000 $\mu g\ L^{-1}$) and measurements were taken. PAH-mix 9 of 16 polycyclic aromatic hydrocarbon compounds in standard contented NAP, ACE, ACY, FLU, PHN, ANT, FLR, PYR, BaA, CRY, BbF, BkF, BaP, DahA, BghiP and IcdP, and the dilution made out of from standard mixture was prepared standards of 1, 5, 10, and 25 $\mu g\ L^{-1}$. The methods applied in GC–MS for EN and PAHs are given in **Tables 1** and **2**, respectively.

The retention times (RT), SIM fragmentation ions, chromatogram programming times of PAHs are given in **Table 3**. The TICs of EN and PAHs were shown in **Figures 3** and **4**.

GC injection conditions				
Applied method			Splitless	
Injection volume			1 µL	
Temperature			250°C	
He gas flow rate			1 mL/min	
Total flow rate			64 mL/min	
Septum cleaning flow			3 mL/min	
Temperature program			250°C for 1 min	
Cleaning flow for split vent			60 mL/min throughout 2.5 min	
Transfer line temperature program			150°C for 0 min, runtime 30 min 280°C for 0 min, runtime 30 min	
Column oven temperature program			**MS information**	
			Collection mode	**Scan**
Rate (°C/min)	Temperature (°C)	Standby time (min)	Solvent delay (min)	10.00
			Gain factor	5.00
	45 (initial)	1	EM voltage obtained	2271
21	150	5	MS source temperature (°C)	230
4	220	0	MS Quad (°C)	150
	300 (post run)	3	Scanned mass range	100–279

Table 1.
The optimum GC-MS conditions applied for endosulfan.

GC injection conditions				
Applied method			Splitless	
Injection volume			1 μL	
Inlet temperature			70°C	
He gas flow rate			1 mL/min	
Total flow rate			64 mL/min	
Septum cleaning flow			3 mL/min	
Temperature program			300 °C for 1 min	
Cleaning flow for split vent			60 mL/min throughout 2.0 min	
Transfer line temperature program			150°C for 0 min, runtime 42.43 min 280°C for 0 min, runtime 41.43 min	
Column oven temperature program			**MS information**	
			Collection mode	**Scan**
Rate (°C/min)	Temperature (°C)	Standby time (min)	Solvent delay (min)	0.00
			Gain factor	5.00
	70 (initial)	1	EM voltage obtained	2329
10	120	1	MS source temperature (°C)	230
7	270	13	MS Quad (°C)	150
	300 (post run)	3	Scanned mass range	100–279

Table 2.
Optimum GC-MS conditions for PAHs.

Pesticides	**RT (min)**	**Target ion**	**Ion 1**	**Ion 2**	**Programming time (min)**
Naphthalene (NAP)	7.584	128.00	129.00	—	6.00
Acenaphthalene (ACE)	12.394	152.00	153.00	—	8.00
Acenaphtylene (ACY)	13.017	153.00	154.00	—	12.67
Fluorene (FLU)	14.694	166.00	165.00	—	13.50
Phenantherene (PHN)	17.846	178.00	176.00	179.00	15.00
Anthracene (ANT)	17.994	178.00	176.00	179.00	17.91
Fluoranthene (FLR)	21.850	202.00	200.00	203.00	18.10
Pyrene (PYR)	22.556	202.00	200.00	203.00	22.15
Benz(a)anthracene (BaA)	26.650	228.00	226.00	229.00	23.00
Chrysene (CRY)	26.879	228.00	226.00	229.00	26.71
Benzo(b)fluoranthene (BbF)	30.281	252.00	250.00	253.00	26.90
Benzo(k)fluoranthene (BkF)	30.376	252.00	250.00	253.00	30.33
Benzo(a)pyrene (BaP)	37.713	252.00	250.00	253.00	35.65
Benzo(g,h,i)perylene (BghiP)	31.539	276.00	138.00	277.00	31.70
Dibenz(a,h)anthracene (DahA)	38.082	276.00	278.00	279.00	37.95
Indeno(1,2,3,c,d)pyrene (IcdP)	39.436	276.00	274.00	277.00	38.50

Table 3.
The retention times (RT), SIM fragmentation ions, and chromatogram programming times of PAHs.

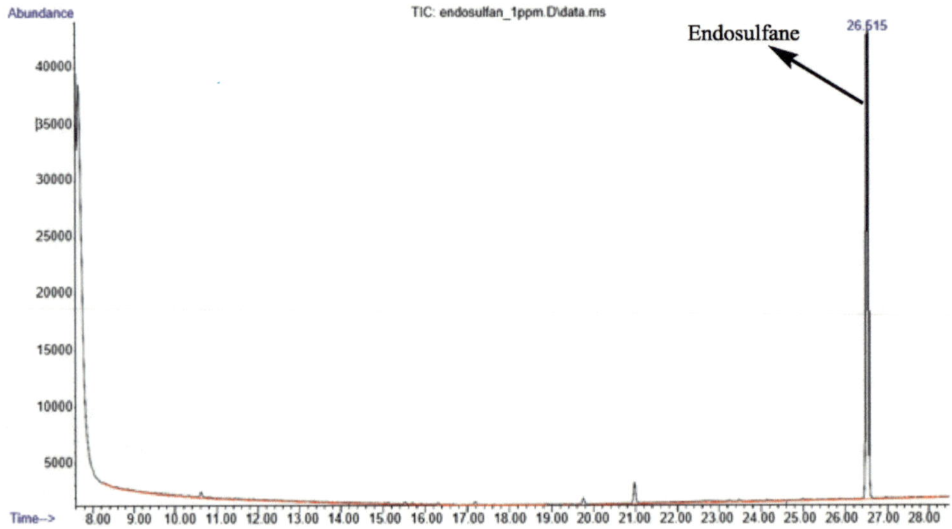

Figure 3.
GC chromatograms of Endosulfan compound.

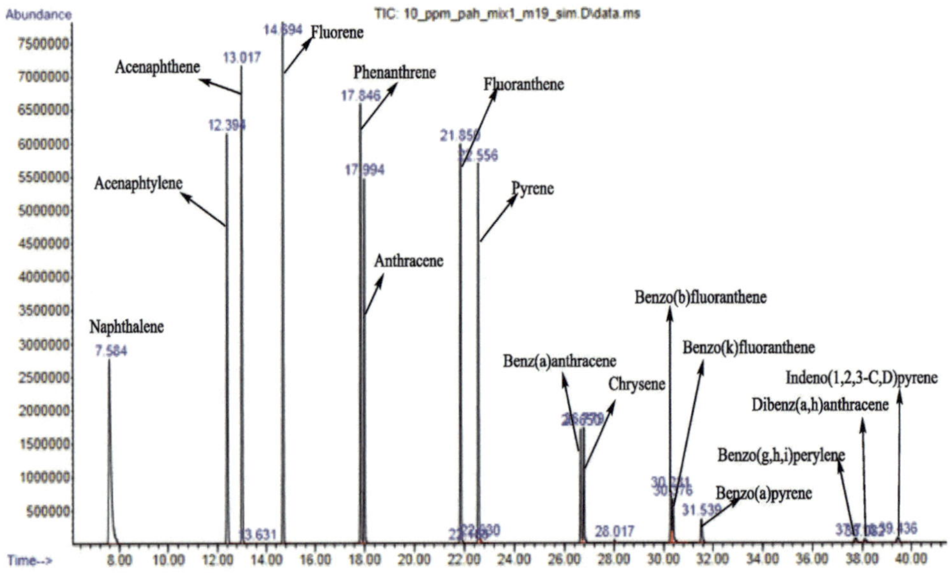

Figure 4.
GC chromatograms of PAH compounds.

The standards (EN and PAHs) were prepared in certain concentrations and readings were made in the device and the calibration graphs were plotted to calculate the amounts in the actual samples (**Figures 5** and **6**).

The linear regression, correlation coefficient, the detection limit (LOD) indicating the performance of the method in the method validation [14], the determination limit (LOQ), relative standard deviation percentage (RSD) and recovery calculations of the pesticides analyzed by GC-MS were shown in **Table 4**.

2.2 Collection of samples and preparation for analysis

For each point determined, twice samples were sampled in the months of July (sowing period) and September (harvesting period) in the periods of paddy sowing

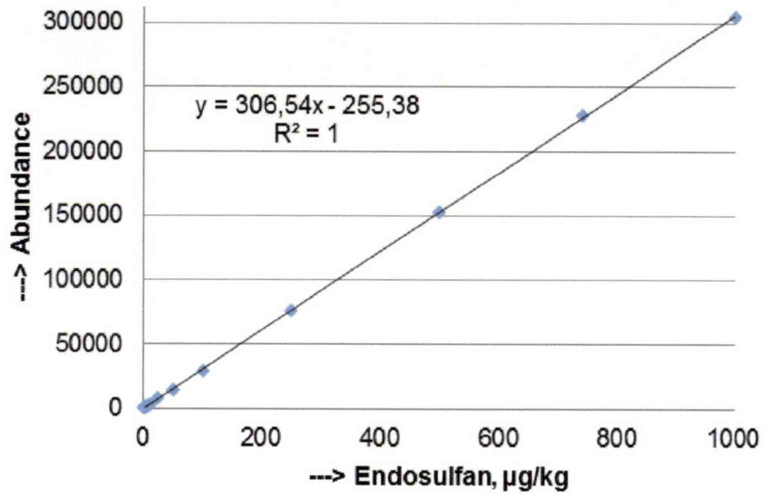

Figure 5.
The calibration graph of Endosulfan compound.

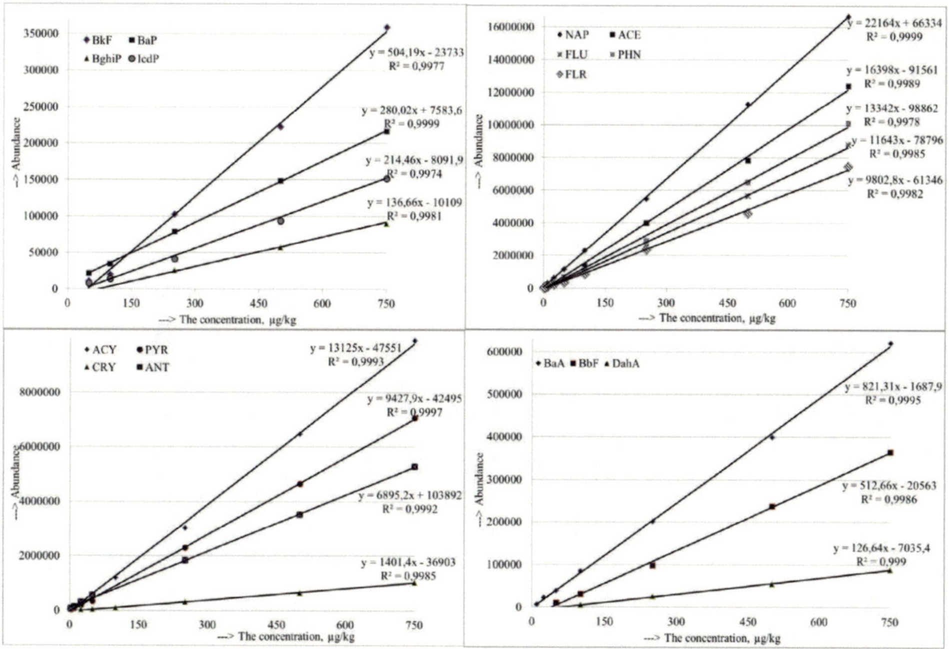

Figure 6.
The calibration graphs of PAHs compounds.

and harvesting. In the paddy pan, selected for sampling, three points were determined to represent each of the ceilings and pan water, soil and plant samples were made from these points. Since there were no paddy in the sowing period, only in the harvesting period of soil and water "at the latest 5 days before the harvest" were done every triple sampling. The pan water, sediment samples, rice grains and rice husk samples from the paddy samples were coded with the abbreviation "w," "sd," "r," and "rh," respectively. The coordinates of the locations, code and sampling dates are given in **Table 5**.

Water samples prepared for analysis of EN and PAHs are enriched in the solid phase extraction system and prepared for analysis by GC-MS. Plant and sediment

	Pesticides	Regression equation (linear range)	R^2	LOD	LOQ	RSD%	Recovery,%
1	EN	305.77x - 245.32 (1–1000 μg/kg)	0.9999	0.79	2.64	7.39	103
2	NAP	23,096x - 6571.3 (0.25–1000 μg/kg)	0.9984	0.03	0.10	1.85	105
3	ACE	16,398x - 91,561 (1–750 μg/kg)	0.9989	0.23	0.76	1.17	98.9
4	ACY	13,125x - 47,551 (0.5–750 μg/kg)	0.9993	0.05	0.17	0.42	101
5	FLU	11,643x - 78,796 (2.5–750 μg/kg)	0.9985	0.43	1.46	1.62	88.1
6	PHN	13,342x - 98,862 (2.5–750 μg/kg)	0.9978	0.64	2.12	2.17	96.5
7	ANT	6895.2x + 103,892 (2.5–750 μg/kg)	0.9992	1.17	3.89	3.65	100
8	FLR	9819.7x - 70,254 (2.5–750 μg/kg)	0.9981	0.64	2.12	2.28	91.6
9	PYR	9427.9x - 42,495 (2.5–750 μg/kg)	0.9997	0.67	2.24	3.42	97.6
10	BaA	821.31x - 1687.9 (25–750 μg/kg)	0.9995	4.05	13.52	4.96	92.9
11	CRY	1401.4x - 36,903 (25–750 μg/kg)	0.9985	6.78	22.6	1.35	86.9
12	BbF	512.66x - 20,563 (50–750 μg/kg)	0.9986	12.3	40.9	4.86	83.2
13	BkF	504.19x - 23,733 (50–750 μg/kg)	0.9977	17.4	58.1	6.8	84.3
14	BaP	280.02x + 7583.6 (100–1000 μg/kg)	0.9999	20.0	66.3	7.87	82.5
15	IcdP	136.66x - 10,109 (250–1000 μg/kg)	0.9981	63.1	210	3.33	86.1
16	NAP	127.67x - 7342 (250–1000 μg/kg)	0.9991	42.2	141	5.05	90.2
17	ACE	214.46x - 8091.9 (50–1000 μg/kg)	0.9974	14.0	46.7	5.72	95.0

Table 4.
The values of linear regression ($y = ax + b$), correlation coefficient (R^2), LOD (μg/kg), LOQ (μg/kg), and RSD% in the pesticides.

Coordinates	Sowing period			Harvest period		
	Code	Sampling date	Watering source	Code	Sampling date	Watering source
40°48′25.0″N 26°18′58.9″E	1-w	20.07.2016	Dam Lake	11-w	22.09.2016	Dam Lake
				11-sd		
	1-sd			11-r		
				11-rh		
41°32′19.7″N 26°36′14.4″E	2-w	27.07.2016	Meriç River	22-w	23.09.2016	Meriç River
				22-sd		
	2-sd			22-r		
				22-rh		
41°20′17.8″N 26°52′31.6″E	3-w	02.08.2016	Ergene River	33-w	21.09.2016	Ergene River
				33-sd		
	3-sd			33-r		
				33-rh		

Table 5.
Sampling's coordinates, codes, and dates.

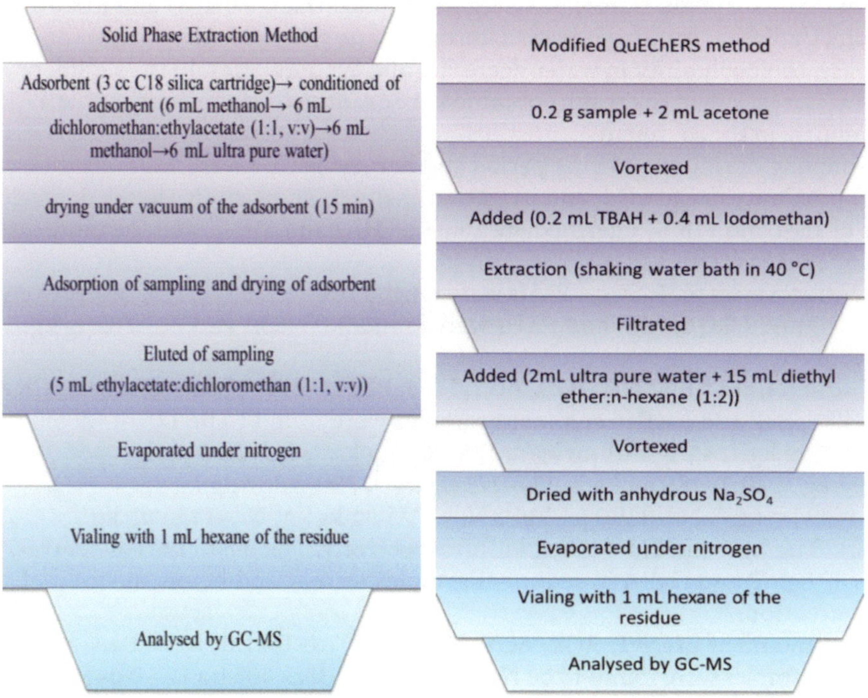

Figure 7.
The enrichment of samples with the solid phase extraction and modified QuEChERS methods.

samples were enriched with modified QuEChERS method and prepared for analysis by GC-MS. The contents of the solid phase extraction and modified QuEChERS methods are given in **Figure 7**.

3. Results and discussion

Along with the developments in industrialization, it has made it necessary to determine organo-contaminant at major and minor levels in the settlements where there is a large amount of factories. In this study, The analysis of EN, NAP, ACE, ACY, FLU, PHN, ANT, FLR, PYR, BaA, CRY, BbF, BkF, BaP, DahA, BghiP and IcdP were performed in paddy, sediment and water samples taken during paddy planting and harvesting from the places such as Yeni Karpuzlu Dam, Muhacirkadı Village (Ergene River) and Üyüklütatar Village (Meriç River). These analyses were performed by using GC-MS. The results of the analysis of EN and PAHs in samples were given in **Tables 6–8**.

Large scale accumulation or pollution of pesticide chemicals or natural chemicals is a source of concern for our global world. Due to the incorporation of these substances into the condensation and evaporation cycle, our natural life creates constant exposure with rain, snow and fog [26]. Soil pollution is closely related to industrial activities, destruction of municipal and industrial waste or environmental accidents. Soil is a complex and heterogeneous matrix with a porous structure containing inorganic and natural organic components [1, 2].

PAH components, which come out from the chemical production factories and vehicles' exhausts along the Ergene River, constitute a serious source of pollution. In the Ergene Basin is located in petrochemical, rubber, plastic, mineral oil, rust oil, paint, leather and other products. Rubber and plastic materials including PAHs are high-risk materials. For this reason, EN and PAHs analyses of the samples collected from rice cultivated areas in the basin selected as clean region (Yeni Karpuzlu Dam), dirty region (Muhacirkadı Village) and less dirty region (Üyüklütatar Village) were determined. As the molecular weights of PAHs increase, their solubility in water decreases and accumulation in the sediment also increases. When the results obtained are examined, it can be seen that PAHs accumulate in the sediment.

Sampling during the sowing period and harvest period were made classifications as for proximity to the road, and irrigation channel, or proximity to the dam with code of 1B-1 and 11B-1, intermediate zones of 1B-2 and 11B-2, and the more distant area of 1B-3 and 11B-3.

The recoveries for EN and PAHs were ranged from 82.5 to 105%, respectively. The LOD and LOQ for EN and PAHs were found 0.03–63.1 and 0.1–210 $\mu g\ kg^{-1}$, respectively.

The amount of EN, NP, ACE, ACY, FLU, PHN, ANT, FLR, PYR, BaA, CRY, BbF, BkF, BaP and IcdP in sediment samples were found to be 17.5–44.9 $mg\ kg^{-1}$, 219–417 $\mu g\ kg^{-1}$, 41.1–64.1 $\mu g\ kg^{-1}$, 176–264 $\mu g\ kg^{-1}$, Nd–70.5 $\mu g\ kg^{-1}$, 66.7–318 $\mu g\ kg^{-1}$, 20.3–485 $\mu g\ kg^{-1}$, 137–273 $\mu g\ kg^{-1}$, 790–1867 $\mu g\ kg^{-1}$, 113–1549 $\mu g\ kg^{-1}$, Nd–1190 $\mu g\ kg^{-1}$, Nd–1775 $\mu g\ kg^{-1}$g, Nd–1436 $\mu g\ kg^{-1}$, Nd–2478 $\mu g\ kg^{-1}$ and Nd–968 $\mu g\ kg^{-1}$, respectively. The pesticide concentrations of BghiP and DahA in sediment samples were found below the limit of determination.

The amount of EN, NP, ACE, ACY, FLU, PHN, ANT, FLR, PYR, BaA, and CRY in water samples were found to be Nd–11.1 $mg\ L^{-1}$, 38.3–105 $\mu g\ L^{-1}$, Nd–10.3 $\mu g\ L^{-1}$, 8.30–14.8 $\mu g\ L^{-1}$, 6.73–10.6 $\mu g\ L^{-1}$, 44.8–104 $\mu g\ L^{-1}$, 18.7–152 $\mu g\ L^{-1}$, 16.2–65.4 $\mu g\ L^{-1}$, 45.0–92.4 $\mu g\ L^{-1}$, 19.4–88.3 $\mu g\ L^{-1}$and Nd–85.4 $\mu g\ L^{-1}$, respectively. The sample chromatograms of EN and PAHs in sediment, water, rice and rice husk are shown in **Figures 8** and **9**. Except for the clean area, EN was determined above the limit of detection in other sampling areas. The amount of EN in the dirty area was determined as ND-11.1 $\mu g\ L^{-1}$. In the Harvest period for the polluted region were found 7.85, 7.99, and 8.64 $\mu g\ L^{-1}$, respectively.

		Sowing period			Harvest period		
		1Bsd1	1Bsd2	1Bsd3	11Bsd1	11Bsd2	11Bsd3
Dam Basin	EN, mg/kg	31.7 ± 1.7	28.0 ± 2.5	21.8 ± 2.6	17.5 ± 1.3	29.0 ± 2.3	22.5 ± 1.2
	NAP, µg/kg	385 ± 30	376 ± 2	325 ± 5	306 ± 8	307 ± 24	314 ± 26
	ACE, µg/kg	61.8 ± 2.6	63.7 ± 5.6	62.5 ± 1.9	59.7 ± 4.1	54.7 ± 5.3	53.0 ± 4.6
	ACY, µg/kg	264 ± 1	254 ± 14	226 ± 9	208 ± 7	223 ± 10	212 ± 16
	FLU, µg/kg	55.6 ± 1.5	70.5 ± 2.9	47.8 ± 2.8	45.1 ± 3.8	53.9 ± 1.3	66.7 ± 2.9
	PHN, µg/kg	318 ± 18	263 ± 23	164 ± 13	124 ± 8	113 ± 3	120 ± 5
	ANT, µg/kg	485 ± 34	383 ± 35	197 ± 2	122 ± 11	102 ± 6	115 ± 8
	FLR, µg/kg	273 ± 10	247 ± 15	236 ± 2	201 ± 14	204 ± 13	204 ± 10
	PYR, µg/kg	1867 ± 126	1617 ± 54	1622 ± 55	1265 ± 99	1245 ± 54	1291 ± 48
	BaA, µg/kg	708 ± 70	1231 ± 110	1244 ± 95	424 ± 42	497 ± 36	113 ± 7
	CRY, µg/kg	1058 ± 75	1190 ± 32	Nd	Nd	Nd	Nd
	BbF, µg/kg	970 ± 57	1775 ± 34	Nd	1293 ± 51	1074 ± 48	1118 ± 30
	BkF, µg/kg	1359 ± 112	1436 ± 47	900 ± 91	1354 ± 44	1124 ± 49	1163 ± 102
	BaP, µg/kg	2462 ± 159	1768 ± 14	2478 ± 121	383 ± 38	1737 ± 119	312 ± 31
	IcdP, µg/kg	867 ± 3	968 ± 59	846 ± 42	921 ± 19	758 ± 53	815 ± 58

		Sowing period			Harvest period		
Meriç-Üyüklütatar Village		**2Bsd1**	**2Bsd2**	**2Bsd3**	**22Bsd1**	**22Bsd2**	**22Bsd3**
	EN, mg/kg	27.0 ± 1.7	24.2 ± 1.9	29.9 ± 2.4	36.9 ± 3.6	44.9 ± 2.7	39.7 ± 2.3
	NAP, µg/kg	336 ± 22	278 ± 18	321 ± 14	296 ± 12	311 ± 33	254 ± 24
	ACE, µg/kg	64.1 ± 1.2	58.5 ± 4.6	58.5 ± 5.1	55.3 ± 4.9	54.6 ± 3.9	50.4 ± 3.1
	ACY, µg/kg	248 ± 17	208 ± 16	244 ± 10	249 ± 28	233 ± 5	258 ± 15
	FLU, µg/kg	44.6 ± 4.4	44.0 ± 1.9	43.8 ± 3.4	50.3 ± 1.1	39.6 ± 3.9	38.9 ± 0.1
	PHN, µg/kg	104 ± 9	126 ± 5	104 ± 12	108 ± 1	69.2 ± 6.1	77.4 ± 3.7
	ANT, µg/kg	85.5 ± 3.1	127 ± 9	67.9 ± 5.2	93.7 ± 1.2	21.0 ± 1.3	36.3 ± 3.5
	FLR, µg/kg	212 ± 1	212 ± 13	196 ± 13	182 ± 5	156 ± 12	149 ± 11
	PYR, µg/kg	1407 ± 61	1415 ± 109	1328 ± 107	1160 ± 17	969 ± 92	881 ± 71
	BaA, µg/kg	1115 ± 108	519 ± 7	341 ± 33	749 ± 53	1093 ± 67	154 ± 15
	CRY, µg/kg	1136 ± 120	Nd	Nd	Nd	777 ± 69	Nd
	BbF, µg/kg	1171 ± 112	1084 ± 99	885 ± 53	1122 ± 109	753 ± 64	661 ± 52
	BkF, µg/kg	1427 ± 123	1011 ± 37	368 ± 16	1391 ± 22	801 ± 67	Nd
	BaP, µg/kg	136 ± 17	Nd	310 ± 25	160 ± 14	Nd	Nd
	IcdP, µg/kg	797 ± 36	777 ± 49	768 ± 95	692 ± 8	Nd	Nd
	IcdP, µg/kg	797 ± 36	777 ± 49	768 ± 95	692 ± 8	Nd	Nd

Muhacirkadı-Ergene Basin		Sowing period			Harvest period		
		3Bsd1	3Bsd2	3Bsd3	33Bsd1	33Bsd2	33Bsd3
	EN, mg/kg	21.4 ± 2.2	23.0 ± 1.6	26.5 ± 1.9	33.7 ± 2.5	21.7 ± 2.3	23.3 ± 2.3
	NAP, μg/kg	322 ± 12	417 ± 21	408 ± 18	219 ± 19	242 ± 23	226 ± 18
	ACE, μg/kg	56.5 ± 3.6	61.5 ± 1.2	56.8 ± 2.6	51.9 ± 2.0	44.1 ± 1.8	57.5 ± 3.4
	ACY, μg/kg	232 ± 18	258 ± 12	237 ± 7	235 ± 14	197 ± 19	176 ± 16
	FLU, μg/kg	44.6 ± 0.7	48.6 ± 3.8	46.0 ± 1.4	Nd	62.4 ± 2.5	50.7 ± 4.1
	PHN, μg/kg	102 ± 6	113 ± 5	77.8 ± 4.7	68.9 ± 6.4	66.7 ± 2.8	101 ± 3
	ANT, μg/kg	81.7 ± 6.4	103 ± 9	99.3 ± 2.5	20.3 ± 1.1	27.9 ± 2.1	79.6 ± 6.2
	FLR, μg/kg	190 ± 1	137 ± 4	196 ± 14	137 ± 7	245 ± 16	179 ± 5
	PYR, μg/kg	1275 ± 72	1314 ± 62	1279 ± 90	790 ± 78	1086 ± 47	1099 ± 52
	BaA, μg/kg	659 ± 66	771 ± 78	1549 ± 143	576 ± 41	664 ± 55	570 ± 50
	CRY, μg/kg	Nd	Nd	989 ± 91	674 ± 32	521 ± 23	Nd
	BbF, μg/kg	1194 ± 28	1014 ± 59	1047 ± 96	Nd	Nd	589 ± 36
	BkF, μg/kg	1243 ± 27	580 ± 12	1098 ± 37	Nd	577 ± 32	732 ± 33
	BaP, μg/kg	Nd	Nd	Nd	62.0 ± 6.2	483 ± 38	Nd
	IcdP, μg/kg	715 ± 28	728 ± 32	708 ± 24	Nd	Nd	Nd

Table 6.
Amount of pesticides in sediment samples in the dam basin, Meriç-Üyüklütatar village, and Muhacirkadı-Ergene Basin (n = 6).

		Sowing Period			Harvest Period		
		1Bw1	**1Bw2**	**1Bw3**	**11Bw1**	**11Bw2**	**11Bw3**
Dam Basin	EN, mg/kg	Nd	Nd	Nd	Nd	Nd	Nd
	NAP, µg/kg	47.6 ± 4.4	47.3 ± 3.4	58.9 ± 2.9	70.1 ± 0.1	47.2 ± 3.2	105 ± 9
	ACE, µg/kg	6.44 ± 0.05	10.3 ± 0.6	Nd	Nd	Nd	Nd
	ACY, µg/kg	8.66 ± 0.87	8.76 ± 0.21	9.68 ± 0.48	12.6 ± 0.7	10.3 ± 1.2	13.8 ± 1.1
	FLU, µg/kg	7.93 ± 0.78	6.73 ± 3.4	8.07 ± 0.27	6.76 ± 0.14	6.74 ± 0.37	7.70 ± 0.61
	PHN, µg/kg	60.6 ± 2.0	104 ± 6	70.4 ± 2.2	44.8 ± 6.1	52.7 ± 1.9	53.8 ± 3.5
	ANT, µg/kg	91.6 ± 3.8	18.7 ± 0.7	110 ± 4	62.0 ± 1.9	76.9 ± 3.6	79.0 ± 6.6
	FLR, µg/kg	18.2 ± 0.6	65.4 ± 4.2	18.3 ± 0.63	16.2 ± 1.6	17.5 ± 0.6	16.5 ± 0.8
	PYR, µg/kg	66.6 ± 4.1	45.0 ± 3.4	62.0 ± 3.1	58.5 ± 0.3	65.8 ± 5.3	55.7 ± 4.2
	BaA, µg/kg	45.5 ± 2.3	63.9 ± 3.2	39.6 ± 3.0	34.9 ± 3.4	38.4 ± 3.3	30.5 ± 1.9
	CRY, µg/kg	62.0 ± 4.3	Nd	Nd	Nd	60.8 ± 0.8	Nd
Meriç-Üyüklütatar Village		**2Bw1**	**2Bw2**	**2Bw3**	**22Bw1**	**22Bw2**	**22Bw3**
	EN, mg/kg	10.6 ± 0.3	Nd	Nd	Nd	Nd	Nd
	NAP, µg/kg	47.4 ± 1.1	53.7 ± 4.6	49.2 ± 2.5	39.5 ± 1.8	43.1 ± 2.2	41.1 ± 1.2
	ACE, µg/kg	Nd	Nd	6.60 ± 0.01	Nd	Nd	Nd
	ACY, µg/kg	8.86 ± 0.71	10.1 ± 0.8	7.86 ± 0.51	9.97 ± 0.13	8.30 ± 0.30	9.10 ± 1.9
	FLU, µg/kg	7.24 ± 0.31	8.26 ± 0.47	7.58 ± 0.72	7.41 ± 0.42	6.84 ± 0.26	7.12 ± 0.59
	PHN, µg/kg	71.0 ± 2.5	79.1 ± 5.8	69.1 ± 1.1	62.1 ± 4.2	60.8 ± 6.3	61.4 ± 5.3
	ANT, µg/kg	111 ± 5	126 ± 10	108 ± 2	94.5 ± 7.9	92.0 ± 7.5	93.3 ± 3.8
	FLR, µg/kg	18.4 ± 0.6	20.4 ± 1.2	20.0 ± 1.4	17.2 ± 0.4	16.9 ± 0.6	17.3 ± 1.2
	PYR, µg/kg	61.8 ± 2.1	71.8 ± 3.2	75.7 ± 6	51.8 ± 3.7	50.8 ± 3.2	54.1 ± 3.8
	BaA, µg/kg	39.3 ± 0.4	44.7 ± 1.9	59.9 ± 4.1	27.4 ± 2.6	26.1 ± 2.1	29.7 ± 2.3
	CRY, µg/kg	Nd	Nd	Nd	Nd	Nd	Nd

		Sowing Period			Harvest Period		
Muhacirkadı-Ergene Basin		**3Bw1**	**3Bw2**	**3Bw3**	**33Bw1**	**33Bw2**	**33Bw3**
	EN, mg/kg	11.1 ± 0.6	7.36 ± 0.63	Nd	8.64 ± 0.32	7.85 ± 0.81	7.99 ± 0.46
	NAP, μg/kg	73.5 ± 2.1	53.0 ± 2.7	58.3 ± 3.9	42.4 ± 1.8	38.3 ± 1.1	46.7 ± 4.5
	ACE, μg/kg	6.82 ± 0.19	Nd	Nd	Nd	Nd	Nd
	ACY, μg/kg	14.8 ± 1.4	14.6 ± 0.69	13.7 ± 1.3	12.7 ± 1.1	10.4 ± 0.8	12.8 ± 1.1
	FLU, μg/kg	10.6 ± 0.8	8.84 ± 0.12	9.09 ± 0.74	7.90 ± 0.58	7.41 ± 0.31	8.54 ± 0.52
	PHN, μg/kg	93.0 ± 1.4	83.3 ± 2.3	76.3 ± 6.2	76.3 ± 0.9	70.5 ± 6.1	88.4 ± 6.6
	ANT, μg/kg	152 ± 9	134 ± 4	121 ± 11	121 ± 2	110 ± 6	144 ± 12
	FLR, μg/kg	24.2 ± 1.5	20.6 ± 0.8	19.0 ± 0.7	18.7 ± 0.2	17.5 ± 1.0	16.9 ± 0.1
	PYR, μg/kg	92.4 ± 1.6	72.8 ± 5.3	63.0 ± 5.1	61.7 ± 1.6	53.7 ± 4.2	48.7 ± 0.4
	BaA, μg/kg	88.3 ± 5.2	43.1 ± 4.1	43.3 ± 2.4	33.3 ± 2.5	30.6 ± 0.9	19.4 ± 0.0
	CRY, μg/kg	85.4 ± 2.4	Nd	Nd	Nd	Nd	Nd

Table 7.
Amount of pesticides in water samples in in the dam basin, Meriç-Üyüklütatar village, and Muhacirkadı-Ergene Basin ($n = 6$).

		Rice Harvest Period			Rice Husk Harvest Period		
Dam Basın		**11Br1**	**11Br2**	**11Br3**	**11Brh1**	**11Brh2**	**11Brh3**
	EN, mg/kg	54.9 ± 4.6	46.4 ± 2.7	54.0 ± 2.8	47.4 ± 0.2	46.7 ± 1.6	41.6 ± 1.7
	NAP, µg/kg	220 ± 12	265 ± 18	280 ± 3	679 ± 27	420 ± 29	327 ± 6
	ACE, µg/kg	58.3 ± 0.4	55.3 ± 0.9	65.1 ± 2.3	246 ± 6	265 ± 16	239 ± 10
	ACY, µg/kg	205 ± 5	246 ± 7	216 ± 2	444 ± 3	374 ± 14	341 ± 6
	FLU, µg/kg	71.9 ± 2.8	76.7 ± 2.3	99.0 ± 4.3	62.3 ± 0.4	59.1 ± 0.3	75.2 ± 0.1
	PHN, µg/kg	204 ± 4	267 ± 7	192 ± 1	259 ± 18	145 ± 2	181 ± 10
	ANT, µg/kg	274 ± 7	396 ± 14	250 ± 2	416 ± 32	157 ± 9	324 ± 7
	FLR, µg/kg	179 ± 3	195 ± 1	194 ± 4	145 ± 4	185 ± 4	229 ± 2
	PYR, µg/kg	1068 ± 25	1148 ± 12	1142 ± 15	779 ± 20	1080 ± 19	1351 ± 21
	BaA, µg/kg	1614 ± 37	1871 ± 86	1770 ± 63	1900 ± 100	2100 ± 17	2760 ± 88
	CRY, µg/kg	1101 ± 22	1247 ± 59	1190 ± 30	1247 ± 59	1317 ± 18	1765 ± 58
	BbF, µg/kg	Nd	796 ± 79	226 ± 25	961 ± 91	945 ± 93	1303 ± 20
	BkF, µg/kg	Nd	Nd	254 ± 24	999 ± 99	1036 ± 103	1253 ± 8
	IcdP, µg/kg	1044 ± 73	1101 ± 24	843 ± 7	2076 ± 66	2134 ± 11	2343 ± 51

		Rice Harvest Period			Rice Husk Harvest Period		
Meriç-Üyüklütatar Village		**22Br1**	**22Br2**	**22Br3**	**22Brh1**	**22Brh2**	**22Brh3**
	EN, mg/kg	53.2 ± 1.4	60.1 ± 3.7	58.4 ± 1.6	43.4 ± 2.8	42.4 ± 3.5	41.8 ± 3.1
	NAP, µg/kg	239 ± 7	356 ± 29	383 ± 18	809 ± 23	628 ± 50	882 ± 25
	ACE, µg/kg	63.4 ± 0.9	70.5 ± 0.9	60.3 ± 1.5	179 ± 12	150 ± 3	153 ± 7
	ACY, µg/kg	194 ± 4	242 ± 7	275 ± 7	403 ± 10	401 ± 12	382 ± 16
	FLU, µg/kg	73.7 ± 3.3	51.0 ± 1.5	208 ± 15	105 ± 4	92.5 ± 3.1	102 ± 1
	PHN, µg/kg	79.3 ± 1.6	139 ± 7	428 ± 16	179 ± 1	193 ± 3	254 ± 5
	ANT, µg/kg	39.6 ± 2.9	172 ± 9	338 ± 26	52.1 ± 5.1	235 ± 6	402 ± 8
	FLR, µg/kg	195 ± 4	213 ± 4	228 ± 3	239 ± 8	297 ± 3	270 ± 7
	PYR, µg/kg	1176 ± 19	1196 ± 16	1349 ± 18	1436 ± 40	1916 ± 33	1673 ± 48
	BaA, µg/kg	1120 ± 52	1478 ± 71	1831 ± 101	2014 ± 79	3788 ± 141	3909 ± 180
	CRY, µg/kg	829 ± 25	969 ± 53	1225 ± 67	1330 ± 42	2375 ± 82	2400 ± 123
	BbF, µg/kg	215 ± 6	202 ± 3	617 ± 8	950 ± 94	294 ± 29	1261 ± 38
	BkF, µg/kg	245 ± 6	261 ± 5	270 ± 26	918 ± 85	537 ± 31	1120 ± 32
	IcdP, µg/kg	893 ± 6	948 ± 24	1395 ± 55	2637 ± 80	1426 ± 21	2058 ± 75

Muhacirkadı-Ergene Basin	Rice Harvest Period			Rice Husk Harvest Period		
	33Br1	33Br2	33Br3	33Brh1	33Brh2	33Brh3
EN, mg/kg	50.2 ± 2.1	17.7 ± 0.5	28.4 ± 1.5	45.6 ± 1.2	47.2 ± 1.6	41.6 ± 3.2
NAP, μg/kg	285 ± 5	445 ± 34	514 ± 42	603 ± 53	524 ± 36	517 ± 31
ACE, μg/kg	77.0 ± 1.4	55.0 ± 0.2	59.9 ± 0.7	74.9 ± 2.7	94.4 ± 2.2	78.7 ± 4.0
ACY, μg/kg	272 ± 8	331 ± 7	336 ± 23	364 ± 19	337 ± 9	313 ± 12
FLU, μg/kg	250 ± 14	176 ± 4	294 ± 16	268 ± 25	168 ± 13	102 ± 3
PHN, μg/kg	269 ± 22	518 ± 9	336 ± 30	197 ± 4	225 ± 5	189 ± 3
ANT, μg/kg	392 ± 33	859 ± 18	519 ± 15	216 ± 2	199 ± 3	205 ± 8
FLR, μg/kg	222 ± 6	191 ± 4	125 ± 6	358 ± 8	378 ± 7	358 ± 9
PYR, μg/kg	1202 ± 30	1078 ± 32	789 ± 2	2293 ± 36	2576 ± 43	2336 ± 55
BaA, μg/kg	1717 ± 6	1560 ± 67	895 ± 61	2415 ± 46	3766 ± 163	4258 ± 169
CRY, μg/kg	1152 ± 10	1077 ± 37	752 ± 5	1526 ± 44	2413 ± 109	2562 ± 52
BbF, μg/kg	901 ± 75	1029 ± 57	1255 ± 115	1256 ± 127	1042 ± 22	1220 ± 37
BkF, μg/kg	949 ± 48	1403 ± 105	599 ± 22	1722 ± 134	1061 ± 31	929 ± 35
IcdP, μg/kg	1397 ± 24	1531 ± 23	1392 ± 41	2332 ± 96	2300 ± 22	1988 ± 52

Table 8.
The amounts of pesticides in the samples of rice and rice husk in the dam basin, Meriç-Üyüklütatar village and Muhacirkadı-Ergene Basin (n = 6).

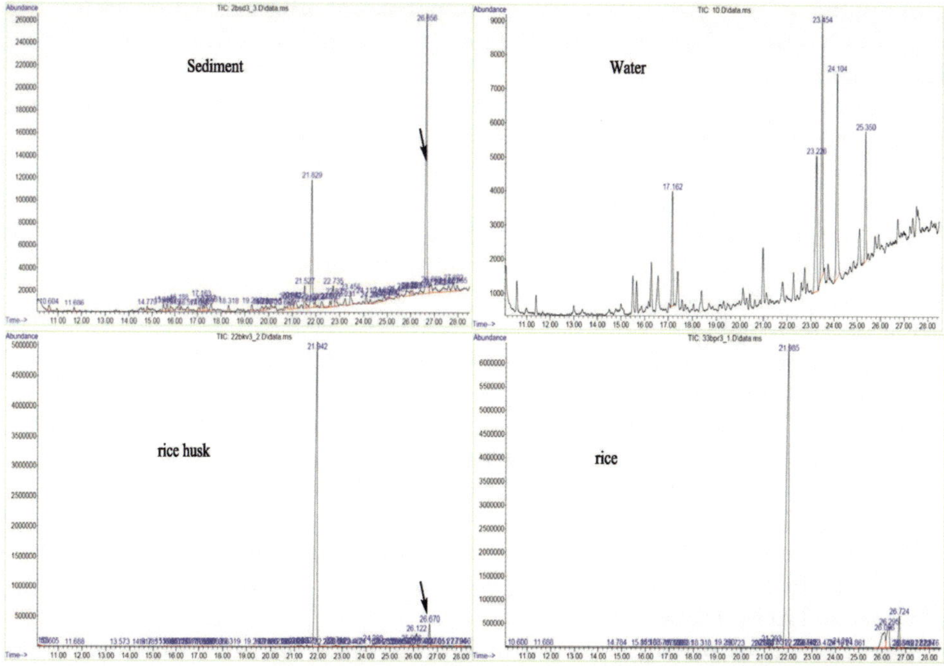

Figure 8.
Endosulfan chromatograms of samples taken from the paddy grown area.

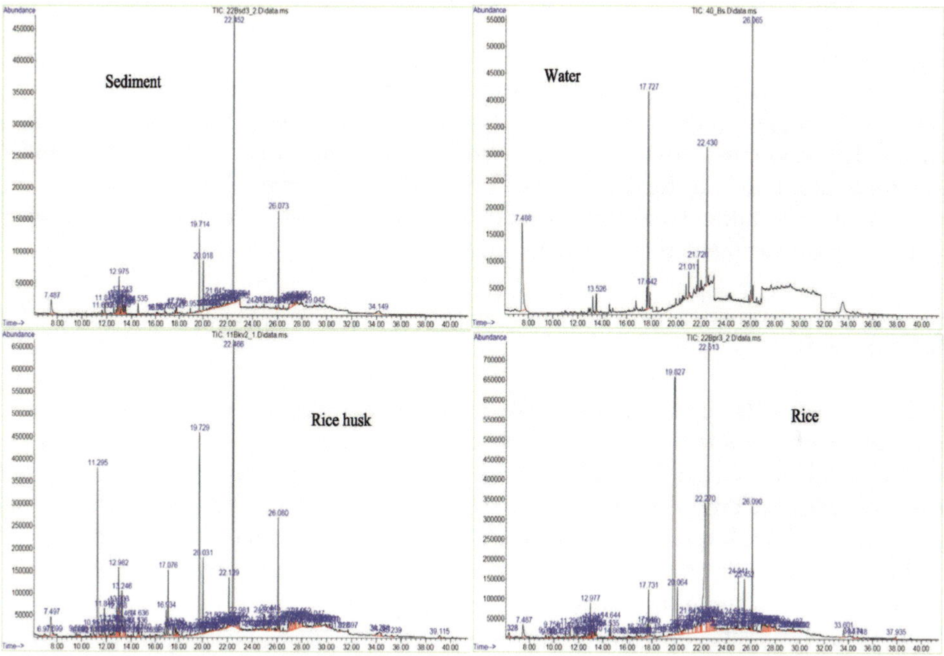

Figure 9.
PAH chromatogram of samples taken from the paddy grown area.

Pesticides strongly absorbed by soil particles are much less likely to evaporate [1, 2, 24]. Since pesticides are more adsorbed in sediment samples, their transition to water decreases. The results were confirmed this. Sodium was found in higher concentrations than sediment samples because of its high solubility in water. This increases the electrical conductivity and reduces the water quality.

The amount of EN, NP, ACE, ACY, FLU, PHN, ANT, FLR, PYR, BaA, CRY, BbF, BkF and IcdP in rice samples were found to be 17.7–60.1 mg kg^{-1}, 220–514 μg kg^{-1}, 55.0–77.0 μg kg^{-1}, 194–336 μg kg^{-1}, 51.0–294 μg kg^{-1}, 79.3–518 μg kg^{-1}, 39.6–859 μg kg^{-1}, 125–228 μg kg^{-1}, 798–1349 μg kg^{-1}, 895–1871 μg kg^{-1}, 752–1247 μg kg^{-1}, Nd–1255 μg kg^{-1}, Nd–1403 μg kg^{-1} and 843–1531 μg kg^{-1}, respectively. The pesticide concentrations of BaP, BghiP, DahA in rice samples were found below the limit of determination. The amount of PAHs in the polluted region was more than twice that of the clean region.

The amount of EN, NAP, ACE, ACY, FLU, PHN, ANT, FLR, PYR, BaA, CRY, BbF, BkF and IcdP in the rice husk samples were found to be 41.6–46.7 mg kg^{-1}, 327–882 μg kg^{-1}, 74.9–265 μg kg^{-1}, 313–444 μg kg^{-1}, 59.1–268 μg kg^{-1}, 145–259 μg kg^{-1}, 52.1–416 μg kg^{-1}, 145–378 μg kg^{-1}, 779–2576 μg kg^{-1}, 1900–4258 μg kg^{-1}, 1247–2562 μg kg^{-1}, 294–1303 μg kg^{-1}, 537–1722 μg kg^{-1} and 1426–2343 μg kg^{-1}, respectively.

The amount of PAHs in the rice husk samples was found twice the amount of rice. Except for rice and water samples, PAHs accumulation was determined in sediment and rice husk samples.

If we summarize briefly, the BghiP and DahA pesticides in the sediment samples in Dam Basin, Meriç-Üyüklütatar Village and the Muhacirkadı-Ergene Basin was be bellowed of limit of detection. It was below the limit of detection of BbF, BkF, BaP, BghiP, DahA and IcdP pesticides content in water samples in the Dam Basin, Meriç-Üyüklütatar Village and the Muhacirkadı-Ergene Basin. It was below the limit of detection the BaP, BghiP and DahA pesticides in the rice and rice husk samples in the Dam Basin, Meriç-Üyüklütatar Village and the Muhacirkadı-Ergene Basin.

When the results were examined, it was determined that the amounts of pesticides were higher in the samples taken near the Ergene river, but the amounts in the edible section were less than in the rice husk. Rice husk has shown a very good adsorbent and reduced the transport of EN and PAHs in of food. EN and PAHs levels in samples taken from river, stream, or near the canal were found to be higher than the samples taken from the inner sides.

4. Conclusion

Ergene River Basin surroundings were selected for this study: one heavily contaminated sites, moderately contaminated sites, and one less contaminated reference sites. The modified QuEChERS method used in this study was practical for mixtures found in environmental samples. This technique performs well, exhibiting good sensitivity, selectivity, and precision in the range of concentrations appropriate for the determination of target analytes. Our study investigated to the Ergene River Basin in sediment, rice, rice husk and water were analyzed for trace organic pollutants. However, the sediment and plant (rice and husk) had measurable and sometimes high levels of PAHs, even though no industrial sources of pollution were known. Other sources of PAH contamination may include runoff from paved roads and exhaust from farm machinery, and factory wastes immediate of the sampling stations. Therefore, in the alives feeding with husk of rice, there may be bioaccumulation of EN and PAHs. Ecological risk assessments for the sediment efficacies concluded that response actions were necessary for the sediment and husk, except for water and rice.

Author details

Barış Can Körükçü[1,2] and Cemile Ozcan[1*]

1 Department of Chemistry, Science and Art Faculty, Kirklareli University, Kirklareli, Turkey

2 Ataturk Soil, Water and Agricultural Meteorology Research Institute, Kirklareli, Turkey

*Address all correspondence to: cemilebal.ozcan@klu.edu.tr

References

[1] Bailey GW, White JL. Review of adsorption and desorption of organic pesticides by soil colloids, with implications concerning pesticide bioactivity. Journal of Agricultural and Food Chemistry. 1984;**12**:324-332. Available from: https://pubs.acs.org/doi/pdf/10.1021/jf60134a007?src=recsys

[2] Soares AA, Albergaria JT, Domingues VF, Alvim-Ferraz MDCM, Delerue-Matos C. Remediation of soils combining soil vapor extraction and bioremediation: Benzene. Chemosphere. 2010;**80**:823-828. DOI: 10.1016/j.chemosphere.2010.06.036

[3] Kurutaş E, Kılınç M. Pestisitlerin Biyolojik Sistemler Üzerine Etkisi. Arşiv Kaynak Tarama Dergisi. 2003;**12**:215-228. Available from: http://dergipark.org.tr/aktd/issue/2231/29481

[4] Monneret C. What is an endocrine disruptor. Comptes Rendus Biologies. 2017;**340**:403-405. DOI: 10.1016/j.crvi.2017.07.004

[5] Kabir ER, Rahman MS, Rahman I. A review on endocrine disruptors and their possible impacts on human health. Environmental Toxicology and Pharmacology. 2015;**40**:241-258. DOI: 10.1016/j.etap.2015.06.009

[6] Chormey DS, Büyükpınar Ç, Turak F, Komesli OT, Bakırdere S. Simultaneous determination of selected hormones, endocrine disrupt or compounds, and pesticides in water medium at trace levels by GC-MS after dispersive liquid-liquid microextraction. Environmental Monitoring and Assessment. 2017;**189**:277-287. DOI: 10.1007/s10661-017-6003-6

[7] Zhao C, Xie H, Mu Y, Xu X, Zhang J, Liu C, et al. Bioremediation of endosulfan in laboratory-scale constructed wetlands: Effect of bioaugmentation and biostimulation. Environmental Science and Pollution Research. 2014;**21**:12827-12835. DOI: 10.1007/s11356-014-3107-1

[8] Kaur R, Rani S, Malik AK, Aulakh JS. Determination of endosulfan isomers and their metabolites in tap water and commercial samples using microextraction by packed sorbent and GC-MS. Journal of Separation Science. 2014;**37**:966-973. DOI: 10.1002/jssc.201301154

[9] ATSDR. Agency for Toxic Substances and Disease Registry (ATSDR), Toxicological Profile for Polycyclic Aromatic Hydrocarbons (PAHs). Atlanta (GA): Department of Health and Human Services, Public Health Service, USA; 1995

[10] Danyi S, Bose F, Brasseur C, Schneider YJ, Larondelle Y, Pussemier L. Analysis of EU priority polycyclic aromatic hydrocarbons in food supplements using high performance liquid chromatography coupled to an ultraviolet, diode array or fluorescence detector. Analytica Chimica Acta. 2009;**633**:293-699. DOI: 10.1016/j.aca.2008.11.049

[11] Shi L, Zheng L, Liu R, Chang M, Huang J, Jin Q, et al. Quantification of polycyclic aromatic hydrocarbons and phthalic acid esters in deodorizer distillates obtained from soybean, rapeseed, corn and rice bran oils. Food Chemistry. 2019;**275**:206-213. DOI: 10.1016/j.foodchem.2018.09.119

[12] Ferrarese E, Andreottola G, Oprea IA. Remediation of PAH contaminated sediments by chemical oxidation. Journal of Hazardous Materials. 2008;**152**:128-139. DOI: 10.1016/j.jhazmat.2007.06.080

[13] EC. Priorty Substances and Certain Other Pollutions According to Annex II of Directing, 2008/105/EC. 2013. Available from: http://ec..europe.eu/

enviromental/water/waterframework/priortysubstances.htm;eylul2013

[14] US. US Food and Drug Administration, Guidance for Industry: Bioanalytical Method Validation. 2018. Available from: https://www.fda.gov/files/drugs/published/Bioanalytical-Method-Validation-Guidance-for-Industry.pdf

[15] WHO. Application of Risk Analysis to Food Standards Issues, Report of the Joint FAO/WHO Expert Consultation, WHO/FNU/FOS/95.3. Geneva, Switzerland: World Health Organization; 1995

[16] WHO. Guidelines for Assessing Quality of Herbal Medicines with Reference to Contaminants and Residues. Switzerland: World Health Organization; 2007. ISBN: 978 92 4 159444 8

[17] Lepom P, Brown B, Hanke G, Lous R, Quevauviller P, Wollgast J. Needs for reliable analytical methods for monitoringe chemical pollutants in surface water under the European Water Framework Directive. Journal of Chromatography. A. 2009;**1216**:302-305. DOI: 10.1016/j.chroma.2008.06.017

[18] WHO. Guidelines for Predicting Dietary Intake of Pesticide Residues (Revised), WHO/FSF/FOS/97.7. Geneva, Switzerland: World Health Organization; 1997

[19] Casida JE, Durkin KA. Pesticide chemical research in toxicology: Lessons from nature. Chemical Research in Toxicology. 2017;**30**(1):94-104. DOI: 10.1021/acs.chemrestox.6b00303

[20] Bulgurcuoğlu AE, Yılmaz B, Chormey DS, Bakırdere S. Simultaneous determination of estrone and selected pesticides in water medium by GC-MS after multivariate optimization of microextraction strategy. Environmental Monitoring and Assessment. 2018;**190**(4):252. DOI: 10.1007/s10661-018-6625-3

[21] Bonansea RI, Amé MV, Wunderlin DA. Determination of priority pesticides in riverr water samples combining SPE and SPME coupled to GC-MS. Chemosphere. 2013;**90**:1860-1869. DOI: 10.1016/j.chemosphere.2012.10.007

[22] Akvan N, Azimi G, Parastar H. Chemometric assisted determination of 16 PAHs in water samples by ultrasonic assisted emulsification microextraction followed by fast high-performance liquid chromatography with diode array detector. Microchemical Journal. 2019;**150**:104056. DOI: 10.1016/j.microc.2019.104056

[23] Schenck FJ, Lehotay SJ, Vega V. Comparison of solid-phase extraction sorbents for cleanup in pesticide residue analysis of fresh fruits and vegetables. Journal of Separation Science. 2002;**25**:883-890. DOI: 10.1002/1615-9314(20021001)25:14 <883::AID-JSSC883>3.0.CO;2-7

[24] Correia-Sa L, Fernandes VC, Carvalho M, Calhau C, Domingues VF, Delerue-Matos C. Optimization of QuEChERS method for the analysis of organochlorine pesticides in soils with diverse organic matter. Journal of Separation Science. 2012;**35**:1521-1530. DOI: 10.1002/jssc.201200087

[25] Kin CM, Huat TG, Kumari A. Application of solid-phase microextraction for the determination of pesticides in vegetable samples by gas chromatography with an electron capture detector. Malaysian Journal of Analytical Sciences. 2008;**12**:1-9. Avaialble from: http://www.ukm.my/mjas/v12_n1/1.pdf

[26] Sheil D. Forests, atmospheric water and an uncertain future: The new biology of the global water cycle. Forest Ecosystems. 2018;**5**:19. DOI: 10.1186/s40663-018-0138-y

Chapter 16

Reducing Emerging Contaminants Ensuing from Rusting of Marine Steel Installations

Karima Hanini, Sameh Boudiba and Merzoug Benahmed

Abstract

Marine steel installations are usually subject to biocorrosion due to their immersing in seawater. Biocorrosion-causing microorganisms, such as bacteria and fungi, often form biofilms on materials, inducing chemical changes in these materials and in the surrounding liquid medium. The formed biofilms resulting from this phenomenon are considered as emerging contaminants. In this work, in addition to the realization of the electrodeposition of zinc on a steel in chloride bath with various concentrations of *Taxus baccata* extracts as additives using a direct courant supply, the study of the corrosion of the obtained substrates was performed in seawater as an aggressive environment. The efficiency against corrosion was evaluated by potentiodynamic polarizations and weight loss measurements. The coated surface morphology was analyzed using brightness meter, thickness meter and adhesion tests. The experimental results showed that all tested extracts performed the quality of the zinc deposits and their efficiency against corrosion indicating that coated samples in the presence of the extracts were more resistant minimizing the emerging contaminants in seawater.

Keywords: *Taxus baccata*, biocorrosion, emerging contaminant, inhibition, electrodeposition

1. Introduction

Emerging contaminants (ECs), also known as emerging organic contaminants [1] and contaminants of emerging concerns (CECs) [2], are a group of natural and synthetic chemicals with their transformation products occurring in water bodies throughout the globe. Actually, these materials, which are also considered as microorganisms (e.i. biofilm) present in the environment, are not monitored and have a high potential of causing environmental damage in addition to be suspected of detrimental effects on ecosystems and human health. It should be noted that as reported in the literature, the exposure to environmental relevant concentrations of ECs caused significant alterations in the microbial community composition and function [3].

The biocorrosion, or microbiologically influenced corrosion (C.I.M), which represent one of the biofilms source is an electrochemical phenomenon of dissolution of a metal affecting all industries, where microorganisms (in particular bacteria) can develop [4]. To minimize the biocorrosion and therefore the emerging contaminants, electrodeposition of zinc is usually used for marine steel installations protection, it is a suitable, fast and cost-effective way to improve the surface hardness, anti-corrosion properties and wear resistance...etc. [5]. The electrodeposition consists principally in the immersion of a steel as an electrode in a container containing an electrolyte and a zinc as a counter electrode. The electrodeposition electrolytes usually contain metal salts, species to increase electrolyte conductivity and additives, added to improve leveling metal and to optimize the properties of the deposited metal (hardness, brightness, corrosion resistance, roughness, porosity, grains size .....) [6]. The clear problem with these additives is that they are all synthetic, toxic, high cost and they have a significant negative impact on the environment, such as furfurale and thiourea [7–11]. To resolve this problem, the new trend in the current research is focused on replacing synthetic additives with less expensive, biodegradable and effective natural products such as extracts obtained from natural products as *Manihot esculenta* [12], *Saccharum officinarum*, *Ananas comosus* [13], *Taxus baccata* [14] and *Daphne gnidium* L. [15].

In this work, two main objectives are targeted. Firstly, we are focused on the use of plant extracts obtained from a plant belonging to Algerian flora (*Taxus baccata*), as additives in chloride baths for the zinc plating of a mild steel (E24–2). Secondly, we will evaluate the effect of the addition of the investigated extracts as additives, namely on the quality of the zinc deposit, on one hand and on the other hand, on the resistance of the deposit against corrosion, therefore the decrease of the emerging contaminants effect.

2. Emerging contaminants

Emerging contaminants are synthetic or naturally occurring chemicals or any microorganisms that are not commonly monitored in the environment but have the potential to penetrate in the environment and thereby causing known or suspected adverse ecological and/or human health effects. These contaminants consist of pesticides, pharmaceuticals, personal care compounds, industrial additives by-products, food additives, water treatment by-products and nanomaterials including metal oxides, zero-valent metals, quantum dots, dendrimers, composites and nanosilver [16].

According to the previous definitions, the products resulting from steel corrosion in marine environment such as biofilms and iron oxides are considered as emerging contaminants.

3. Corrosion

Corrosion is the spontaneous destruction of metals and alloys caused by chemical, biochemical and electrochemical interactions between materials and their environments. These environments can be humidity, oxygen, inorganic and organic acids, high pressures, temperature, chlorides, etc. During corrosion, metals tend to convert to more thermodynamically stable compounds [17].

4. Zinc electroplating

Zinc electroplating provides corrosion resistance by acting as a barrier and sacrificial coating. Because zinc is more reactive that steel, the zinc coating corrodes first, protecting the steel substrate. The rate of corrosion of zinc is at least 10 times slower than that of steel, thus a thin coating of zinc can protect steel for long time.

Electroplating is an electrochemical process, which takes place according to a certain number of elementary stages of electrolytic crystallization leading to the reduction of the dissolved species M^{+n} into metal atoms M, and to the incorporation of these atoms into the crystal lattice of the substrate. In general, for a coating, we have to make the cathode the object to be coated, the anode is then made of the metal, which we want to coat and the electrolyte will contain a solution of a salt of this metal. Species to increase electrolyte conductivity and additives are added to improve leveling metal and to optimize the properties of the deposited metal [11].

5. Additives

The development of coatings requires when using a direct current, the addition in small quantities of certain organic or inorganic additives to the electrolysis bath. These additives will modify the mechanisms of electrodeposition by selectively affecting the rate of germination and/or growth of deposited species. The use of additives in electrolysis baths is expressed by a reduction in surface roughness, an increase in hardness and an improvement in both deposits gloss and morphology [6].

6. Experimental

6.1 Plant extracts

The stems and leaves of *Taxus baccata* were harvested in April 2016 in Babor (altitude 2000 m), in little kabylia (northern end of the wilaya of Sétif, in Algeria). The aerial parts were cut into small pieces, they were then subjected to an extraction procedure by maceration in a lukewarm methanol–water mixture (7/3:V/V) for 24 hours. This operation was repeated three times. The various recovered fractions were then combined and evaporated under reduced pressure at a temperature below 70° C until a syrupy residue was obtained. The latter phase was taken up by boiling water, decanting for one nights to eliminate the chlorophyll. The filtered mixture was subjected to liquid–liquid extraction using several solvents separately in a sequence of increasing polarity, starting from dichloromethane (MDE), ethyl acetate (EAE) and n-butanol (BE), where the organic phase was recovered for each solvent. The latter solutions were evaporated to dryness using a rotary evaporator to obtain the desired extracts [18].

6.2 Materials

The tests were carried out on a mild steel of grade E24–2, having a chemical composition in % by weight (C, 0.17; Mn, 0.6; P, 0.035; S, 0.035; Si, 0.04; Al, 0.02 to 0.05; N, 0.01; Fe remaining). Plates of 2 cm x 6 cm x 0.1 cm in dimension were coated with epoxy resin leaving a single exposure surface. The substrates were plated with a pure zinc (99.91%). Prior to each experiment, the surfaces of all samples were mechanically abraded using different types of silicon carbide emery paper, cleaned with acetone, rinsed with distilled water then dried with air.

6.3 Electrodeposition bath

The chloride solution for the coating was a mixture of $ZnCl_2$ (65 g/l), KCl (200 g/l), H_3BO_3 (20 g/l) and different concentrations of *Taxus baccata* extracts (1 g/l to 1.6 g/l) [19].

6.4 Experimental conditions

All experiments were carried out in an aerated medium, pH = 5 and current of 0.04 A. The steel sample represents the anode of the electrochemical cell, while the zinc plate represents the cathode leaving a distance of 1 cm between them. The zinc plating of the steel was carried out for 30 min, with gentle stirring, by partially immersing the steel sample and the zinc electrode in the chloride bath. To determine the weight of deposited zinc on surfaces, all substrates were weighed before and after electroplating. At the end of the process, the samples were removed from the bath, cleaned with distilled water and air dried [14].

6.5 Quality of the deposited zinc layer

The thickness of the deposited zinc layer was measured with an Elektro-Physik (eXacto) apparatus and the adhesion of the coated zinc to the substrate was examined by the ASTM D3359 method [20]. For the adhesion test, an "X" was etched on the film and an attached adhesive tape was applied to the samples, and then removed strongly. This test is macroscopic and more qualitative. The gloss of the zinc deposits was measured using a Poly Gloss meter with a large beam of white light at a measuring angles of 20°, 60° and 85°. Calibration was performed automatically using a highly polished black standard built into the gloss meter. The final gloss values were the average of three measurements taken for each coating.

6.6 Potentiodynamic polarization measurements

Potentiodynamic polarization measurements were performed in seawater with the coated samples as working electrode, a platinum rod as counter electrode and a saturated calomel electrode (SCE) as a reference electrode. A controlled computer (Voltalab PGZ 301) instrument with Voltamaster 4 software was employed for this purpose. The measurements were applied in the potential range of ±1500 mV at a sweep rate of 1 mV/s.

6.7 Gravimetric measurements

To evaluate the corrosion resistance of the electrodeposited substrates, weight loss measurements were made. Each sample coated in chloride baths containing different concentrations of the extracts was partially immersed in seawater (corrosive medium). Measurements were collected every five days for a month and the Eq. (1) was used to determine the corrosion rate [21]:

$$CR = \frac{w}{At} \tag{1}$$

w: average weight loss.
A: total area of one mild steel specimen.
t: immersion time.

7. Results and discussion

7.1 Quality of electrodeposited zinc

The quality of the electrodeposited zinc was observed through the deposit parameters as the brightness, the strength adhesion and the thickness. The obtained results are presented in **Table 1** for MDE, **Table 2** for EAE and **Table 3** for BE. Where: + adhesion is strong, ++ adhesion is very strong.

Via **Tables 1–3** we noted that there is an increase in the deposited mass, thickness and adhesion when increasing the concentration of MDE (1.6 g/l). However, for EAE and BE, these parameters reached a maximum value at 1.2 g/l and then decreased, indicating that the better deposit (nucleation and growth) is found in these concentrations. The obtained results may be related to two considerations: the first one is that the adsorption of additives on the surface, leads to a partial coating of the steel, thus blocking the active sites and causing a decrease in the nucleation rate. The second consideration is that the additive will complex with one of the electroactive species in the solution, therefore the step of dissociation of the complex introduces a new kinetic constant before the redox reaction of the electroactive species at the electrode surface [22]. Furthermore, and according to ASTM D 523 [23] regulations, we observed that the deposits obtained with the addition of the extracts were matt. However, when adding different concentrations (1.2; 1.4 and 1.6 g/l) of BE, the deposits were semi-gloss. In addition, all measured thicknesses were in agreement with ASTM A879 and ASTM B633 [24].

7.2 Potentiodynamic polarization measurement

The corrosion resistance of the electrodeposited mild steel was tested in seawater at 298 K to evaluate the effect of adding extracts to the chloride baths. The electrochemical parameters such as E_{corr}, i_{corr} and CR are collected in **Table 4** for MDE, **Table 5** for EAE and **Table 6** for BE.

From **Tables 4–6**, it can be seen that the addition of the investigated extracts as additives gave rise to significant decreases in current densities as well as the corrosion rate compared to the sample obtained without extracts addition. This indicates that the studied extracts strongly modified the quality of the deposit producing coatings more resistant to corrosion and therefore lessening the formation of biofilms, which represent one of emerging contaminants. It is also noted that the dependence of the

C (g/l)	Mass deposited (g)	Thickness (μm)	Adhesion	Brightness (GU)
Without extracts	0.0423	15	+	Matt 05.30
1	0.0248	8.67	+	Matt 14.15
1.2	0.0395	13.92	+	Matt 02.30
1.4	0.0424	15.12	++	Matt 01.75
1.6	0.0481	17.15	++	Matt 19.10

Table 1.
Mass, thickness, brightness and strength adhesion of the deposited zinc layer in the presence of MDE.

C (g/l)	Mass deposited (g)	Thickness (μm)	Adhesion	Brightness (GU)
Without extracts	0.0423	15	+	Matt 05.30
1	0.0450	15.99	++	Matt 08.60
1.2	0.0551	19.60	++	Matt 14.65
1.4	0.0477	16.20	++	Matt 19.40
1.6	0.0366	12.11	+	Matt 09.70

Table 2.
Mass, thickness, brightness and strength adhesion of the deposited zinc layer in the presence of EAE.

Concentration (g/l)	Mass deposited (g)	Thickness (μm)	Adhesion	Brightness (GU)
Without extracts	0.0423	15	+	Matt 05.30
1	0.0467	16.10	++	Matt 13.40
1.2	0.0685	24.83	++	Semi bright 32.85
1.4	0.0405	14.36	++	Semi bright 31.45
1.6	0.0393	13.40	+	Semi bright 31.75

Table 3.
Mass, thickness, brightness and strength adhesion of the deposited zinc layer in the presence of BE.

MDE extract	C (g/l)	$-E_{corr}$ (V/SCE)	i_{corr} ($mAcm^{-2}$)	CR (mm/y)
Without extract	/	923.8	0.6514	9.791
With extract	1	1064.4	0.4653	5.442
	1.2	1176.4	0.3220	4.840
	1.4	1095.3	0.1427	2.145
	1.6	1136.4	0.1411	1.650

Table 4.
Polarization parameters for corrosion of electroplated mild steel without and with different concentrations of MDE at 293 K.

EAE extract	C (g/l)	$-E_{corr}$ (V/SCE)	i_{corr} ($mAcm^{-2}$)	CR (mm/y)
Without extract	/	923.8	0.6514	9.791
With extract	1	787.8	0.000967	0.01131
	1.2	159.2	0.000184	0.002158
	1.4	699.2	0.000347	0.005219
	1.6	618	0.005706	0.06674

Table 5.
Polarization parameters for corrosion of electroplated mild steel without and with different concentrations of EAE at 293 K.

density of the corrosion current and the additive concentration in the electroplating bath was not linear, probably due to the fact that in each electroplating process there is an optimum additive concentration, whereby the deposit quality is the best [25].

7.3 Gravimetric measurements

Figure 1 shows the corrosion rate curves of the mild steel in the absence and in the presence of optimal concentrations of *Taxue baccata* extracts used as additives during their immersion in seawater for one month.

Table 7 gathers the values of the corrosion rate in the absence and in the presence of optimal concentrations of *Taxue baccata* extracts tested separately as additives in zinc baths.

The examination of **Table 7** and **Figure 1** displayed a better corrosion resistance for all plated samples in the presence of optimal concentration of extracts as additives than those plated in their absences. Furthermore, the plated sample without additives recorded a corrosion rate value of 0.0052 ($mg\ cm^{-2}\ h^{-1}$) during one month of immersion time in seawater. In contrary, the corrosion rate values for plated specimens with additives were in the range of 0.0025 and 0.0018 $mg\ cm^{-2}\ h^{-1}$.

BE extract	C (g/l)	$-E_{corr}$ (V/SCE)	i_{corr} ($mAcm^{-2}$)	CR (mm/y)
Without extract	/	923.8	0.6514	9.791
With extract	1	1107.6	0.2283	3.431
	1.2	1076.4	0.1287	1.933
	1.4	1143.6	0.2631	3.955
	1.6	1124	0.3574	5.371

Table 6.
Polarization parameters for corrosion of electroplated mild steel without and with different concentrations of BE at 293 K.

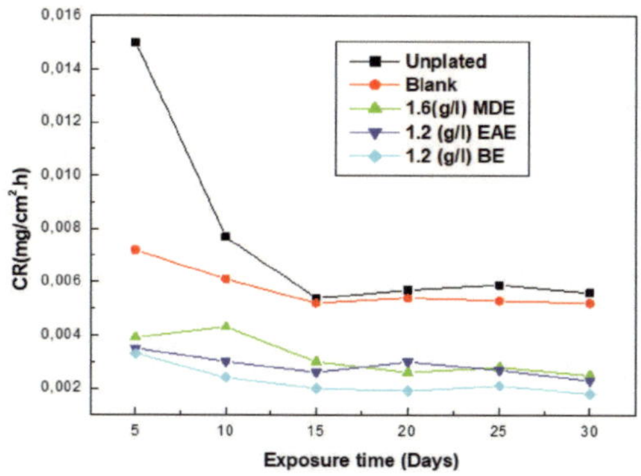

Figure 1.
Variation of the corrosion rate versus exposure time for unplated and zinc plated mild steel samples without and with optimal concentration of MDE, EAE and BE.

Substrate	C (g/l)	Time (Days)	5	10	15	20	25	30
Unplated	/	CR (mg/ cm².h)	0.015	0.0077	0.0054	0.0057	0.0059	0.0056
Blank	/		0.0072	0.0061	0.0052	0.0054	0.0053	0.0052
Plated with MDE	1.6		0.0039	0.0043	0.003	0.0026	0.0028	0.0025
Plated with EAE	1.2		0.0035	0.003	0.0026	0.003	0.0027	0.0023
Plated with BE	1.2		0.0033	0.0024	0.002	0.0019	0.0021	0.0018

Table 7.
Corrosion parameters obtained from weight loss measurements of the electroplated mild steel using optimal concentrations of MDE, EAE and BE.

The obtained values of corrosion rate for the specimen plated in the presence of 1.6 g/l of MDE and 1.2 g/l of EAE and BE were lower than the others. This ascertainment is in good agreement with that obtained from potential polarization.

8. Conclusion

As part of the challenge against emerging contaminants, the use of three extracts obtained from *Taxus baccata* as additives in electrdeposition of zinc was evaluated. The obtained results leads to the following points:

1. Electroplating results showed that all extracts can be used as wetting and leveling additives.

2. In addition to being wetting and leveling the BE can also be regarded as brighteners at a well-defined concentration.

3. Samples coated in the presence of different concentrations of *Taxus baccata* extracts, in particular the EAE, exhibited a higher corrosion resistance than that coated in their absence, and therefore they can minimize the emerging contaminants in seawater.

Acknowledgements

The authors gratefully acknowledge the support of Tebessa University. We also acknowledge the ANABIB gas pipes unit in Tebessa (Algeria) and the laboratory of Fat Factory ENAP Soug-Ahras (Algeria) for the provision of research facilities for this work.

Author details

Karima Hanini[1], Sameh Boudiba[2*] and Merzoug Benahmed[1]

1 Laboratory of Active Molecules and Applications, Tebessa University, Tebessa, Algeria

2 Laboratory of Organic Materials and Heterochemistry, Tebessa University, Tebessa, Algeria

*Address all correspondence to: boudibasameh@gmail.com; sameh.boudiba@univ-tebessa.dz

References

[1] Jurado, A., et al., *Emerging organic contaminants in groundwater in Spain: a review of sources, recent occurrence and fate in a European context.* Science of the total environment, 2012. **440**: p. 82-94.

[2] Dulio, V., et al., *Emerging pollutants in the EU: 10 years of NORMAN in support of environmental policies and regulations.* Environmental Sciences Europe, 2018. **30**(1): p. 1-13.

[3] Geissen, V., et al., *Emerging pollutants in the environment: a challenge for water resource management.* International soil and water conservation research, 2015. **3**(1): p. 57-65.

[4] Feugeas, F., et al., *Corrosion influencée par les micro-organismes: influence du biofilm sur la corrosion des aciers, techniques et résultats récents.* Journal de physique III, 1997. 7(3): p. 631-663.

[5] Alizadeh, M. and H. Safaei, *Characterization of Ni-Cu matrix, Al2O3 reinforced nano-composite coatings prepared by electrodeposition.* Applied Surface Science, 2018. **456**: p. 195-203.

[6] Kanani, N., *Electroplating: basic principles, processes and practice*. 2004: Elsevier.

[7] Arts, J.H., et al., *Subacute (28-day) toxicity of furfural in Fischer 344 rats: a comparison of the oral and inhalation route.* Food and chemical toxicology, 2004. **42**(9): p. 1389-1399.

[8] Gupta, G.D., A. Misra, and D.K. Agarwal, *Inhalation toxicity of furfural vapours: an assessment of biochemical response in rat lungs.* Journal of applied toxicology, 1991. **11**(5): p. 343-347.

[9] Smith, P.B. and C. Crespi, *Thiourea toxicity in mouse C3H/10T12 cells expressing human flavin-dependent monooxygenase 3.* Biochemical pharmacology, 2002. **63**(11): p. 1941-1948.

[10] DIEKE, S.H. and C.P. RICHTER, *Acute toxicity of thiourea to rats in relation to age, diet, strain and species variation.* Journal of Pharmacology and Experimental Therapeutics, 1945. **83**(3): p. 195-202.

[11] Gamburg, Y.D. and G. Zangari, *Theory and practice of metal electrodeposition*. 2011: Springer Science & Business Media.

[12] Loto, C., A. Olofinjana, and R. Loto, *Effect of Manihot esculenta c. leaf extract additive on the zinc electroplating on mild steel in acid chloride solution.* International Journal of Electrochemical Science, 2014. **9**: p. 3746-3759.

[13] Loto, C.A., *Synergism of Saccharum officinarum and Ananas comosus extract additives on the quality of electroplated zinc on mild steel.* Research on Chemical Intermediates, 2014. **40**(5): p. 1799-1813.

[14] Hanini, K., et al., *Influence of different polyphenol extracts of Taxus baccata on the corrosion process and their effect as additives in electrodeposition.* Sustainable Chemistry and Pharmacy, 2019. **14**: p. 100189.

[15] Belbah, H., et al., *Effect of a Natural Brightener, Daphne gnidium L. on the Quality of Nickel Electroplating from Watts Bath.* Journal of New Materials for Electrochemical Systems, 2016. **19**(2).

[16] Naidu, R., et al., *Emerging contaminants in the environment: risk-based analysis for better management.* Chemosphere, 2016. **154**: p. 350-357.

[17] Fontana, M.G. and N.D. Greene, *Corrosion engineering*. 2018: McGraw-hill.

[18] S. Boudiba, K. Hanini, I. Selatnia, A. Saouane, S. Hioun, and M. Benahmed, "Experimental, theoretical and mathematical studies of *Echium italicum* L. extract as a corrosion inhibitor for carbon steel in acidic medium," *Materials Research Express*, vol. 6, p. 086546, 2019.

[19] Loto, C., et al., *Effect of Allium Cepa (onion) extract additive on the morphology of zinc electroplated mild steel in acid chloride solution.* Can. J. Pure Appl. Sci, 2015. **9**(1): p. 3267-3277.

[20] Astm, D., *3359-02: standard test methods for measuring adhesion by tape test.* ASTM International, West Conshohocken, PA, 2002.

[21] Loto, C. and R. Loto, *Effects of Nicotiana tobaccum extract additive on the quality of electroplating of zinc on mild steel.* Polish Journal of Chemical Technology, 2013. **15**(1): p. 38-45.

[22] Youssef, K.M., C. Koch, and P. Fedkiw, *Influence of additives and pulse electrodeposition parameters on production of nanocrystalline zinc from zinc chloride electrolytes.* Journal of the Electrochemical Society, 2004. **151**(2): p. C103.

[23] ASTM, D., *523-89, Standard Test Method for Specular Gloss.* ASTM International, 1999.

[24] Anchors, H.E., *Anchor bolt and sleeve assembly of material indicated below with capability to sustain, without failure, a load equal to 6 times the load imposed when installed in unit masonry assemblies and equal to 4 times the load imposed when installed in concrete as determined by testing per ASTM E 488 conducted by a qualified independent testing and inspecting agency. 1.* Material: Carbon-steel components, zinc plated to comply with ASTM B, 2010. **633**: p. 061053-6.

[25] Muresan, L.M. and S.C. Varvara, *Leveling and brightening mechanisms in metal electrodeposition.* M. Nunez, Nova Science Publishers, Inc. New York, 2005.

Chapter 17

Chemical Mechanical Planarization-Related to Contaminants: Their Sources and Characteristics

Jihoon Seo

Abstract

Chemical mechanical planarization (CMP) process has been widely used to planarize a variety of materials including dielectrics, metal, and semiconductors in Si-based semiconductor devices. It is one of the most critical steps to achieve the nanolevel wafer and die scale planarity. However, various contaminants are observed on the wafer surfaces after the CMP process, and they become the most critical yield detractor over many generations of rapidly diminishing feature sizes because they have the most direct impacts on device performance and reliability. This book chapter provides (1) CMP consumables-induced contaminants such as residual particles, surface residues, organic residues, pad debris and metallic impurities, pad contamination, watermark, etc., (2) brush-induced cross-contamination during post CMP cleaning, (3) post-CMP cleaning for removing these contaminants. Fundamental understanding of the formation of various types of CMP contaminants and their characteristics will significantly benefit the development of next-generation CMP slurries and post-CMP cleaning solutions.

Keywords: semiconductor manufacturing process, chemical mechanical planarization (CMP), defects, contaminants, cleaning, post-CMP

1. Introduction

Chemical mechanical planarization (CMP) is a critical and enabling process to achieve nanolevel local and global planarization across 300 mm wafer in integrated circuit (IC) manufacturing [1–3]. There are three main applications of the CMP process in the semiconductor device manufacturing: the formation of the transistors (front-end-of-line, FEOL), the local connections between transistors (middle-of-line, MOL), and the interconnect structures (back-end-of-line, BEOL). FEOL processes form the transistors and build the device architecture. One of the important FEOL CMP processes is the shallow trench isolation (STI) CMP. STI CMP uniformly polishes the step height of SiO_2, formed by the gap-filling process, and stops on an underlying Si_3N_4 film [4–6]. MOL CMP processes include the necessary steps to connect the individual transistors by mainly polishing W contact metal/liner and interlayer dielectric (ILD) layer [7, 8]. BEOL processes enable the multilevel interconnect network where Cu lines are isolated by the dielectric materials [9].

CMP process is mainly utilized to achieve the desired removal rates, rate selectivity between exposed materials, uniformity, etc. by the synergistic interplay of chemical and mechanical interactions. During this process, the wafer is pressed against a polishing pad under the applied down pressure. The slurry is applied onto the center of the pad and is transported into the pad/wafer gap through the pores and grooves of the polishing pad while rotating the pad at high speed, which generates various chemical and mechanical actions at the slurry/pad-wafer interface (**Figure 1**) [1]. Many factors including CMP consumables (slurry, pad, wafer, conditioner, retainer ring, etc.) and their process/tool conditions can have an influence on the polishing performances (**Figure 1**) [1, 10, 11].

As CMP technology has grown by leaps and bounds over the past several decades, considerable progress has been made. However, the presence of CMP-induced defects that can cause device failure and the severe yield loss has become a major concern in the modern semiconductor manufacturing process [3, 12, 13]. Removable defects (residual particles, organic residues, foreign materials, metallic impurities, etc.) and non-removable defects (scratches, corrosion, dishing, erosion, delamination, etc.) are presented on the polished wafer surfaces. The removable defects, also known as CMP-related to contaminants, should be completely removed in the subsequent cleaning process while minimizing the further formation of non-removable defects [14, 15]. CMP consumables themselves can be the source of the contaminants during polishing and cleaning [13]. CMP slurries for the dielectric process are mainly composed of abrasive particles, pH adjuster, dispersant, passivation agent for high selectivity, and deionized water (DIW) [16] Metal CMP slurries contain the additional chemical reagents like oxidizer, chelating agent, corrosion inhibitor, etc. to control the electrochemical behaviors of metal films during polishing [16, 17]. Typical CMP slurry components are listed in **Table 1** [1].

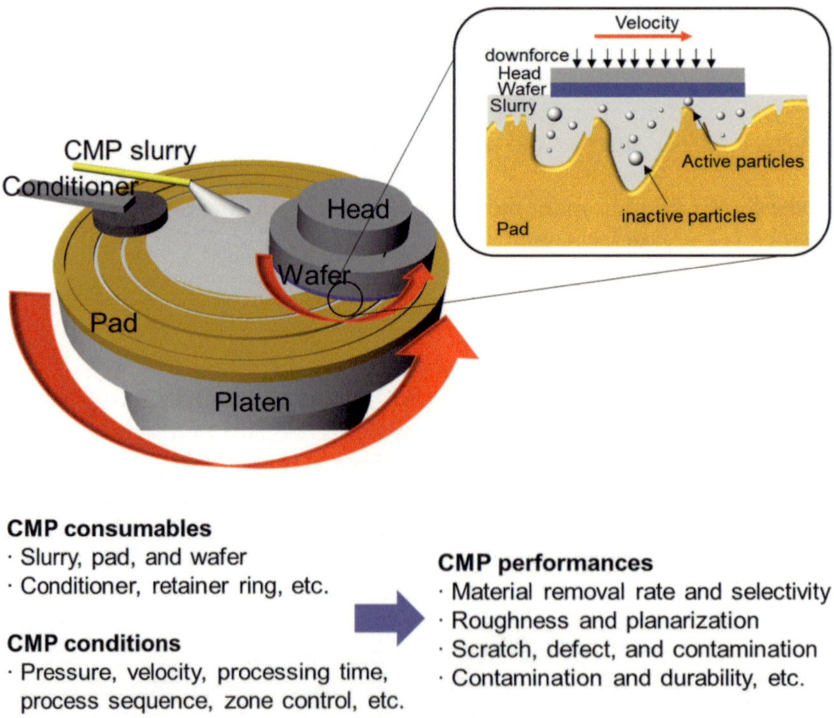

Figure 1.
Schematic of a typical CMP process and CMP consumables and conditions associated with CMP performances. Reprinted from Ref. [1] with permission from the Journal of Materials Research.

Applications	Slurry components
Dielectric CMP: SiO_2, Si_3N_4, Poly-Si, SiC, SiCN, etc.	Abrasive: Ceria/Silica
	pH adjusting agents such as HNO_3, KOH, NH_4OH, etc. and buffers
	Dispersant & Surfactants.
	Chemical agents for high selectivity (PAA, amino acids, etc.)
	Other chemical agents for high polish rates and better performance.
Metal CMP: Cu, W, Co, Ta, TaN, Ti, Ru, etc.	Abrasive: Silica/Alumina.
	pH adjusting agent such as HNO_3, KOH, NH_4OH, etc. and buffers.
	Oxidizer: H_2O_2, KIO_3, $Fe(NO_3)_3$, $K_2S_2O_8$, etc.
	Complexing agents: glycine, citric acid, etc.
	Corrosion inhibitors: Azole and its derivatives.

Table 1.
Typical CMP slurry components. Reprinted from Ref [1] with permission from the Journal of Materials Research.

Considerable effort has been devoted to minimizing the formation of contaminants during polishing by optimizing CMP consumables and their process/tool conditions. As the minimum feature size has shrunk below 7 nm and beyond, the devices require more stringent conditions to achieve a smooth defect-free wafer surface. Thus, the demand for understanding of the origin of CMP-related to contaminants and their characteristics is increasing in both industrial and academic research. This chapter provides an overview of the origin and characteristics of various CMP-related to contaminants that can be generated or presented on the wafer surfaces after CMP and post-CMP cleaning process. It also provides important insights into the cleaning chemistry to remove these contaminants during post-CMP cleaning. The challenges related to post-CMP cleaning solutions are highlighted.

2. CMP consumables-induced contaminants

Some of the CMP related to contaminants, such as residual particles, surface residues, organic residues, and metallic impurities, are common to most CMP processes, which are directly associated with CMP consumables [3, 12, 13]. Various types of CMP-related to contaminants and their impacts in the semiconductor manufacturing process were summarized in **Table 2** [18]. These contaminants are presumably attributed to the chemical reactions of slurry components at the slurry/pad-wafer interface. The sources and characteristics of the contaminants listed in **Table 2** will be discussed in this section.

2.1 Residual abrasive particles

Abrasive particle is not only one of the main components in CMP slurries (**Table 1**) [16, 17], but also a common contaminant observed after all CMP processes (**Figure 2a**) [13]. Silica and ceria have been widely employed as abrasive particles for CMP processes [16]. The adsorption of silica abrasives on the films is driven by the electrostatic attractive forces between abrasives and films in a certain

Contaminants	Dielectric CMP	Metal CMP	Effects
Particulate	Silica or ceria, fine fragments of film or pad, etc.	Silica or alumina, metal hydroxide precipitates, fine fragments of film or pad, etc.	1. Cause local roughness an block photolithography 2. Pinholes in new grown films: metal precipitates leads to metallic contamination 3. Shorts by conductive particles
Organic	Buffers, surfactants, etc.	Buffers, surfactants, inhibitors, etc.	1. Affects wettability and cleanability 2. Outgassing 3. Poor adhesion of deposited layers
Metallic	Na^+, K^+,Ca^{2+}, etc.	W_xO_y, Cu^{2+}, Al^{3+}, Fe^{3+}, IO^{4-}, I^-/I_2, $Fe(CN)_6{}^{3-}$, $Fe(CN)_6{}^{4-}$, etc.	1. Alkali metal ions: high mobility influences electrical characteristics 2. Copper: fast diffuser in Si 3. Many metals can form silicide, and/or affect the oxidation 4. Noble metal ions cause etching of Si

Table 2.
Some CMP-related to contaminants and their effects in the semiconductor manufacturing process. Modified and reprinted with permission from Ref. [18], American Vacuum Society.

pH range. The pH_{IEP} of silica abrasive is about pH 2.5 [22], so the silica particles show a negative surface charge at above pH 2.5 and lead to the contamination of positively charged films that have higher pH_{IEP} values. The preferential adsorption of silica abrasives on Cu and Co films was observed after the Cu CMP process when Co is used as the liner (**Figure 2b**) [20]. As expected, the IEPs of Cu and Co species are much higher than those of TaN and SiO_2 films (**Table 3**). W films are covered with a passivation layer in acidic pH range according to the Pourbaix diagram [32]. So, the silica abrasives can remain on the polished W films due to their electrostatic attraction (The pH_{IEP} of WO_x is 0.5 as listed in **Table 3**) [24]. In some cases, the alumina particles ($pH_{IEP} \sim 7$) are used as the abrasive for W CMP, and they are observed on the W films after polishing due to its positive charge in the acidic medium [13, 24]. Co films and other metal films can also be contaminated with the silica abrasive during polishing [14, 15]. These particle contaminants can be controlled by the chemical reactions between slurry components and films being polished. Moreover, silica abrasives are weakly bound to the films and can be easy to be removed by under-cut and particle lift-off or their combination during cleaning [15].

Ceria-based slurry has been widely used for STI CMP to uniformly polish the step height of SiO_2, formed by the gap-filling process, and stop on an underlying Si_3N_4 film [4–6]. Residual ceria abrasives are discovered after STI CMP process (**Figure 2c**) [21, 33]. In contrast with a silica abrasive, ceria abrasive is more strongly coupled with the dielectric materials (in particular, SiO_2 film) via the formation of strong Ce-O-Si bonding [4, 34]. It is well known that the surface Ce^{3+} species are the active sites for the formation of strong Ce-O-Si bonds with SiO_2 films during polishing [4, 21]. Various ceria abrasives such as smaller particles with higher surface Ce^{3+} concentrations [35], the core/shell type Ce^{3+} rich ceria [36], and metal-doping or coated ceria abrasive [37] have been investigated to improved SiO_2 removal rates, but making their removal during cleaning more difficult. Since the pH_{IEP} of the ceria abrasive, SiO_2, and Si_3N_4 films are 7.3, 2.5, and ~ 5.0 (**Table 3**)

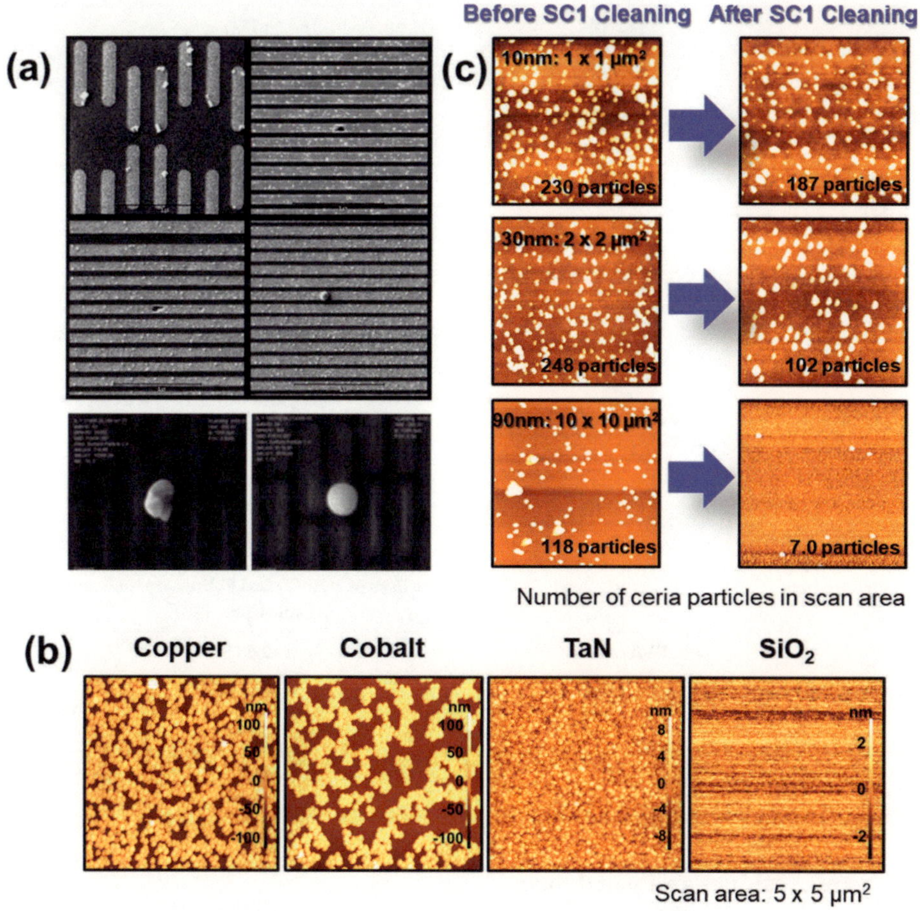

Figure 2.
(a) Residual abrasive particles on the wafer surfaces after the CMP process. (b) Atomic force microscopy (AFM) images of adsorbed three different sized ceria particles on the SiO_2 films and the corresponding number of particles before and after SC1 cleaning. (c) Topographic AFM images of Cu, Co, TaN, and SiO_2 films contaminated with silica slurry at pH 10. Reprinted with permission from Ref. [3]. Copyright 2010 American Chemical Society. Used with the permission of HongJin Kim [19]. Reproduced with permission from Refs. [20, 21]. Copyright 2019 IOP Publishing.

[38], respectively, the particles can effectively interact with the SiO_2 films due to the electrostatic attractions between them. The surface charges of ceria abrasive are different depending on the nature of additives (e.g., dispersant, passivation agent for high selectivity, etc.) and the slurry pH [33]. Positively charged ceria particles, dispersed with amino acid, led significant contamination of negatively charged SiO_2 films while negatively charged ceria particles, dispersed with a weak organic acid or poly(acrylic acid), showed a higher level of contamination of Si_3N_4 films [33]. Thus, cleaning of ceria particles from the wafer surfaces has become more challenging. The pH_{IEP} of abrasive particles, films to be polished, CMP consumables, and organic residues are listed in **Table 3**.

These residual particles cause not only an increase in local roughness but also poor photolithography results by blocking the UV light (**Table 2**) [18]. Residual particles on the wafer surfaces can also lead to pinholes in the subsequently deposited film [18]. In some cases, just two residual abrasives on the surfaces can make the device bad [9]. The particle larger than ½ the minimum feature size becomes a "killer defect" [39].

	Materials	pH_{IEP}
Abrasive particles	Silica	2.5 [22]
	Ceria	7.3 [23]
	Alumina	~7.0 [24]
Dielectric CMP	SiO_2	2.5 [25]
	Si_3N_4	~5.0 [26]
	Poly-Si	~3.3 [27]
Metal CMP	Cu	The IEPs of CuO and $Cu(OH)_2$ are 8.5 and 9.5, respectively [24].
	Co	The IEPs of CoO, Co_3O_4, and $Co(OH)_2$ particles are 9.2, 9.5, and 11.4, respectively [24].
	W	~0.5 [24] (WOx)
	TaN/TiN	~4.0 [27]/~3.6 [28]
	Ru	4.2-5.2 [29] (RuO_X)
Consumables	Polishing pad	The IEPs of IC1000 and Politex are ~ 3.2 and 4, respectively [30].
	PVA brush	~2.5 [15]
Organic residues	Cu-BTA	~10 [20]
	Co-BTA	~10 [31]

Table 3.
The pH_{IEP} of abrasive particles, films to be polished, CMP consumables, and organic residues.

2.2 Organic residues and pad debris

The polished wafers are also contaminated with organic residues (**Figure 3a**), which are originated from the slurry components such as dispersants, additives for the selectivity, complexing agents, corrosion inhibitors, etc. (**Table 1**). One of the main sources of organic residues is insoluble metal complexes. Azole derivatives (more specifically, benzotriazole (BTA)) have been widely used as corrosion inhibitors for metal films during polishing. BTA can strongly chemisorb onto the metal film by forming a chemical bond with a surface metal ion through the nitrogen lone pair electrons [41]. For example, each Cu^+ ion can coordinate with two nitrogen ligands of BTA^- during the Cu CMP process, which forms a polymeric product with the BTA^- acting as bridging ligands (**Figure 3b**) [40, 41]. The third nitrogen atom in BTA^- of Cu-BTA complexes can bind to Cu surfaces, resulting in the polymeric protecting layer on the Cu films [41]. Recently, Seo et al. [20] reported that Cu and Co ions dissolved from Cu and Co films, respectively, can react with BTA and form 4-15 nm Cu-BTA/Co-BTA particles when Co is used as the liner in Cu interconnect structures (**Figure 3c**). These particles adsorb on only Cu surface, not Co film at pH 10 [20]. Since the zeta-potentials of both particles at pH 10 are close to ~0 mV (**Table 3**), there is a negligible electrostatic interaction of Cu-BTA/Co-BTA particles with Cu and Co films. They suggested that the adsorption of these particles on Cu film is not only attributed to the hydrophobic interaction between particles and Cu film but also the chemisorption via the lone pair electrons on the nitrogen atoms in the Cu-BTA/Co-BTA particles [20, 40]. Other organic additives can also be adsorbed on the films via van der Waals and hydrophobic interactions. In some cases, these may convert hydrophilic to hydrophobic of the film surfaces.

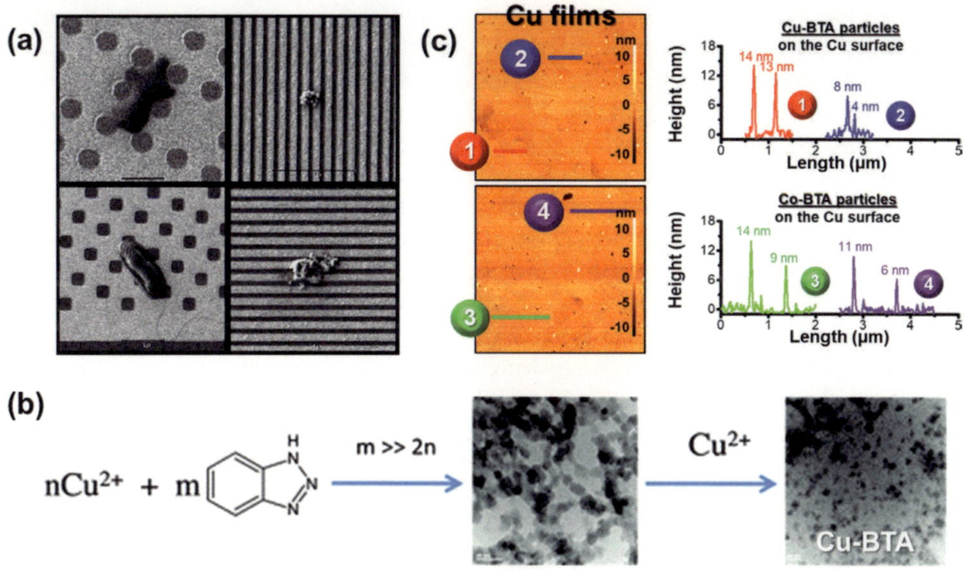

Figure 3.
(a) Organic residues generated from pad materials and others during polishing. (b) The formation of Cu-BTA complexes during polishing; TEM images of samples collected from the wafer surfaces after the exposure to the slurry components containing hydrogen peroxide, glycine, and BTA. (c) Topographic AFM images of Cu films contaminated with Cu-BTA (upper figure) and Co-BTA complexes (lower figure) at pH 10 in a scan area of 5 × 5 μm². AFM profiles show the height of Cu-BTA/Co-BTA particles adsorbed on the Cu films. Reprinted with permission from Refs. [3, 40]. Copyright 2010 and 2009 American Chemical Society. Reproduced with permission from Ref. [20]. Copyright 2019 IOP Publishing.

The hydrophobic nature of the film surfaces can attract water droplets containing organic contaminants, leading to the watermarks and more organic residues [42]. These adsorbed organic contaminants affect the wettability and cleanability of the wafer surface, resulting in the poor adhesion of subsequently deposited layers (**Table 2**).

Most polishing pads are made of polymeric materials such as polyurethane. During polishing, the pad is conditioned with a diamond conditioner to regenerate the pad asperities and remove the accumulated particles on the pad, but generating 0.2 to 300 μm pad debris [43]. Although in-situ conditioning enables a higher removal rate and better planarity by maintaining stable pad surface properties, it can cause more pad debris compared to ex-situ conditioning [43]. Most of the pad debris is in the range of 0.2 to 0.3 μm. Some of the large pad debris (20-300 μm) are not only very irregular shapes, but also covered by abrasive particles [44]. This pad debris is known as a source of micro-scratches, and it should be completely removed during cleaning. Both hybrid clean (i.e., acidic plus alkaline cleans) and alkaline-clean processes are effective in removing pad debris from the wafer surfaces by the electrostatic repulsion between them in the alkaline medium [45].

2.3 Metallic impurities

The CMP process leaves metallic impurities in the concentrations of 10^{11}-10^{12} atoms/cm^2. These contaminants may originate from the abraded metal lines, metal ions in the slurries, the environment of the CMP tool [15]. During the metal CMP process, chelating agents are able to form a metal complex with metal ions on metal surfaces (Cu, W, Co, Ta, TaN, Ti, Ru, etc.). Metal ions dissolved from metal surfaces or metal residues may be the main source for metallic contaminants (**Table 2**). These metallic cations not only are affected by the surface charge, but also can be precipitated on the surface of Si devices, which is expressed by ≡Si-OH(s) + Me^{n+}(aq) ↔ ≡ $SiOMe^{(n-1)+}$ (s) + H^{+}(aq). Heavy metals (Cu, Fe, Ni, Cr, Co, and Mo) that deposited on the wafer surface by the galvanic reaction can diffuse into the Si devices during heat treatments and cause excessive leakage currents, resulting in the device degradation and reliability problems [46]. Other metals (Al, group II metals, and Ti) may have much lower diffusivities and may not diffuse significantly into the Si devices [39]. Metal ions such as Cu, Co, Fe, Al, Zn, and Mg can hydrolyze in the alkaline based cleaning solution and form insoluble metal hydroxides that are remained on the wafer surfaces [15]. Cu electromigration occurs through the movement of Cu atoms or Cu ions when there is a strong electrical current [47]. The undesirable metallic particles can cause short circuits between metal lines, whereas the metal hydroxides may cause open circuits [39].

Mobile ions such as alkaline metals (Na^{+} and K^{+}) originated from the slurry components such as salts and NaOH/KOH (pH adjuster) [18] (**Table 2**) can cause flatband shifts and surface-related leakage currents due to their electrical characteristics of high mobility [39]. Fe ions have been used as a catalyst for W CMP slurry [8]. Fe ions (Fe^{3+}, $Fe(CN)_6^{3-}$, $Fe(CN)_6^{4-}$, etc.) and FeO_x caused from W CMP slurries are observed on the polished wafers (**Table 2**) [18, 48]. Acidic cleaning solutions are useful for removing metallic impurities and suppressing the adsorption of metallic species. Critical metallic impurities on the Si device continue to decrease as the device feature shrinks down. For the current technology nodes, the acceptable metallic contaminants are less than 10^8 atoms/cm^2 and approach the limit of detection [46].

Some metallic contaminants directly come from the metal interconnect lines. After the Cu CMP process, pyramid-shaped Cu particles (Cu, CuO, and CuOH) detached from the Cu films are discovered on the surface [49] Metal flakes such as Ti or W-Ti on the top of the replace metal gate (RMG) after W RMG CMP process are observed [50, 51]. Metals at partially filled can be broken during the RMG CMP process, and they are a source of metal flake. These metal flakes are trapped inside the brush and re-deposit to the wafer surface by the cross-contamination process. In some cases, the delamination of metal films is occurred at the wafer edge due to the edge over erosion or a poor adhesion between metal and barrier film, which is another source of metal flakes [50, 51]. These metal flakes are known as a potential killer defect in the current RMG technologies.

2.4 Pad contamination

In some cases, the by-products are generated during the metal CMP process, and they are discovered on the pad surface [52, 53]. Han et al. observed the large stain on the pad after the polishing of Cu films [52]. The brown-colored by-products are formed and accumulated on the pores and grooves of the polishing pad, which is able to disturb the slurry transportation during polishing. These contaminants are caused by the chemical reactions between the slurry components and Cu films. They suggested that an additional pad cleaning step will be required to remove these by-products from the polishing pad and improve the pad lifetime [52]. Later, Lu et al. reported the pink by-products remained on the polishing pad after Co CMP process [53]. They compared Raman spectroscopy of by-products with that of the precipitates (Co-BTA particles) made from a mixture of $Co(NO_3)_2$ and BTA. Both samples showed the same Raman peaks, indicating that by-products observed on the polishing pad are Co-BTA particles.

2.5 Watermarks

When water evaporates from the hydrophobic surfaces, it leaves the residues containing organic residues, particles, and metallic impurities that were present in the evaporation water layer, which is known as "watermark". Watermarks are observed in hydrophobic regions or Mixed hydrophobic and hydrophilic areas [32, 54]. During the Si CMP and cleaning process, the oxidation of Si occurs in the presence of O_2 in the water ($Si + O_2 \rightarrow SiO_2$), and it is dissolved into water ($SiO_2 + H_2O \rightarrow H_2SiO_3 \rightarrow H+ HSiO_3^-$). The dissolved species may precipitate to form the residues containing Si and O. Watermarks that may contain organic residues and Cu oxide particles have also been one of the challenges for Cu CMP and cleaning process. Such watermarks tend to cause significant degradation in device performance [15]. IPA-based Marangoni drying process was proposed and used to eliminate watermarks [55]. The addition of surfactants that can convert hydrophobic to hydrophilic of the films will prevent the formation of watermarks after drying [42].

3. Brush-induced cross-contamination during post CMP cleaning

The abrasive particles are removed from the wafer surfaces by the direct contact between wafers and PVA brushes during post CMP cleaning. The brush is compressed to the wafer surfaces, and then the particle contaminants are removed by the physical force of the compressed brush. However, the surface and inside the pore structure of PVA brushes are contaminated with the particles, organic residues, and pad debris (**Figure 4a**), which can be transported to the next wafers and cause cross-contamination of the wafers during the brush scrubbing [56, 58]. More cross-contamination is observed on the wafer surfaces when the contact pressure and contact area between the brush and the wafer increase [59]. Also, the longer brush contact time (lower brush rotation speed) results in more cross-contaminated particles on the wafers. Before brush scrubbing, brush soaking treatment and break-in and their optimized process may be useful to reduce the cross-contamination and improve the cleaning efficiency [58]. Also, the ultrasonication method with DIW was very effective in removing the contaminants from the PVA brushes without damage [56].

The ring-shaped CuO residue is rarely observed at the wafer center region after the Cu barrier CMP process with acid-based slurries [60]. Chelating agents in the acidic medium are able to effectively form water-soluble complexes with Cu ions and pull them into the slurries. More polymers or corrosion inhibitors are added at lower pH slurry, which may lead to conductive organic residues during polishing. These organic residues can be dissolved in the cleaning solutions and move between the brush and wafer surfaces during cleaning when there is a direct solid–solid contact between them, making an electrical circuit [60]. Cu^{2+} ions released from the Cu films during cleaning can transfer to the brush, and they react with oxygen in the ambient environment to convert to CuO residue where the electro circuit is provided by the organic residues [60].

Particle contaminants at the backside surface of wafers are also reported (**Figure 4b**) [57]. The wafer backside surface contacts with the slurries during polishing and cleaned with brush scrubbing and nozzle. Cleaning solutions are dispensed from an overhead nozzle onto the wafer backside. The locations of the wafer backside ring signature are well-matched with the inner ring, outer ring, and clean nozzle, which means that the polishing and the downstream surface cleaning process make the wafer backside ring signature [57].

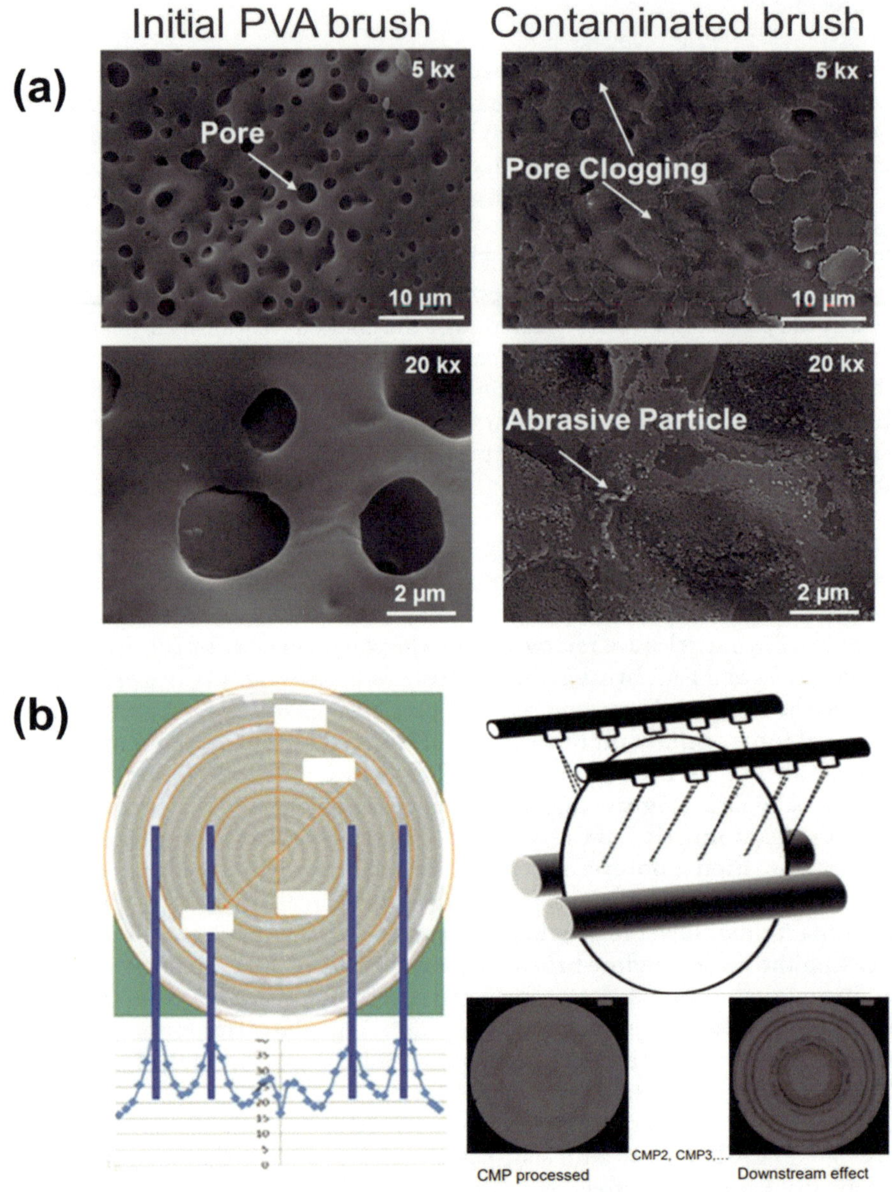

Figure 4.
(a) Scanning electron microscopy (SEM) images of initial PVA brush and contaminated PVA brush. (b) Wafer backside signature after CMP and cleaning process. Reproduced with permission from Ref. [56]. Copyright 2019 IOP Publishing. Used with the permission of HongJin Kim [57].

4. Post-CMP cleaning for removing CMP-related to contaminants

Table 4 shows traditional post-CMP cleaning solutions that have been widely used to remove CMP-related to contaminants over the past several decades [46]. SC-1 solution is a mixture of NH_4OH (29 wt%)/H_2O_2 (30 wt%)/DIW, which is very useful for removing particles, organic residues, and some metallic contaminants from the water surface through under-cut and particle lift-off or their combination [46]. SC-2 solution consisting of HCl (37%)/H_2O_2 (30 wt%)/DIW is very effective in removing metallic contaminants via the formation of soluble metal complexes

Cleaning solution	Compositions and conditions	Contaminant removal
SC-1	NH_4OH (29 wt%)/H_2O_2 (30 wt%)/H_2O, 1:1:5-1:1:100 at 40-75°C	Particles, organics, and some metallic contaminants
SC-2	HCl (37%)/H_2O_2 (30 wt%)/H_2O, 1:1:6-1:1:50 at 40-75°C	Metallic contaminants
SPM	H_2SO_4 (96 wt%)/H_2O_2 (30 wt%), 2:1-4:1 at 90-40°C	Organic residues
DHF	HF, 1:10-1:200 at 25°C	Sacrificial oxide removal and native oxide removal

Table 4.
Traditional cleaning solutions and their conditions.

with Cl^- ions [46]. Sulfuric acid-peroxide mixture (SPM) of H_2SO_4 (96 wt%)/H_2O_2 (30 wt%) is able to remove photoresist and other organic residues by forming a very strong oxidizing agent, which can be expressed as $H_2SO_4 + H_2O_2 \rightarrow H_2SO_5$ (Caro's acid) + H_2O [61]. Caro's acid can easily dissolve the organic residues during cleaning. Hydrofluoric acid (HF) diluted with H_2O (DHF) is useful to remove the oxide layer from the wafers. These traditional cleaning solutions have been modified to meet the post-CMP cleaning process requirements for advanced technology nodes.

Some equimolar solutions containing NH_4OH and H_2O_2 (4.20 and 4.13 mol/L, respectively) to remove even 10 nm ceria particles (~99% cleaning efficiencies) from SiO_2 films were proposed by Seo et al. [21]. Non-traditional cleaning solutions composed of a bond-breaking reagent, complexing reagent, cleaning additive and pH adjuster were proposed for post STI CMP cleaning process [62, 63]. For post-metal CMP cleaning, the contaminants need to be completely removed while minimizing the individual corrosion of metals, localized pitting, and bimetallic/galvanic corrosion. Many cleaning compositions consisting of oxidizers, complexing agents, cleaning agents, and pH adjuster have been developed so far [20, 31]. However, there are still several cleaning challenges for the future technology nodes, while considerable progress has been made [57]; (1) improvement of cleaning efficiency, (2) the removal of smaller particles from the films, (3) the prevention of cross-contamination by brush scrubbing, (4) the removal of new-types contaminants-very thin metal flake, (5) wafer backside cleaning, (6) universal cleaning solution, (7) environmentally friendly post-CMP cleaning, (8) TMAH-free cleaning solution.

5. Toxicity issues and safety considerations

Toxic gases (e.g., PH_3, AsH_3) and the III–V containing liquid (in particular, As) can be generated during the polishing of III–V materials such as GaAs, InGaAs, InAs, and InP, which poses an environment, health, and safety (EHS) risk. Exposure to inorganic As can cause irritation of the stomach and intestines, decrease in the production of red and white blood cells, skin changes and lung irritation [64, 65]. Hence, one of the goals of the CMP processes of III–V materials is to achieve high planarity without generating toxic by-products. Also, Ru film can be converted to highly volatile RuO_4 (a toxic gas) in the acidic pH during polishing. Compounds of Ru stain the skin very strongly, and the ingested Ru is retained strongly in bones. The addition of chelating agents may help to reduce the formation of highly volatile RuO_4 during polishing. The formation of toxic by-products during polishing must be avoided by controlling slurry chemistry such as oxidizer, complexing agent, corrosion inhibitor, and the slurry pH.

6. Conclusions

As the performances of devices at 7 nm node and beyond have become very sensitive to contaminants and defects, there has been a growing interest in understanding the sources and characteristics of CMP-related contaminants. An overview of various contaminants generated during the CMP process and their characteristics were discussed in this book chapter. There has been significant progress in understanding the fundamental science and technology of the sources of CMP-induced contaminants and their characteristics for the past several decades. Nevertheless, a more fundamental understanding of various chemical and mechanical reactions that occur between slurry components, polishing pad, and wafer surfaces will help us to propose new strategies and novel concepts of CMP slurries to minimize the formation of the contaminants during polishing. As expected, Post-CMP cleaning has become more important than ever to eliminate many of these CMP-induced contaminants. A close collaboration between the consumables manufacturers and the fab end-users is required to overcome many challenges and some issues related to the contaminants in the CMP and post-CMP cleaning process.

Acknowledgements

The author gratefully acknowledges Prof. S.V. Babu for many useful discussions, valuable comments, and suggestions.

Author details

Jihoon Seo
Department of Chemical and Biomolecular Engineering and Center for Advanced Materials Processing, Clarkson University, Potsdam, New York, United States of America

*Address all correspondence to: jseo@clarkson.edu

References

[1] Seo, J., A review on chemical and mechanical phenomena at the wafer interface during chemical mechanical planarization. *Journal of Materials Research* **2020**, 1-23.

[2] Suryadevara, B., *Advances in chemical mechanical planarization (CMP)*. Woodhead Publishing: 2016.

[3] Krishnan, M.; Nalaskowski, J. W.; Cook, L. M., Chemical mechanical planarization: slurry chemistry, materials, and mechanisms. *Chemical reviews* **2010,** *110* (1), 178-204.

[4] Srinivasan, R.; Dandu, P. V.; Babu, S., Shallow trench isolation chemical mechanical planarization: A review. *ECS Journal of Solid State Science and Technology* **2015,** *4* (11), P5029-P5039.

[5] Seo, J.; Moon, J.; Moon, S.; Paik, U., Interpolymer complexes of poly (acrylic acid) and poly (ethylene glycol) for low dishing in STI CMP. *Applied Surface Science* **2015,** *353*, 499-503.

[6] Seo, J.; Moon, J.; Bae, J.-Y.; Yoon, K. S.; Sigmund, W.; Paik, U., Control of adhesion force between ceria particles and polishing pad in shallow trench isolation chemical mechanical planarization. *Journal of nanoscience and nanotechnology* **2014,** *14* (6), 4351-4356.

[7] You, K.; Seo, J.; Kim, P. J. H.; Song, T., Control of Tungsten Protrusion with Surface Active Agent during Tungsten Chemical Mechanical Polishing. *ECS Journal of Solid State Science and Technology* **2017,** *6* (12), P822-P827.

[8] Seo, J.; Kim, J. H.; Lee, M.; You, K.; Moon, J.; Lee, D.-H.; Paik, U., Multi-objective optimization of tungsten CMP slurry for advanced semiconductor manufacturing using a response surface methodology. *Materials & Design* **2017,** *117*, 131-138.

[9] Krishnan, M.; Lofaro, M., Copper chemical mechanical planarization (Cu CMP) challenges in 22 nm back-end-of-line (BEOL) and beyond. In *Advances in Chemical Mechanical Planarization (CMP)*, Elsevier: 2016; pp 27-46.

[10] Khanna, A. J.; Kakireddy, V. R.; Fung, J.; Jawali, P.; Yamamura, M.; Kenchappa, N. B.; Hariharan, V.; Redfield, D.; Bajaj, R., High-Performance Pad Conditioning (HPPC) Arm for Augmenting CMP Performance. *ECS Journal of Solid State Science and Technology* **2020,** *9* (6), 064012.

[11] Khanna, A. J.; Yamamura, M.; Kakireddy, V. R.; Chockalingam, A.; Jawali, P.; Kenchappa, N. B.; Hariharan, V.; Redfield, D.; Bajaj, R., Investigation of the Impact of Pad Surface Texture from Different Pad Conditioners on the CMP Performance. *ECS Journal of Solid State Science and Technology* **2020,** *9* (6), 064011.

[12] Lefevre, P., Defects Observed on the Wafer after the CMP Process. *Microelectronic Applications of Chemical Mechanical Planarization* **2007**, 511-561.

[13] Tseng, W.-T., Approaches to defect characterization, mitigation, and reduction. In *Advances in Chemical Mechanical Planarization (CMP)*, Elsevier: 2016; pp 433-462.

[14] Alety, S. R.; Lagudu, U. R.; Popuri, R.; Patlolla, R.; Surisetty, C. V.; Babu, S., Cleaning Solutions for Ultrathin Co Barriers for Advanced Technology Nodes. *ECS Journal of Solid State Science and Technology* **2017,** *6* (9), P671.

[15] Chen, Y.; Mikhaylichenko, K.; Brown, B.; Redeker, F., Post-CMP Cleaning. In *Handbook of Silicon Wafer Cleaning Technology*, Elsevier: 2018; pp 253-301.

[16] Seo, J.; Paik, U., Preparation and characterization of slurry for chemical mechanical planarization (CMP). In *Advances in Chemical Mechanical Planarization (CMP)*, Elsevier: 2016; pp 273-298.

[17] Lee, D.; Lee, H.; Jeong, H., Slurry components in metal chemical mechanical planarization (CMP) process: A review. *International Journal of Precision Engineering and Manufacturing* **2016,** *17* (12), 1751-1762.

[18] Zhang, L.; Raghavan, S.; Weling, M., Minimization of chemical-mechanical planarization (CMP) defects and post-CMP cleaning. *Journal of Vacuum Science & Technology B: Microelectronics and Nanometer Structures Processing, Measurement, and Phenomena* **1999,** *17* (5), 2248-2255.

[19] Kim, H. J. In *Process Optimization in Post W CMP In-situ Cleaning*, NCCAVS 2015 CMP User Group Spring Meeting, Albany, NY, Albany, NY, 2015.

[20] Seo, J.; Vegi, S. H.; Babu, S., Post-CMP Cleaning Solutions for the Removal of Organic Contaminants with Reduced Galvanic Corrosion at Copper/Cobalt Interface for Advanced Cu Interconnect Applications. *ECS Journal of Solid State Science and Technology* **2019,** *8* (8), P379.

[21] Seo, J.; Gowda, A.; Babu, S., Almost Complete Removal of Ceria Particles Down to 10 nm Size from Silicon Dioxide Surfaces. *ECS Journal of Solid State Science and Technology* **2018,** *7* (5), P243-P252.

[22] Seo, J.; Kim, J. H.; Lee, M.; Moon, J.; Yi, D. K.; Paik, U., Size-dependent interactions of silica nanoparticles with a flat silica surface. *Journal of colloid and interface science* **2016,** *483*, 177-184.

[23] Seo, J.; Gowda, A.; Khajornrungruang, P.; Hamada, S.; Song, T.; Babu, S., Trajectories, diffusion, and interactions of single ceria particles on a glass surface observed by evanescent wave microscopy. *Journal of Materials Research* **2020**, P1-11.

[24] Kosmulski, M., Compilation of PZC and IEP of sparingly soluble metal oxides and hydroxides from literature. *Advances in colloid and interface science* **2009,** *152* (1-2), 14-25.

[25] Penta, N. K.; Veera, P. D.; Babu, S., Charge density and pH effects on polycation adsorption on Poly-Si, SiO2, and Si3N4 films and impact on removal during chemical mechanical polishing. *ACS applied materials & interfaces* **2011,** *3* (10), 4126-4132.

[26] Alety, S. R.; Sagi, K. V.; Babu, S., Role of Ce3+ Ions in Achieving High Silicon Nitride Polish Rates. *ECS Journal of Solid State Science and Technology* **2017,** *6* (12), P898-P903.

[27] Shih, C.-J.; Lung, B.-H.; Hon, M.-H., Colloidal processing of titanium nitride with poly-(methacrylic acid) polyelectrolyte. *Materials chemistry and physics* **1999,** *60* (2), 150-157.

[28] Guo, Z.; Xiong, J.; Yang, M.; Xiong, S.; Chen, J.; Wu, Y.; Fan, H.; Sun, L.; Wang, J.; Wang, H., Dispersion of nano-TiN powder in aqueous media. *Journal of Alloys and Compounds* **2010,** *493* (1-2), 362-367.

[29] Luxton, T. P.; Eick, M. J.; Scheckel, K. G., Characterization and dissolution properties of ruthenium oxides. *Journal of colloid and interface science* **2011,** *359* (1), 30-39.

[30] Penta, N. K.; Matovu, J. B.; Veera, P. D.; Krishnan, S.; Babu, S., Role of polycation adsorption in poly-Si, SiO2 and Si3N4 removal during chemical mechanical polishing: Effect of polishing pad surface chemistry.

Colloids and Surfaces A: Physicochemical and Engineering Aspects **2011,** *388* (1-3), 21-28.

[31] Seo, J.; Vegi, S. H.; Ranaweera, C.; Baradanahalli, N.; Han, J.-H.; Koli, D.; Babu, S., Formation of Cobalt-BTA Complexes and Their Removal from Various Surfaces Relevant to Cobalt Interconnect Applications. *ECS Journal of Solid State Science and Technology* **2019,** *8* (5), P3009-P3017.

[32] Li, Y., *Microelectronic applications of chemical mechanical planarization*. John Wiley & Sons: 2007.

[33] Gowda, A.; Seo, J.; Ranaweera, C. K.; Babu, S., Cleaning Solutions for Removal of ~ 30 nm Ceria Particles from Proline and Citric Acid Containing Slurries Deposited on Silicon Dioxide and Silicon Nitride Surfaces. *ECS Journal of Solid State Science and Technology* **2020,** *9* (4), 044013.

[34] Cook, L. M., Chemical processes in glass polishing. *Journal of non-crystalline solids* **1990,** *120* (1-3), 152-171.

[35] Seo, J.; Moon, J.; Kim, J. H.; Lee, K.; Hwang, J.; Yoon, H.; Yi, D. K.; Paik, U., Role of the oxidation state of cerium on the ceria surfaces for silicate adsorption. *Applied Surface Science* **2016,** *389*, 311-315.

[36] Kim, K.; Seo, J.; Lee, M.; Moon, J.; Lee, K.; Yi, D. K.; Paik, U., Ce 3+-enriched core–shell ceria nanoparticles for silicate adsorption. *Journal of Materials Research* **2017,** *32* (14), 2829-2836.

[37] Cheng, J.; Huang, S.; Li, Y.; Wang, T.; Xie, L.; Lu, X., RE (La, Nd and Yb) doped CeO2 abrasive particles for chemical mechanical polishing of dielectric materials: Experimental and computational analysis. *Applied Surface Science* **2020,** *506*, 144668.

[38] Kosmulski, M., pH-dependent surface charging and points of zero charge. IV. Update and new approach. *Journal of colloid and interface science* **2009,** *337* (2), 439-448.

[39] Bearda, T.; Mertens, P. W.; Beaudoin, S. P., Overview of Wafer Contamination and Defectivity. In *Handbook of silicon wafer cleaning technology*, Elsevier: 2018; pp 87-149.

[40] Li, Y.; Gong, M.; Ramji, K.; Li, Y., Role of Cu- Benzotriazole Nanoparticles in Passivation Film Formation. *The Journal of Physical Chemistry C* **2009,** *113* (42), 18003-18013.

[41] Xue, G.; Ding, J.; Lu, P.; Dong, J., SERS, XPS, and electroanalytical studies of the chemisorption of benzotriazole on a freshly etched surface and an oxidized surface of copper. *The Journal of Physical Chemistry* **1991,** *95* (19), 7380-7384.

[42] Mikhaylichenko, K. In *Cleaning Efficiency Improvement Solutions for FEOL CMP*, Surface Preparation and Cleaning Conference (SPCC 2019), Portland, Oregon, USA, Portland, Oregon, USA, 2019.

[43] Prasad, Y. N.; Kwon, T.-Y.; Kim, I.-K.; Kim, I.-G.; Park, J.-G., Generation of pad debris during oxide CMP process and its role in scratch formation. *Journal of the Electrochemical Society* **2011,** *158* (4), H394.

[44] Yang, J. C.; Kim, H.; Oh, D. W.; Won, J.-H.; Lee, C.-G.; Kim, T., Experimental evaluation of the effect of pad debris size on microscratches during CMP process. *Journal of electronic materials* **2013,** *42* (1), 97-102.

[45] Tseng, W.-T.; Rill, E.; Backes, B.; Chace, M.; Yao, Y.; DeHaven, P.; Ticknor, A.; Devarapalli, V.; Khojasteh, M.; Steber, D., Post Cu CMP cleaning of polyurethane pad debris. *ECS Journal of*

Solid State Science and Technology **2013,** *3* (1), N3023.

[46] Kern, W., Overview and evolution of silicon wafer cleaning technology. In *Handbook of silicon wafer cleaning technology*, Elsevier: 2018; pp 3-85.

[47] Lloyd, J.; Clement, J., Electromigration in copper conductors. *Thin solid films* **1995,** *262* (1-2), 135-141.

[48] White, D.; Parson, T.; Lieten, R. R.; Fu, B.; White, M. In *Understanding interfacial surface interactions*, Surface Preparation and Cleaning Conference (SPCC), Portland, Oregon, USA, Portland, Oregon, USA, 2019.

[49] Hong, J.; Niu, X.; Liu, Y.; He, Y.; Zhang, B.; Wang, J.; Han, L.; Yan, C.; Zhang, J., Effect of a novel chelating agent on defect removal during post-CMP cleaning. *Applied Surface Science* **2016,** *378*, 239-244.

[50] Kim, H. J.; Lawyer, A.; Egan, B.; Ng, M.; Huang, C.-C.; Han, J.-H., Study on the mechanism of Nano-Flake defect during tungsten contact chemical mechanical polishing. *ECS Journal of Solid State Science and Technology* **2018,** *7* (4), P175.

[51] Kim, H. J.; Govindarajulu, V.; Bohra, G.; Wang, H.-M.; Koli, D., Metal flake defect and its formation mechanism during replacement metal gate CMP process. *ECS Journal of Solid State Science and Technology* **2016,** *5* (10), P637.

[52] Han, J.-H.; Hah, S.-R.; Kang, Y.-J.; Park, J.-G., Effect of polish by-products on copper chemical mechanical polishing behavior. *Journal of the Electrochemical Society* **2007,** *154* (6), H525.

[53] Lu, H.-S.; Zeng, X.; Wang, J.-X.; Chen, F.; Qu, X.-P., The effect of glycine and benzotriazole on corrosion and polishing properties of cobalt in acid slurry. *Journal of The Electrochemical Society* **2012,** *159* (9), C383-C387.

[54] Gale, G. W.; Cui, H.; Reinhardt, K. A., Aqueous cleaning and surface conditioning processes. In *Handbook of silicon wafer cleaning technology*, Elsevier: 2018; pp 185-252.

[55] Tang, J. S.; Lu, W.; Xi, B.; Martinez, E.; Li, F.; Ko, A.; Todd, C.; Lee, J. T. In *Marangoni dryer integrated high performance cleaner for Cu/low k post strip clean for 45nm technology node and beyond*, Solid state phenomena, Trans Tech Publ: 2008; pp 337-340.

[56] Lee, J.-H.; Ryu, H.-Y.; Hwang, J.-K.; Yerriboina, N. P.; Kim, T.-G.; Hamada, S.; Wada, Y.; Hiyama, H.; Park, J.-G., A breakthrough method for the effective conditioning of PVA brush used for post-CMP process. *ECS Journal of Solid State Science and Technology* **2019,** *8* (6), P307.

[57] Kim, H. J. In *HOLISTIC APPROACH TO THE UNDERSTANDING CMPINDUCED DEFECTS* NCCAVS 2020 CMP User Group Spring Meeting, April 2020; 2020.

[58] Kim, H. J., Effect of brush treatment and brush contact sequence on cross contaminated defects during CMP in-situ cleaning. *Tribology and Lubricants* **2015,** *31* (6), 239-244.

[59] Kim, H. J.; Bohra, G.; Yang, H.; Ahn, S.-G.; Qin, L.; Koli, D., Study of the cross contamination effect on post CMP in situ cleaning process. *Microelectronic engineering* **2015,** *136*, 36-41.

[60] Kim, H.; Hong, S.; Jin, Y.; Lim, D. H.; Kim, J.-y.; Hwang, H.; Kim, T., Investigation of copper oxide ring formation during post chemical mechanical polishing cleaning of Cu interconnect. *ECS Journal of Solid State Science and Technology* **2017,** *6* (8), n542.

[61] Jones, C. W., *Applications of hydrogen peroxide and derivatives*. Royal Society of Chemistry: 2007.

[62] Thomas, E.; White, M.; White, D.; Das, A. K., Post cmp cleaning compositions for ceria particles. Google Patents: 2020.

[63] White, D. In *High Performance, Ceria Post-CMP Cleaning Formulations for STI/ILD Dielectric Substrates*, Surface Preparation and Cleaning Conference (SPCC 2017), Austin, Texas, USA, Austin, Texas, USA, 2017.

[64] Babu, S. V., CMP: an overview, the good and not so good. In *ICPT 2019*, Hsinchu, Taiwan, 2019.

[65] Ong, P.; Teugels, L., CMP processing of high mobility channel materials: Alternatives to Si. In *Advances in Chemical Mechanical Planarization (CMP)*, Elsevier: 2016; pp 119-135.